Reappraising Oppenheimer

Berkeley Papers in History of Science

1 *Max Planck: A bibliography of his non-technical writings.* Compiled by Henry Lowood.
2 *William Henry Bragg and William Lawrence Bragg: A bibliography of their non-technical writings.* Compiled by Henry Lowood.
3 *A bibliography of quantitative studies on science and its history.* Compiled by Roger Hahn.
4 *Ernest Rutherford: A bibliography of his non-technical writings.* Compiled by Henry Lowood.
5 *Literature on the history of physics in the 20th century.* Compiled by J.L. Heilbron and Bruce R. Wheaton.
6 *An inventory of published letters to and from physicists, 1900-1950.* Compiled by Bruce R. Wheaton and J.L. Heilbron.
7 *A bibliography of early modern algebra, 1500-1800.* Compiled by Robin E. Rider.
8 *Calendar of the correspondence of Pierre Simon Laplace.* Compiled by Roger Hahn.
9 *Werner Heisenberg: A bibliography of his writings.* Compiled by David C. Cassidy and Martha Baker.
10 *Guide to sources in Northern California for history of science and technology.* Compiled by Robin E. Rider and Henry E, Lowood.
11 *The Nobel population 1901-1937: A census of the nominators and nominees for the prizes in physics and chemistry.* Compiled by Elisabeth Crawford, J.L. Heilbron, and Rebecca Ullrich.
12 *Eighteenth-century chemistry as an investigative enterprise.* Frederic L. Holmes.
13 *Weighing imponderables and other quantitative science around 1800.* J.L. Heilbron.
14 *Between biology and medicine: The formation of intermediary metabolism.* Frederic L. Holmes.
15 *Biology and social thought: 1850-1914.* Peter J. Bowler.
16 *The new calendar of the correspondence of Pierre Simon Laplace.* Compiled by Roger Hahn.
17 *A scientific correspondence during the chemical revolution: Louis-Bernard Guyton de Morveau and Richard Kirwan, 1782-1802.* Edited by Emmanuel Grison, Michelle Goupil, and Patrice Bret.
18 *Laboratories, workshops, and sites: Concepts and practices of research in industrial Europe, 1800-1914.* Robert Fox and Anna Guagnini.
19 *The structure of knowledge: Classifications of science and learning since the Renaissance.* Edited by Tore Frängsmyr.
20 *Michael Frayn's* Copenhagen *in debate.* Edited by Matthias Dörries.
21 *Reappraising Oppenheimer: Centennial studies and reflections.* Edited by Cathryn Carson and David A. Hollinger.

Reappraising Oppenheimer
Centennial Studies and Reflections

Edited by
Cathryn Carson and David A. Hollinger

Berkeley Papers in History of Science, Volume 21

Office for History of Science and Technology
University of California, Berkeley

Berkeley, California

Printed in the United States of America

Office for History of Science and Technology
543 Stephens Hall # 2350
University of California
Berkeley, CA 94720-2350

Library of Congress Control Number: 2005936057

Contents

J. Robert Oppenheimer and Ernest O. Lawrence on an excursion, 1930s.
Courtesy The J. Robert Oppenheimer Memorial Committtee and Mary Prudhomme.

Oppenheimer entertaining at Los Alamos during the war.
Courtesy Los Alamos National Laboratory.

J. Robert Oppenheimer, Enrico Fermi, and Ernest O. Lawrence at the University of California Radiation Laboratory, probably late 1930s.
Courtesy Lawrence Berkeley National Laboratory.

Ernest O. Lawrence, Glenn T. Seaborg, and J. Robert Oppenheimer in early 1946, at the controls of the magnet of Lawrence's 184-inch cyclotron.
Courtesy Lawrence Berkeley National Laboratory.

Kitty and Robert Oppenheimer visiting Japan, 1960.
Courtesy The J. Robert Oppenheimer Memorial Committee.

Oppenheimer receiving the Fermi Award from President Lyndon B. Johnson, 1963.
Courtesy Lawrence Berkeley National Laboratory.

Preface

J. Robert Oppenheimer, that protean figure, still holds fascination today. So an observer would conclude upon surveying his centennial: old biographies and new ones, treatments in film, literature, and television, even an opera. About this man, scientist and intellectual, advisor to the powerful and political martyr, this man who stood at the center of his century—about Oppenheimer there is something simultaneously engrossing and tantalizing. Just when we think we understand him, he always, somehow, shows a new side.

The present volume brings together the best of the new historical studies on many aspects of Oppenheimer's life. It offers cutting-edge scholarship, fresh insights, and deep contextualization, as well as an entrée into the fierce debates still surrounding the man. It aims to deliver all this in a publicly accessible format. An introduction to an ongoing discussion, it brings important new findings not just to the specialist, but to other interested readers and to students and teachers. In this it complements the online exhibit "J. Robert Oppenheimer: A Life," winner of a *Scientific American* web award (http://ohst.berkeley.edu/oppenheimer/exhibit/).

This volume originated in a conference in 2004, at a moment when the community of historians writing on Oppenheimer had reached critical mass. The conference had the advantage of a historic location, the University of California, Berkeley, where Oppenheimer in the 1930s built up a school of theoretical physics, and where he and a small group of colleagues in 1942 carried out the first calculations fixing the design parameters of an atomic bomb.

In our own deliberations, a hundred years after Oppenheimer's birth, fifty years after his security hearing, we found wide areas of agreement—and also contention. The chapters that follow reflect the balance at the conference. They are grouped under rubrics representing the most active areas of research today:

Oppenheimer as Physicist
The Communist Question
The Soviet Comparison: New Insights

Postwar Politics: The Oppenheimer Case in Context
Cultural Resonances
The Public Oppenheimer

Each section begins with a short overview highlighting and contextualizing its essays. At the end of the book, the selective bibliography will assist readers who want to pursue matters further, into the popular as well as the scholarly literature.

It is our pleasure to thank the people and organizations who made this volume possible. Along with the contributors, the conference benefited from the participation of several people not represented here, including commentators Charlie Weiner and Jessica Wang, as well as Chancellor Robert M. Berdahl. We were honored by Charles Oppenheimer's presence and comments at our conference. The larger Oppenheimer Centennial at Berkeley gained much from the excellent exhibits created by the Bancroft Library (David Farrell and Eric Vettel) and the Library (Diane Fortner), and from the participation of the Department of Physics (coordinated by Virginia Rapp). The staff of the Office for History of Science and Technology—Diana Wear, Kate Spohr, Alex Wellerstein, Ellen Bales, and Kevin Saavedra—not only put on a remarkable centennial program, but have also seen the conference volume through to completion. Without their efforts, often above and beyond the call of duty, none of this would have been possible. Finally, this material is based upon work supported by the National Science Foundation under Grant No. SES-0332474. We gratefully acknowledge this support. Any opinions, findings, and conclusions or recommendations expressed in this material are those of the author(s) and do not necessarily reflect the views of the National Science Foundation.

Cathryn Carson
David A. Hollinger

INTRODUCTION

Cathryn Carson

Robert Oppenheimer all but sensed that his life would be lived as an example. A self-profiled intellectual and a quantum theorist who built up a coterie of admiring students at Berkeley, his manners, charisma, and (by the later 1930s) his politics stamped the founding generation of American theoretical physicists. However unexpected his colleagues found his appointment at Los Alamos, its spectacular accomplishments came to stand for the Manhattan Project at large. Then his postwar apotheosis epitomized the physicists' entry into positions of power, just as the McCarthy-era stripping of his security clearance defined their political bounds.

Oppenheimer's story, his person, his significance and symbolism have fascinated commentators since he first appeared on the world's stage. If his career encapsulates the transformation of science in the twentieth century, then it is understandable that popular understandings of him have experienced wild swings. As his trajectory first reached its postwar heights, then passed through its lowest of lows, his public representations have reflected the era and moment as much as the cipher himself.

Historical scholarship has taken some time to come to grips. In recent years, however, several excellent scholarly biographies have appeared, now complemented by studies of particular issues and broader accounts of his context. It has become possible to reappraise Oppenheimer—the purpose of the present collection of centennial studies and reflections.

Oppenheimer (1904-1967): A Life

The outlines of Oppenheimer's life have never been in question. Of late, the details have come better into focus, however. Born on April 22, 1904, to Julius and Ella Oppenheimer, young Robert spent a precocious childhood in the cocoon of his well-off New York family: mother, father, and younger brother Frank. Recent scholarship by S.S. Schweber and others has filled out the formative influence upon Robert of the Ethical Culture School—the individual teachers he grew to admire, first and foremost, and the progressive social philosophy of Ethical Culture.[1] The intense but withdrawn, diffident but demanding young man came to know the attractions of the New Mexico landscape in a formative summer on horseback. He did not lack ambition either. His three years at Harvard (A.B., summa cum laude, Chemistry, 1925) were spent oscillating among French literature, philosophy (reading the *Principia mathematica* with Whitehead), and independent graduate-level study in thermodynamics, quantum physics, and physical chemistry.

As his Harvard letters make plain, in Smith and Weiner's exceptional edition, alongside intellectual triumphs Robert as an undergraduate suffered intense episodes of emotional turmoil.[2] Depression shadowed his postgraduate studies in Europe, as he struggled with experiments at Cambridge's Cavendish Laboratory before finding his métier as a theoretical physicist. But Oppenheimer returned to the United States a brilliant success, his German Ph.D. in hand, obtained in 1927 under the aegis of Göttingen's quantum theorist, Max Born. The young American's mastery of the mathematical formalism of the new quantum mechanics would launch his scientific career. As David Cassidy's biography explores with exemplary care, Oppenheimer moved to the front of the research field with his contributions to relativistic quantum mechanics and quantum field theory.[3] Beginning in 1929, he occupied a joint position as a professor at the California Institute of Technology and the University of California in Berkeley. Over the course of the 1930s, Oppenheimer built up a school of theoretical physics in Berkeley, making it nearly as much of a mecca for rising young theorists as his colleague Ernest Lawrence's Radiation Laboratory would be for cyclotroneers. This decade brought his deepest engagement with theoretical physics, even if large-impact results were relatively few.

Oppenheimer's politics took a turn to the left in the 1930s; on this all commentators agree. His commitments during the Popular Front years—the Spanish Loyalist cause, San Francisco Bay Area labor organizing—set him into close friendships with like-minded compatriots. These centered on a circle of Berkeley professors and other Bay Area intellectuals, including one woman, Jean Tatlock, whom Oppenheimer nearly married, and another, Kitty Puening, who would become his wife. The same commitments brought him into proximity to the Communist Party—to what degree is still fiercely debated, of course. Jean, Kitty, Robert's brother Frank (a physicist too), and several close colleagues had been CP members. Was Robert as well? If so, what could that mean? And how long did he remain close to the Party? The evidence from the 1930s has been sifted and weighed in diverging assessments, most conspicuously in the recent

1. S.S. Schweber, *In the shadow of the bomb: Bethe, Oppenheimer, and the moral responsibility of the scientist* (Princeton, 2000).
2. Alice Kimball Smith and Charles Weiner, eds., *Robert Oppenheimer: Letters and recollections* (Cambridge, 1980).
3. David Cassidy, *J. Robert Oppenheimer and the American century* (New York, 2005). See also Jeremy Bernstein, *Oppenheimer: Portrait of an enigma* (Chicago, 2004).

biographies by Gregg Herken on the one hand, Kai Bird and Marty Sherwin on the other.[4] However the controversy is resolved, it *is* certain that Oppenheimer's magnetic embodiment of the leftist intellectual—scintillating in conversation, facile in manner, at ease with Sanskrit, Donne, or *Das Kapital*—powerfully attracted student-acolytes who took up his habits, tastes, and poses, in politics as in style of life.

When nuclear fission burst onto the scene at the turn of 1939, the possibilities were scientifically exciting and politically threatening. Oppenheimer's involvement with what would become the Manhattan Project began in 1941, even before American entry into the Second World War. In 1942 he headed a Berkeley study group on fast neutron research that metamorphosed into Los Alamos's theoretical core. His graduate students followed him into the project; nearly every American physicist was mobilized for military research of some stripe. In particular, Army General Leslie Groves selected the bohemian theorist without administrative experience as scientific director of the bomb-design laboratory—located, on Oppenheimer's proposal, on a remote New Mexico mesa not far from the physicist's favored vacation retreat. The driven pace at Los Alamos, the enormous technical advances, the site's wonderland quality, and its scientific community life have all been recounted in memoirs and explored by scholars. The most compelling portrayal may be Jon Else's milestone documentary *The day after Trinity*.[5] In accounts of Los Alamos, naturally, Oppenheimer takes a central place: an inspiring genius with a view for the whole. Recent scholarship has sought to understand his individual role against the backdrop of the much larger Manhattan Project, giving space, alongside the accomplishments of theoretical physicists, to the contributions of engineers, chemists, and metallurgists, army personnel and practitioners of the trades.[6] In all these different treatments, still, there is no question about the individual view of Los Alamos's director regarding the bomb's eventual use: against counterarguments from his scientists, Oppenheimer fully backed the atomic bombing of Hiroshima and Nagasaki, on August 6 and 9, 1945.

4. Gregg Herken, *Brotherhood of the bomb: The tangled lives and loyalties of Robert Oppenheimer, Ernest Lawrence, and Edward Teller* (New York, 2002); Kai Bird and Martin J. Sherwin, *American Prometheus: The triumph and tragedy of J. Robert Oppenheimer* (New York, 2005).
5. E.g., Richard Rhodes, *The making of the atomic bomb* (New York, 1986); Lawrence Badash, Joseph O. Hirschfelder, and Herbert P. Broida, eds., *Reminiscences of Los Alamos, 1943-1945* (Dordrecht, 1980); Lillian Hoddeson, Paul W. Henriksen, Roger A. Meade, and Catherine Westfall, *Critical assembly: A technical history of Los Alamos during the Oppenheimer years, 1943-1945* (Cambridge, 1993); Jon Else, David Webb Peoples, Janet Peoples, Pyramid Films & Video, KTEH-TV, *The day after Trinity: J. Robert Oppenheimer and the atomic bomb* (San Jose, CA, 1981).
6. For a sampling of perspectives, see Jeff Hughes, *The Manhattan Project: Big science and the atom bomb* (New York, 2003); Robert S. Norris, *Racing for the bomb: General Leslie R. Groves, the Manhattan Project's indispensable man* (South Royalton, VT, 2002); Jon Hunner, *Inventing Los Alamos: The growth of an atomic community* (Norman, OK, 2004); Charles Thorpe and Steven Shapin, "Who was J. Robert Oppenheimer? Charisma and complex organization," *Social studies of science, 30* (2000), 545-590.

Come the end of the war, Oppenheimer was the man of the hour. His Los Alamos stature opened onto an extraordinary presence in the public arena and the corridors of power. In that transformation he personified the sudden revaluation of physics writ large. Immediately after the war, making common cause with the grass-roots "Scientists' Movement," he spoke and lobbied for international control of atomic energy. Whether or not he felt guilt for the atomic bombings (a sense of "blood on his hands," as Harry S. Truman's dismissive report had it) and whatever he meant by the evocative metaphor that "the physicists have known sin," he was politically persuaded that international cooperation was the only way to secure global post-Hiroshima survival. The extent of his insider status in certain Washington circles is plain, putting his stamp on the Acheson-Lilienthal report proposing an international Atomic Development Authority and moving at ease through high-level national advisory bodies. Quickly he was knotting a network of elite contacts—and consolidating the shift in his political views. In 1947, when the Army's Manhattan Project changed over to the civilian Atomic Energy Commission, he was placed on the AEC's powerful General Advisory Committee. The GAC unanimously elected him its chairman. He moved that same year from California to Princeton to take up the directorship of the Institute for Advanced Study and became a fixture of the East Coast advisory scene.

In that charmed circle of scientific advisors whose activities are now becoming clear, Oppenheimer had a front-row seat as the Cold War settled in.[7] International hopes to steer off a nuclear arms race dissolved; the Acheson-Lilienthal report was buried, and the Soviet Union tested its own nuclear device in 1949. Oppenheimer became known as a Cold War liberal, deeply skeptical of the U.S.S.R. Still, in the internal debate of 1949-50 over the hydrogen bomb, he and his scientist colleagues on the GAC recommended against pursuing the new weapon. If the "Super" worked, it would be many times more destructive than the now-familiar atomic bomb. Instead Oppenheimer backed assembling an extensive arsenal of smaller nuclear weapons. Yet in January 1950, in a move that historians have analyzed in great depth, President Truman overrode his official scientific advisors, ordering a crash program of thermonuclear development.[8] Over the next years, Oppenheimer continued to argue for candor in weapons policy, as well as alternative nuclear strategies to simply building

7. See, e.g., Gregg Herken, *Cardinal choices: Presidential science advising from the atomic bomb to SDI* (New York, 1992); James Hershberg, *James B. Conant: Harvard to Hiroshima and the making of the nuclear age* (New York, 1993); complementarily, Allan A. Needell, *Science, Cold War and the American state: Lloyd V. Berkner and the balance of professional ideals* (Amsterdam, 2000); Martin J. Collins, *Cold War laboratory: RAND, the Air Force, and the American state, 1945-1950* (Washington, D.C., 2002); Sharon Ghamari-Tabrizi, *The worlds of Herman Kahn: The intuitive science of thermonuclear war* (Cambridge, 2005).

8. Starting from Herbert F. York, *The advisors: Oppenheimer, Teller, and the Superbomb* (San Francisco, 1976); and Peter Galison and Barton Bernstein, "In any light: Scientists and the decision to build the Superbomb, 1942-54," *Historical studies in the physical and biological sciences, 19:2* (1989), 266-347.

more and bigger bombs as fast as possible. In these appeals he found some admirers, but scarcely undiluted success.

Clearly, Oppenheimer's position was not secure—and that in more ways than one. The H-bomb's advocates (including politicians and their aides, scientists like Edward Teller and Ernest Lawrence, and powerful figures in the military, especially the Air Force) noted his dissent as a sign of bad judgment or worse. His persuasive eloquence made him dangerous, and his sharp tongue earned him enemies, including AEC Commissioner Lewis Strauss. While Oppenheimer still rode the wave of postwar power, his seeming influence-seeking made his opponents suspect his ambition. More than that, his political past left him vulnerable. What other high-level advisor had such a spectacular array of left-wing commitments to answer for? Although his one-time affiliations had not deterred General Groves, Army counterintelligence had kept a close eye on the Los Alamos director during the war. An ambiguous encounter with an old Berkeley friend from his leftist days, Haakon Chevalier, had raised concerns about Soviet attempts at penetrating the Manhattan Project. Oppenheimer's inconsistent reports on the "Chevalier incident" would continue to hover over him. J. Edgar Hoover's FBI had already opened a file on the physicist in 1941; the file grew in bulk after the war, collecting data from informants, wiretaps, and other surveillance. The FBI also targeted Frank Oppenheimer as a part of a supposed Rad Lab communist conspiracy during the war. In 1949, Robert Oppenheimer effectively testified against his own graduate students before a closed session of the House Un-American Activities Committee. Senator Joe McCarthy could not touch him directly, but McCarthyism put him on notice, and public outcry over atomic espionage made the atmosphere more threatening. Recent scholarship has greatly filled out our understanding of the loyalty-security regime.[9]

In 1954 Oppenheimer would be brought down in an "inquiry" before a special AEC panel challenging his suitability for a security clearance. The course of the hearings has been reconstructed in detail in the new Oppenheimer biographies, in part with the help of the wiretaps that the FBI still had on his phone.[10] Openly or tacitly, much was at issue: Oppenheimer's past political affiliations, his misrepresentations to security officers, his stance on the H-bomb and his commentary on the nuclear arms race, his personal indiscretions—and the vitriolic hatred of now-AEC chairman

9. Jessica Wang, *American science in an age of anxiety: Scientists, anticommunism, and the Cold War* (Chapel Hill, NC, 1999); Lawrence Badash, "Science and McCarthyism," *Minerva, 38* (2000), 53-80.

10. See also Richard Polenberg, ed., *In the matter of J. Robert Oppenheimer: The security clearance hearing* (Ithaca, NY, 2002), and the full transcript from the United States Atomic Energy Commission, *In the matter of J. Robert Oppenheimer: Transcript of hearing before Personnel Security Board and texts of principal documents and letters* (Cambridge, 1971), as well as Priscilla McMillan, *The ruin of J. Robert Oppenheimer and the birth of the modern arms race* (New York, 2005), and Philip M. Stern with Harold P. Green, *The Oppenheimer case: Security on trial* (New York, 1969).

Lewis Strauss. Facing the charges, Oppenheimer refused to resign as a consultant. More than that, he insisted on asserting his loyalty. But he failed miserably in defending himself. In this extrajudicial setting, a string of scientific and political statesmen (even intelligence officials) speaking for him could not undo the damage. Prosecutor Roger Robb's citations from the FBI file, critical testimony from one-time colleagues like Teller, and Oppenheimer's own tactical mistakes sealed the Gray Board's decision. On action of the AEC commissioners in June 1954, Oppenheimer was definitively stripped of his security clearance.

With the news (and publication of the hearing transcript), Oppenheimer was turned into an instant martyr on the left. On the right, however, his case was seen as a justified payback and a reassertion of control over stunning security lapses. As for the man himself, no longer a government consultant, Oppenheimer moved ever more into the role of the cultural commentator. Although historians have only begun to pay attention, the scientist-intellectual in a transatlantic mold was a piquant figure on the American scene.[11] Oppenheimer's reflections on science, society, and philosophy entranced hearers and readers—yet pinning down his meanings could be hard. The same held true of his comments on nuclear weapons: while working to contain nuclear arms and deploring their effects, he expressed little public regret at his decisive involvement in bringing them to practical use. Politically, Oppenheimer's semi-official rehabilitation came in 1963, when President Lyndon B. Johnson honored him with the Fermi Award. But he seemed resigned to playing a hollowed-out version of his former persona. Oppenheimer died of cancer on February 18, 1967, at the age of sixty-two.

Orientations

A good dramatic story, Oppenheimer's life has attracted commentators of every sort. Their accounts have become overlaid in historical memory, leading to an unusually dense complex of interpretations with which today's analysts have to contend. Early postwar portraits played on the image of the abstruse theorist: the ivory-tower intellectual, posed on the cover of *Time* against a blackboard of formulae, now a nuclear mastermind and advisor on the national scene. Oppenheimer's charisma was part of the picture—for this was the press's simplification of a deeply complicated personality whose other, more troubled manifestations were less frequently discussed. Also part of the picture was his political past. The physicist was too savvy about the anti-Communist crusade of the late 1940s to try to push his one-time left-wing commitments out of sight. In his own account, these were excusable as the folly of youth. The full-throated admiration that he found immediately after the war, however, started yielding to more ambivalent tones. By 1953, mass media outlets had been turned partly against him; unflattering portrayals of his devious influence seeded a more equivocal public presence.

11. Especially Charles Thorpe, *Oppenheimer: The tragic intellect* (Chicago, 2006).

With the security hearing, of course, the narrative changed entirely: it became a story of rise *and* fall. In different tellings, the revised Oppenheimer figure was sometimes brought low by distinctly personal flaws; sometimes his fate stood in for that of scientists or intellectuals in general. What was interesting about Oppenheimer was now often his presumptive Faustian pact. Some versions of the story gave more room to unattractive aspects of his character. Dramatization began in earnest with the journalist Robert Jungk's reportage *Brighter than a thousand suns: A personal history of the atomic scientists.* Following on Jungk came Heinar Kipphardt's restaging of the security hearing.[12] The same theme of rise and fall structured Michel Foucault's famous comment upon the appearance of a new kind of intellectual:[13]

> Perhaps it was the atomic scientist (in a word, or rather a name: Oppenheimer) who acted as the point of transition between the universal and the specific intellectual. It's because he had a direct and localised relation to scientific knowledge and institutions that the atomic scientist could make his intervention; but, since the nuclear threat affected the whole human race and the fate of the world, his discourse could at the same time be the discourse of the universal. Under the rubric of this protest, which concerned the entire world, the atomic expert brought into play his specific position in the order of knowledge. And for the first time, I think, the intellectual was hounded by political powers, no longer on account of a general discourse which he conducted, but because of the knowledge at his disposal: it was at this level that he constituted a political threat.

Foucault all but made *l'affaire Oppenheimer* the pivot of the history of the twentieth-century intellectual.

As our accounts of the man have grown more complex, we have not lost the fascination with puzzling out his character. A certain de-heroization has conspicuously taken hold. At the same time, recent scholarship has oriented itself to different sorts of questions. A huge amount of effort has been invested in getting past the clichés and reconstructing the details of his story; here the foci have been the Berkeley milieu, his postwar policy conflicts, and the security hearing itself. Curiously, the Los Alamos years have been less intensely reexamined—perhaps because they received overwhelming attention from the start. On occasion, other analytical approaches have been explored.[14] However, they seem to have made little impact on the main line of

12. Robert Jungk, *Brighter than a thousand suns* (New York, 1958); Heinar Kipphardt, *In the matter of J. Robert Oppenheimer: A play freely adapted on the basis of the documents* (New York, 1968).
13. Michel Foucault, "Truth and power," *Power/knowledge: Selected interviews and other writings 1972-1977,* Colin Gordon, ed. (New York, 1980), 109-133, on 127-128.
14. E.g., Rachel L. Holloway, *In the matter of J. Robert Oppenheimer: Politics, rhetoric, and self-defense* (Westport, CT, 1993); Brian C. Taylor, "The politics of the nuclear text: Reading Robert Oppenheimer's *Letters and recollections,*" *Quarterly journal of speech, 78* (1992), 429-449;

Oppenheimer studies. That mainline scholarship has focused intensely on ever-deeper historical contextualization: more knowledge about not just the life, but the times. The essays in this volume share in that trend.

The state of the field also opens up further questions, many of them pushing for yet more attention to context. However, the context in question is typically less centered upon Oppenheimer's experience as an individual; rather, it highlights the broader trends in which he may have had a small part. An excellent example is the scientific progressivism of the 1930s.[15] Other than marking Oppenheimer as a man of the Left, it is still hard to know how to situate him (and his students) in relation to that movement. And because the scientific Left of the 1930s was far broader than nuclear physicists, its bearing on his concerns calls out for reconsideration—especially as it fed into early post-Hiroshima scientific activism, which he viewed from a peculiarly ambivalent position. In this respect at least, the atomic scientists were *not* the pivot of modern intellectuals' history—other scientists got there first—and Oppenheimer, suitably recontextualized, may help us explore that as well.

In other ways, too, we can deepen our understanding of Oppenheimer and his context in their mutual relations. Before Oppenheimer, American physicists had never assumed the intellectual's persona. His unique ambitions only highlighted his separateness in a discipline dominated by experimenters and machine-builders like his colleague Ernest Lawrence. Yet it was the theorist who came to be called the father of the bomb, and Oppenheimer's ascendancy marked a critical turn in American conceptions of the promise and power of intellectuals and science. An emerging story of physical theory in the Cold War can be paired with the rise of "theory" and the influence of a self-consciously intellectual elite in the social sciences, for instance.[16] These considerations point us, finally, toward assessing the iconic work that Oppenheimer has done in contemporary culture. And here he cannot be analyzed alone, instead

---

Charles Robert Thorpe, "Disciplining experts: Scientific authority and liberal democracy in the Oppenheimer case," *Social studies of science, 32* (2002), 525-562.

15. Peter J. Kuznick, *Beyond the laboratory: Scientists as political activists in 1930s America* (Chicago, 1987); Katherine Pandora, *Rebels within the ranks: Psychologists' critiques of scientific authority and democratic realities in New Deal America* (Cambridge, 1997); Loren Butler, "Robert S. Mulliken and the politics of science and scientists, 1939-1946," *Historical studies in the physical and biological sciences, 25:1* (1994), 25-45; also Elizabeth Hodes, "Precedents for social responsibility among scientists: The American Association of Scientific Workers and the Federation of American Scientists, 1938-1948," Ph.D. dissertation, History, University of California, Santa Barbara, 1982; and Andrew John Jewett, "To make America scientific: Science and democracy in American public culture, 1900-1950," Ph.D. dissertation, History, University of California, Berkeley, 2002.

16. E.g., Nils Gilman, *Mandarins of the future: Modernization theory in Cold War America* (Baltimore, 2003); S.M. Amadae, *Rationalizing capitalist democracy: The Cold War origins of rational choice theory* (Chicago, 2003); Jennifer S. Light, *From warfare to welfare: Defense intellectuals and urban problems in Cold War America* (Baltimore, 2003).

taking his place within a small cadre of other highly public scientists.[17] The scientist's bearing, his mode of speech and his concerns, are rich in their ambiguity. If we can see Oppenheimer in his complexity as something more than a singular individual—as a man mirroring and responding to a panoply of contradictory forces—then we may come closer to capturing his sense that the life he lived had significance beyond its individual confines.

17. E.g., Peter Goodchild, *Edward Teller: The real Dr. Strangelove* (Cambridge, 2004); and beyond the United States, Mary Jo Nye, *Blackett: Physics, war, and politics in the twentieth century* (Cambridge, 2004); Gennady Gorelik and Antonina W. Bouis, *The world of Andrei Sakharov: A Russian physicist's path to freedom* (Oxford, 2005); Cathryn Carson, *Heisenberg in the atomic age: Science and the public sphere* (in preparation).

PART ONE

Oppenheimer as Physicist

"I didn't start to make a school," Oppenheimer once remarked of the cadre of students who had clustered around him. "I started really as the propagator of the theory which I loved, about which I continued to learn more, and which was not well understood and which was very rich." Oppenheimer's delight in quantum mechanics, the new physics of the 1920s, carried him forward into some of the next decade's most exciting domains. Relativistic quantum field theory, nuclear physics, and astrophysics were the fields in which he and his students would work. In seminars and colloquia Oppenheimer's gifts shone through: broad mastery of topics both theoretical and empirical, and amazing quickness of understanding. His own research in the 1930s ran in many directions at once, but—due in part to the state of the field, in part to the character of his interventions—he made no great breakthroughs in any.

The essays in this section offer complementary accounts of Oppenheimer's science. David C. Cassidy gives a broad overview of Oppenheimer's "school" of theoretical physics. Along with the school's internal dynamics, Cassidy reveals the external forces that made it possible. He examines the style of physics that the leader conveyed to his students and postdocs, and he traces its ramifications into the Manhattan Project and the postwar era. While Cassidy paints Oppenheimer's physics against the broad backdrop of the "American century," Karl Hufbauer focuses on the scientist's work on black holes. Carried out collaboratively with his students, Oppenheimer's theoretical studies from the late 1930s are sometimes put forward as Nobel-worthy material. Hufbauer firmly situates Oppenheimer's work in the context of contemporary boundary-crossing between physics and astrophysics. He explains how Oppenheimer came to the topic—and why he dropped it, and why almost no one cared.

CHAPTER ONE

From Theoretical Physics to the Bomb

J. Robert Oppenheimer and the American School of Theoretical Physics

David C. Cassidy

More than any other events of their time the two world wars that engulfed the globe during the twentieth century shaped the historical and political contours of that century. It was a century that encompassed America's rise onto the world stage to become a dominant economic, cultural, and military power in world affairs. An important contributor to the nation's success was, of course, the success of its physics community. And a key element of that success was the rise of the American school of theoretical physics under the impetus in California of J. Robert Oppenheimer. How did he do it? How did such an unlikely figure as J. Robert Oppenheimer, a man known more for his unfocussed, even dilettantish, intellectual interests than for his breakthrough research or teaching, become, in Hans Bethe's rather hyperbolic obituary assessment, the head of "by far the greatest school of theoretical physics that the United States has ever known"?[1]

As with most historical questions the answers lie in the times, the circumstances, and the people themselves. A great deal has been written about Oppenheimer, research schools, and the Oppenheimer school in particular. My aim is to examine the success of the school by looking briefly at Oppenheimer and the internal dynamics of

This essay is based on papers delivered to the history of science colloquium, Magnus-Haus, Berlin, 17 Mar 2003, and at the J. Robert Oppenheimer Centennial at Berkeley, 22-24 Apr 2004. It is adapted from D.C. Cassidy, *J. Robert Oppenheimer and the American century* (New York, 2005). I am indebted to Cathryn Carson, Jürgen Renn, and the other participants on both occasions for their valuable comments.

The following abbreviations are used: *BHPD*, Raymond T. Birge, *History of the physics department*, 5 vols., bound typescript (Berkeley, 1965); *HSPS*, *Historical studies in the physical and biological sciences*; IEB, International Education Board; JRO, J. Robert Oppenheimer; NBL, Niels Bohr Library, Center for History of Physics, College Park, MD; NRC, National Research Council; RAC, Rockefeller Archive Center, Tarrytown, NY; TBL, The Bancroft Library, University of California, Berkeley; DPUCB, Department of Physics records, University of California, Berkeley, CU-68, TBL.

1. Hans Bethe, "J. Robert Oppenheimer 1904-1967," Royal Society of London, *Biographical memoirs of fellows* (1968), vol. 14, 391-416, on 396.

his school, as well as the external forces promoting their success. These forces derived from the deliberate efforts of powerful American experimentalists to bring American physics, especially American theoretical physics, to the forefront of world research, of which Oppenheimer and his school were the direct beneficiaries.

In the context of American practical universities early in the century, theoretical studies lagged far behind experimental research; and American theorists lagged far behind their European counterparts. The disparity was made obvious at the end of the First World War in the public sensation attending the confirmation of Einstein's theory of general relativity. In 1921 Einstein made a widely publicized visit to the United States, where Princeton University conferred upon him an honorary doctorate. Four years later German and other European theorists astonished the world with the second of the two great revolutions in twentieth-century physics – the breakthrough to quantum mechanics. American physicists could only look with envy as recognition for such achievements went to their former enemy and current competitor, Germany.

Even though the theoretical physics of relativity and the quantum seemed to hold little practical application, these theories carried an enormous international prestige value. German scientists were well aware of this. As the postwar German inflation accelerated exponentially, leading German scientists began to funnel precious foundation money into postdoctoral fellowships for promising young theorists in atomic and quantum physics in order to keep them in the profession.

Americans pursued a similar policy at that time and for similar nationalistic reasons. As Daniel Kevles, Silvan S. Schweber, David Hollinger, and others have shown, in the depths of the world war American chemists and physicists set out to do something about their second-rate status. Many of the promoters of theoretical and other forms of physical science were members of the recently founded National Research Council of the National Academy of Sciences. As the war ended, the NRC, which had been instrumental in setting war-time science policy, took the lead in setting post-war policy, and theoretical physics was among its top priorities.[2]

Even as the NRC, led by experimentalist Robert A. Millikan and astronomer George Ellery Hale, looked to the prestige value of theoretical physics, developments within physics were demanding greater numbers of physicists skilled in theoretical work. As elsewhere, experimental research in America was focusing more directly upon the properties of atoms, electromagnetic fields, and the atomic structure of matter. Some of this work did impact the economically beneficial fields of electrical engineering and materials science. But such work was also requiring sophisticated mathematical and analytical skills that were not within the expertise of most experimentalists. Theoretical

2. Daniel J. Kevles, *The physicists: The history of a scientific community in modern America* (New York, 1971); Silvan S. Schweber, "The empiricist temper regnant: Theoretical physics in the United States 1920-1950," *HSPS, 17:1* (1986), 55-98; David A. Hollinger, *Science, Jews, and secular culture* (Princeton, 1996); Robert E. Kohler, *Partners in science: Foundations and natural scientists 1900-1945* (Chicago, 1991); Stanley Coben, "The scientific establishment and the transmission of quantum mechanics to the United States, 1919-1932," *American historical review, 76* (1971), 442-466.

physicists adept in mathematical manipulations yet closely tied to empirical data were now essential to the future progress and competitive ability of American physics.

With the financial backing of the Rockefeller Foundation, the National Research Council began promoting the nation's science through the funding of young researchers, especially in atomic physics. In 1919 Millikan told the public: "In a few years we shall be in a new place as a scientific nation and shall see men coming from the ends of the earth to catch the inspiration of our leaders and to share in the results which have come from our developments in science."[3]

Millikan was right. By the early 1930s American work in quantum theoretical physics, as well as in experimental research, was within the first ranks of world research, and foreign physicists were coming to the United States to learn the latest developments. John Slater recalled how impressed he was by the American papers given during an APS meeting in June 1933: "not so much the excellence of the invited speakers as the fact that the younger American workers on the program gave talks of such high quality on research of such importance that for the first time the European physicists present were here to learn as much as to instruct."[4]

The quality of American physics further accelerated after 1933 with the arrival of European theorists and experimentalists. Although these refugees from persecution now faced discrimination and unemployment in the United States, they contributed significantly to the rise and success of American physics during the 1930s and thereafter, providing important preconditions for the success of the Manhattan Project and other ventures during the next world war.

One indication of the ascent of American physics in this period is the increase in the number of doctorates. In 1921 the United States awarded only 31 doctorates in physics. By 1930 that number had more than tripled to 99; and in 1940 it was 170.[5] As military funding flowed into physics after the Second World War, the growth of physics doctorates was nearly exponential during the early years of the Cold War.[6]

The NRC's strategy to build up American theoretical work entailed several other components. Two of these involved sending Americans to Europe to learn the new quantum physics at the source, and inviting Europeans to teach the new physics in the United States as visiting and permanent professors. In 1926 Max Born, then visiting the Massachusetts Institute of Technology, reported to the dean at Göttingen that he had already received two job offers. "While in the United States the experimental physics is in high blossom," he wrote, "the theoretical physics is entirely undeveloped. Europe's lead must be recognized for the moment; but at the same time the ambition

3. Robert A. Millikan, "The new opportunity in science," *Science, 50* (1919), 292-297, on 297.

4. John C. Slater, "Quantum physics in America between the wars," *Physics today, 21:1* (1968), 43-51, on 43.

5. Kevles (ref. 2), 200; and M. Lois Marckworth, *Dissertations in physics: An indexed bibliography of all doctoral theses accepted by American universities, 1861-1959* (Stanford, 1961).

6. David Kaiser, "Cold war requisitions, scientific manpower, and the production of American physicists after World War II," *HSPS, 33:1* (2002), 131-159, esp. graph, 135.

of the Americans is directed toward the goal of filling this gap, initially by attracting European assistance."[7]

In addition to attracting Europeans, Millikan and the NRC pursued a domestic policy of grant elitism, what they called "best science." Emphasizing speed, their post-doctoral fellowships went only to a small number of the very best young American physicists, and they directed the fellows and their research funds only to a small number of carefully selected American universities that they intended to transform quickly into elite research institutions. Naturally, their own students and universities were among those selected. Even after it succeeded, the elitism continued for decades thereafter. As a result of it, during the 1930s only ten universities produced over 50% of the nation's physics doctorates. During the 1950s, the same universities formed the top ten institutions, producing nearly 40% of the doctorates, while the top two recipients of cold-war military funding—Berkeley and MIT—produced over one third of the doctorates among the top ten. (See Table 1.) Most of these universities are still the leading producers of doctorates in physics and other sciences in the United States.

Although he was the direct beneficiary of his elders' strategy, J. Robert Oppenheimer made his own way in physics. After graduating in chemistry and physics from Harvard in 1925, Oppenheimer headed for the Cavendish Laboratory, where he intended to obtain a doctorate with Ernest Rutherford as a theorist assisting in experimental work. But he gravitated instead toward R.H. Fowler and Fowler's former student Paul Dirac. Dirac was then making important contributions to the emergence of the new quantum

Table 1. The top ten American institutions producing doctorates in physics during the 1930s and 1950s.

1930-1939		1950-1959	
Total for United States:	1,269	Total for United States:	4,175
Institution	Number	Institution	Number
1. Caltech	89	1. UC Berkeley	339
2. U Michigan	85	2. MIT	303
3. U Chicago	72	3. Harvard	178
4. UC Berkeley	70	4. U Connecticut	188
5. Cornell	67	5. U Illinois	158
6. MIT	67	6. Yale	149
7. Harvard	52	7. Cornell	147
8. U Connecticut	50	8. U Chicago	141
9. Yale	48	9. Caltech	137
10. U Illinois	41	10. U Michigan	134
Total:	641		1,863
% of national total	50.5%		39.5%

Source: Derived from data in Marckworth (ref. 5).

7. Born to Kurator, Univ. Göttingen, 13 Apr 1926, Universitätsarchiv, Göttingen, Personalakte Prof. Dr. Max Born.

mechanics, and Oppenheimer began working on applications of the new formalism to old or previously unresolved atomic problems. He received his doctorate in theoretical physics in 1927 at arguably the premier school in quantum theory at the time, Max Born's institute in Göttingen. There Oppenheimer helped to produce, among other applications, the well-known Born-Oppenheimer approximation in the quantum mechanics of molecules. Back home, the NRC was offering postdoctoral fellowships. With encouragement from NRC board member Karl T. Compton, who was then scouting for recruits in Göttingen, Oppenheimer applied to work at the newly founded California Institute of Technology with Paul Epstein, a mathematical physicist whom Millikan, the head of Caltech, had recently imported from Munich. Born recommended Oppenheimer shortly after he submitted his dissertation. In December 1926 he wrote to the NRC: "I would like to point out here only one who rises far above the average. He is Mr. Robert Oppenheimer, a young American who is extraordinarily good in mathematics, has good physical understanding and promises to become an exceptional scholar."[8]

Oppenheimer was one of only fourteen NRC fellows selected in physics for 1927/28.[9] He divided his time that year between Harvard and Caltech, working on applications of quantum mechanics to metal electrons and emission spectra. Fresh from Europe with a Göttingen doctorate and several German and British publications to his credit, the twenty-three-year-old Oppenheimer was suddenly a hot commodity. By that spring he had received as many as ten university teaching offers, including two from Europe and one each from Harvard and Caltech. But Oppenheimer wanted another stint abroad. In 1928 he was one of only four Americans newly selected by the NRC to work abroad in physics under the auspices of the Rockefeller Foundation's International Education Board.[10]

In hindsight Oppenheimer's second tour of Europe proved vital to American physics. He worked with Fowler in Cambridge and Ehrenfest in Leiden, and attended Heisenberg's lectures in Leipzig on ferromagnetism, where he first met another IEB fellow, Isidor I. Rabi. But the high point was his work with Wolfgang Pauli at the Swiss Federal Institute of Technology (Eidgenössische Technische Hochschule, ETH) in Zurich. There he met Rudolf Peierls and Felix Bloch and contributed to the relativistic quantum electrodynamics then under construction by Pauli and Heisenberg. To a certain extent Oppenheimer acquired from these European theorists, especially Pauli, precisely what the Americans still lacked—maturity and imagination. Although Americans like Oppenheimer were technically proficient in the new quantum formalism, they needed what Rabi reportedly called "a taste, style, and sense of what was

8. Born to Trowbridge, IEB, 26 Dec 1926, RAC, IEB series 1.3, folder 1095: JRO, 1926-1945. Oppenheimer submitted a publishable version of his dissertation on 24 Dec 1926, "Zur Quantentheorie kontinuierlicher Spektren," *Zeitschrift für Physik, 41* (1927), 268-293.
9. NRC, minutes of meeting on 23 Apr 1927, excerpt, RAC, IEB series 1.3, folder 1095: JRO, 1926-1945; and NRC, "Preliminary report of the fellowship board in physics, chemistry, mathematics 1919-1930," RAC, RG 1.1, series 200, box 173, folder 2100.
10. Ibid.

important in science."[11] Rabi recalled that, during their European sojourn, young Americans like himself and Oppenheimer were so embarrassed at their performance compared with the Europeans that they vowed to remedy this deficiency upon their return to the States. Said Oppenheimer: "Americans who had studied in Europe and came back...decided that there was no reason that things like that shouldn't be going in this country, and went to work on it."[12] Their elders had the same goal in mind, and they had carefully selected these young men to achieve it.

The University of California, Berkeley, was at that time one of the universities that Millikan and the NRC had targeted for rapid transformation into elite research institutions. As part of the effort, in 1928 Raymond T. Birge, then assistant physics chair at Berkeley, managed to entice an outstanding NRC fellow from Yale to Berkeley—Ernest O. Lawrence. Birge wrote to Lawrence of his plans for the second open position: "As you may know, we are gradually getting quite a strong department, and are constantly strengthening it. There are two positions open for next year, and besides that being offered to you, we will probably try to get a really able mathematical physicist of the productive kind for the other position."[13] Two months later Birge hired another proven NRC fellow for that position—J. Robert Oppenheimer. At first Birge shared Oppenheimer with Millikan at Caltech, but the equal sharing lasted only several years, until Berkeley changed its schedule and demanded more of Oppenheimer's time.[14] Oppenheimer later said of the arrangement: "I visited Berkeley and I thought I'd like to go to Berkeley because it was a desert. There was no theoretical physics and I thought it would be nice to try to start something. I also thought it would be dangerous because I'd be too far out of touch so I kept the connection with Caltech."[15] By 1929, when Oppenheimer arrived on campus, Berkeley and the American community had pinned their hopes on Lawrence and Oppenheimer to bring the university into the top ranks of American physics, and on the twenty-five-year-old Oppenheimer to help bring the nation into successful competition with European theoretical research.

To what extent did Oppenheimer and his school meet these expectations? Did it really achieve the huge success that Bethe attributed to it? Upon closer look, it is evident that Oppenheimer was neither a popular teacher nor an excellent bureaucrat. Students avoided his courses because of his high demands and poor teaching style. Nor did Oppenheimer produce a large number of doctorates—only 22 in the period from

11. Reported by James R. Killian, *The education of a college president: A memoir* (Cambridge, 1985), 17.

12. JRO, interview with Charles Weiner, Nov 1966, NBL; I.I. Rabi, interview with Thomas S. Kuhn, 8 Dec 1963, NBL.

13. Birge to Lawrence, 23 Feb 1928, Lawrence papers, TBL, BANC MSS 72/117c, or microfilm 2248, roll 004.

14. Elmer H. Hall (Berkeley physics chair) to Millikan, 26 Mar 1931, Millikan Papers, Archives, California Institute of Technology, box 27, folder 11; JRO to Millikan, 5 Mar 1933, folder 12; and course bulletins for the period.

15. JRO, interview with Thomas S. Kuhn, 20 Nov 1963, NBL.

1930 until he left for Los Alamos in 1943, and only six through 1939. During the same period, 1930-39, the entire department produced 70 doctorates. (See Appendix.) Oppenheimer was not listed as an examiner of any of the Caltech graduates.

But Oppenheimer learned fast, and with capable postdocs able to fill in the gaps in the teaching process, the school proved an inspiration to those who managed to follow the research program. Because of his first-hand knowledge of quantum mechanics, the Depression-era economic conditions, and the continued support of the NRC into the late 1930s, Oppenheimer was able to gather about him a group of good and, in some cases first-rate, doctoral and postdoctoral researchers at both Berkeley and Caltech. By one count, in 1938 Oppenheimer's Berkeley research group consisted of 14 graduate students and postdocs. This was small by today's standards, and certainly much smaller than the nearly 200 people then employed in Lawrence's heavily-funded Radiation Laboratory.[16] But for theoretical physics at the time, it was quite impressive. A school of this size compared favorably with the powerful European schools of a decade earlier, such as those of Born in Göttingen, Sommerfeld in Munich, and Bohr in Copenhagen; and it far exceeded in size the other American groups at the time.

In addition to the deliberate federal policies promoting Oppenheimer and his school, economic and social factors worked to the school's advantage. Most of Oppenheimer's students and postdocs were funded by NRC fellowships, university grants, and teaching assistantships. But the onset of the Depression forced a gradual decline in the department's annual budget until, in 1933/34, it dropped by over 21%.[17] Faculty salaries fell by as much as 8.4% that year. But within a year the faculty, including Oppenheimer, had forced its salaries back even higher than before.[18] Unfortunately, such largesse came at the expense of student funding at a time when the number of physics students was growing rapidly. The economic bubble of the twenties had produced a burgeoning middle class eager for education leading to professional careers. By the mid-1930s American universities and graduate schools were overflowing. The numbers of Berkeley graduate students in physics jumped from 65 in 1932 to 80 in 1938/39.[19] Since job prospects in industry and academe remained scarce, most students naturally stayed as long as possible, and most postdocs remained in place until their funds ran out. Every spring, as Oppenheimer headed to Pasadena to offer an advanced theory course, an entourage of Berkeley students and postdocs made the trek south for the summer. Robert Serber, one of Oppenheimer's last NRC fellows, was married. "We thought nothing of giving up our houses or apartments in Berkeley," he wrote,

16. DPUCB, box 6, folder 6:22.
17. DPUCB, box 1, folder 1:6: budget 1921-1937; Birge, faculty meeting of 23 Aug 1933, DPUCB, box 2, folder 2:4; and *BHPD*, vol. 4, XII-1 to 2, and XIII-6 to 7. In 1933/34, the number of teaching assistants was cut by 25% and the salaries of those remaining by 30.7%.
18. *BHPD*, vol. 4, XII-2; DPUCB, box 1, folder 1:6. JRO's salary rose from $3,300 in 1931/32 to $3,368 in 1934/35; it went to $4,000 upon promotion to full professor in 1935.
19. DPUCB, box 2, folders 2:4, 2:7; *BHPD,* vol. 4, XIV-34.

"confident that we could find a garden cottage in Pasadena for twenty-five dollars a month. We didn't own more than could be packed in the back of a car."[20]

Aside from attracting the right people and obtaining sufficient funds, the most important factor in the success of the Oppenheimer school was Oppenheimer himself. "His mind was so quick and his speech so fluent," wrote Serber, "that he dominated nearly every gathering." A man who carefully cultivated his public persona with the mannerisms, cultural trappings, and attire of a European intellectual, his influence within the geographically isolated world of California physics was all the greater. His students adored him, clinging to every new idea he offered, every cultural preference he expressed. "These were post-depression days when students were poor," wrote Serber, "...and Oppie was introducing them to an unfamiliar way of life. We acquired something of his tastes. We went to concerts together and listened to chamber music."[21] They made of him "something like a god," wrote one contemporary observer, thereby creating the devotion and identity that small group dynamics require for the achievement of a common outlook that in science might be called a research school.[22] And this, to a great extent, defined the school's success. Owing to the policies of the profession's leaders, the social-economic environment of the times, and the professional expertise and charismatic persona of its leader in a geographically isolated location, the school achieved in terms of numbers and its cohesive internal dynamic a status that could well justify calling it one of the nation's greatest schools of theoretical physics until recent time. This, Cathryn Carson has pointed out, "is a school in the sense of establishing a new discipline in a new country."[23]

But what about the physics it produced? Did the members of this school produce the kinds of breakthrough achievements that we often associate with the greatest of theoretical physics schools, such as Born's, Sommerfeld's, or Landau's? The answer might begin with the school's dominant figure. In order to keep abreast of the latest work occurring outside their world, Oppenheimer established a Journal Club modeled after Cambridge's Kapitza Club. The club met on Monday evenings around the fireplace in nearby Stephens Hall (now, of course, home to the Office for History of Science and Technology). The members heard a student or postdoc report on an assigned article, followed by extensive discussion and critical assessment. Lawrence, McMillan, and others from the Rad Lab joined the theorists, as the work of the two groups converged. Oppenheimer's intellect shone forth in such meetings. Birge declared in his history of the department: "No matter what field of physics was being discussed, Oppenheimer had more numerical facts in his head and ready for instant application

20. Robert Serber, "The early years," I.I. Rabi et al., *Oppenheimer* (New York, 1969), 11-20, on 19.

21. Ibid., and Robert Serber with Robert P. Crease, *Peace and war: Reminiscences of a life on the frontiers of physics* (New York, 1998), 25.

22. Robert Nisbet, *Teachers and scholars: A memoir of Berkeley in depression and war* (New Brunswick, NJ, 1992), 136.

23. Cathryn Carson, session commentary, J. Robert Oppenheimer Centennial at Berkeley, 23 Apr 2004.

than did any experimental physicist present, even those working in the particular field being discussed."[24] Lawrence once dragged a reluctant colleague to one of the club's meetings just to marvel at the intellectual display.

During his pre-Los Alamos years in California, from late 1929 to early 1943, Oppenheimer published 36 research papers and 16 abstracts of conference papers. All but two of these appeared in the *Physical review*, by then among the world's leading physics journals. Four of these papers, especially the ones on astrophysics, still generate citations in the research literature.[25] Of these 36 research papers, only 12 appeared with Oppenheimer as a single author. He coauthored the remaining two thirds with his students, postdocs, and an occasional colleague. Serber proved by far his closest collaborator, publishing eight of the 24 coauthored papers with Oppenheimer. Even if students could not afford the travel costs to APS conferences, half of Oppenheimer's conference abstracts listed a coauthor, an indication of his esteem for his students' work and his willingness to promote their interests.

A review of these papers indicates, however, that, however facile his mind, Oppenheimer did not produce the most innovative and brilliant research, nor, of course, did he ever receive a Nobel Prize. But is the brilliance that leads to a Nobel really needed or even advantageous for a school leader? After all, Sommerfeld never received a Nobel, his work was only briefly brilliant, and he was best known for his textbook, *Atombau und Spektrallinien*; Born and Landau received Nobels only late in their careers (Bohr, however, received his in 1922). Instead, as these others seem to suggest, what makes a brilliant school leader is the ability to lead—to select the most promising problems for his group and to guide his students to the forefront of research in those areas. Bethe provides another glowing assessment of Oppenheimer in this regard: "Probably the most important ingredient he brought to his teaching was his exquisite taste. He always knew what were the important problems, as shown by his choice of subjects. He truly lived with these problems, struggling for a solution, and he communicated his problems to his group."[26]

Many of the group's problems arose directly from the new discoveries arising from experimental work at the time. But the difficulty for Oppenheimer's group in producing breakthrough work arose from the circumstance that the theoretical physics of the day was not always able to account for the new experimental results. As Silvan S. Schweber has put it, those were years in which the empiricist temper was regnant in American physics: the close connection of theoretical research with new and exciting discoveries at the forefront of experimental research. This was precisely the combination that enabled the greatest successes in those areas of physics in which the match of theory to experiment was the closest. The experimental forefront was opening new territory

24. *BHPD*, vol. 3, IX-31.
25. *Science citation index* (Philadelphia, annual). A bibliography of JRO's scientific papers is provided by Alice Kimball Smith and Charles Weiner, eds., *Robert Oppenheimer: Letters and recollections* (Cambridge, 1980), 359-362.
26. Bethe (ref. 1), 396.

in fields that we now categorize as nuclear physics, high-energy research, fundamental quantum physics, and astrophysics. As before, theorists were needed to help the experimentalists understand their data.[27] Oppenheimer's choice of problems for his group closely reflected the reigning empiricist temper, but because of the mismatch between theory and experiment in some areas, the problems Oppenheimer picked were not always the most promising of solution, nor did he (or they) display the single-minded devotion or the Nobel-quality brilliance required to solve them.

For problems to solve, Oppenheimer and his group often turned to Lawrence and his accelerator group at Berkeley and to Anderson, Millikan, and Lauritsen at Caltech. Oppenheimer's star postdoc Robert Serber, one of the last NRC fellows, noted that most of his friends at Berkeley during the late thirties were experimentalists then working with Lawrence at the Rad Lab—Luis Alvarez, Edwin McMillan, and Milton White. "They would bring up the things they were working on and ask us questions. Then of course there was Oppenheimer and Lawrence. It wasn't a separate community; we were all living together."[28] There were plenty of new developments to keep everyone busy, some of them occurring in California. In the annus mirabilis of 1932, Anderson discovered the positron; Chadwick at the Cavendish Laboratory discovered the neutron; Urey at Columbia discovered deuterium and the deuteron; Blackett and Occhialini in UK discovered the existence of cosmic-ray showers, the creation of streams of particles when a high-speed cosmic ray strikes the upper atmosphere or a lead plate in a cloud chamber; Heisenberg presented the neutron-proton theory of the nucleus; Pauli invented the neutrino to help retain energy conservation in beta decay; and at the Rad Lab Lawrence's new cyclotron was accelerating protons to the then-record energy of one million electron volts.[29] By 1937 Lawrence's 37-inch machine was the largest in the world. Two years later he received Berkeley's first Nobel Prize in physics for his invention.

Lawrence's accelerators were then sufficient only for nuclear studies, while the high energies of what we now call high-energy physics were then found only in cosmic-ray interactions, a specialty of the Caltech researchers. As the neutron and two new nuclear-scale forces began to join the theorist's toolkit of applications, the theorists of the Oppenheimer group were gradually able to account for some of the Rad Lab's nuclear data, but important puzzles remained. Similarly, relativistic quantum elec-

27. Schweber (ref. 2).
28. Robert Serber, interview with Charles Weiner, Gloria Lubkin, and Barry Richman, 10 Feb 1967, NBL.
29. For Lawrence's cyclotrons, see J.L. Heilbron and Robert W. Seidel, *Lawrence and his laboratory: A history of the Lawrence Berkeley Laboratory* (Berkeley, 1989), vol. 1; Laurie Brown, "Nuclear forces, mesons, and isospin symmetry," Brown et al., *Twentieth century physics* (London and New York, 1995), vol. 1, 357-419, esp. 371-372; Laurie Brown and Lillian Hoddeson, eds., *The birth of particle physics* (New York, 1983), with papers by the original players; Helge Kragh, *Quantum generations: A history of physics in the twentieth century* (Princeton, 1999), chapt. 13; and Steven Weinberg, *The discovery of subatomic particles*, 2nd edn. (New York, 2003).

trodynamics was plagued at that time with such internal difficulties that it seemed practically useless for the high-energy cosmic-ray shower events. Because of this and other factors, Heilbron and Seidel conclude from their study of the Oppenheimer-Lawrence collaboration, Lawrence's Rad Lab did not develop "the sort of theory-driven experiments that mark the big-machine physics of the postwar era."[30]

By the same token, the Oppenheimer school displayed in some nuclear and most high-energy work the empirically driven character of American physics at the time. This is evident from a comparison of the Oppenheimer group's work with that of Europeans, notably Pauli, Heisenberg, and their students. In contrast to the Americans, Pauli and Heisenberg, as founders of quantum mechanics, followed the more radical and speculative approach of European veterans of the quantum revolution. Both sides were struggling at that time with infinities appearing in theoretical predictions of high-energy events. Oppenheimer had confirmed these infinities while in Zurich in a famous paper published in 1930.[31] This was followed by work on Dirac's hole theory, quantum electrodynamic theories of cosmic-ray showers and their absorption in matter, and the meson theory of nuclear forces and cosmic rays in connection with what was later found to be, in hindsight, the wrong meson. Eager to provide applications of relativistic quantum mechanics to empirical data, yet blocked by infinities, Oppenheimer and his students invoked the empiricist temper in an effort to force the theory into agreement with the data by introducing artificial cutoffs of the infinite results into their equations. The effort did yield the contemporary theory of what we now call lower energy "cascade showers," and it renewed hope then of a similar theory for high-energy events. Heisenberg and Pauli took the opposite approach: they essentially abandoned the available theory in search of alternative theories involving hypothetical minimum lengths and even the existence of a "lattice world" on the order of the Planck length.[32] Oppenheimer at times reacted with such hostility toward Heisenberg's radical suggestions that he seemed to be reflecting a lingering inferiority complex of American physicists toward their German counterparts.

As it turns out, despite Oppenheimer's radical politics at the time, the more conservative and empirical approach of the Oppenheimer school eventually did yield the way forward with the postwar invention of "renormalization." Ironically, the school's most speculative and least empirically driven work, that on the collapse of dying stars, was quickly abandoned by its authors and generally ignored by their contemporaries.

30. Heilbron and Seidel (ref. 29), 259.

31. JRO, "Note on the theory of the interaction of field and matter," *Physical review, 35* (1930), 461-477.

32. The divergent attempts to account for cosmic-ray showers are explored by, among others, Peter Galison, "The discovery of the muon and the failed revolution against quantum electrodynamics," *Centaurus, 26* (1983), 262-316; Galison, *How experiments end* (Chicago, 1987), chapt. 3; and D. Cassidy, "Cosmic-ray showers, high energy physics, and quantum field theories," *HSPS, 12* (1981), 1-39.

Yet, from the "whiggish" perspective of recent discoveries using the Hubble Space Telescope, such work came closest in hindsight to Nobel Prize quality.[33]

Eleven of Oppenheimer's 36 research papers produced in that period dealt with less problematic nuclear collision phenomena arising from accelerator experiments at Berkeley and Caltech. Among these more successful papers was one providing the now well-known Oppenheimer-Phillips process in nuclear disintegration. In 1937 Niels Bohr came to Berkeley to lecture on the liquid-drop model of the nucleus, a model that became the basis two years later of the Bohr-Wheeler theory of nuclear fission.

The discovery of fission and the outbreak of war in Europe in 1939 led to only a slow mobilization of science in the United States. But it did induce physicist entrepreneurs such as Ernest Lawrence to seize the opportunity to tap military funds for basic research—including theoretical work. In early 1941 a small fission research committee had already assigned Lawrence the task (and the funds) of investigating the electromagnetic separation of fissionable U-235 from uranium ore. Oppenheimer, like many others at the time, regardless of their political orientation, readily joined the military research effort. Oppenheimer worked with Lawrence on the electromagnetic separation of uranium isotopes. After he displayed his command of fission theory during an early conference on bomb physics, Arthur Compton appointed him head of fast-neutron theory at Berkeley (on Lawrence's recommendation). In May 1942, Oppenheimer succeeded Gregory Breit as director of all fission theory in the United States. By then, with the graduation of thirteen more doctorates in just three years, the Oppenheimer school had trained and prepared a critical mass of young theorists at Berkeley and Caltech who were now willing and able to form the nucleus of research for what would become the Manhattan Project. That group, augmented by others, laid the foundations and set the agenda for the Manhattan Project and beyond. Even more to the point, the project and its need for theorists helped to catapult Oppenheimer into a position of national leadership and helped to transform his California school into an American school of theoretical physics.

As the coordinator of American nuclear bomb theory in 1942, Oppenheimer's task was again closely connected to experimental data. He needed to compare theoretical calculations on the operation of a fission bomb with known data on nuclei in order to assess the feasibility of a national effort to build the bomb. During the summer of 1942, Oppenheimer secretly assembled a small research group in his office in Le Conte Hall to explore the prospects. The nine participants represented a cross-section of the American theoretical physics community at that time: two American-born leaders —Oppenheimer and John Van Vleck; four European émigrés—Hans Bethe, Edward Teller, Felix Bloch, and Emil Konopinski; and three members of Oppenheimer's school —Robert Serber, Eldred Nelson, and Stanley Frankel.

The Oppenheimer team thoroughly examined every phase of bomb development —the critical masses of uranium and plutonium needed to sustain an explosive chain

33. See the essay by Karl Hufbauer in this volume.

reaction, the speeds with which the sub-critical masses must be brought together in order to trigger each type of bomb, the size and weight of the actual bombs, and the destructive energy released.[34]

Theoretically at least, the achievement of fission bombs of uranium and plutonium seemed to the researchers completely unproblematic. Only high technical hurdles, especially regarding the plutonium bomb, remained to its actual construction. Having settled that issue, the Berkeley bomb theorists spent most of their summer research exploring an even more powerful weapon—the hydrogen or thermonuclear bomb powered by the fusion process. As the result of astrophysical work during the 1930s, this process was already well known as the source of power for the sun and stars. It was now an interesting and challenging theoretical problem to discover how to unleash these enormous energies on earth. Because the amount released was more than one thousand times the destructive force of a fission bomb, such a bomb was then called the "Super." Teller was instantly and steadfastly intrigued. Creating such a device became his life's project almost from the very start. But the obstacles to a fusion bomb were immense. Because of this, the Berkeley researchers (except for Teller) concluded that the Super should be put aside until after the war.

Meanwhile, Oppenheimer's positive report to the joint military and civilian authorities on the fission bomb was enough to justify a full-scale effort. In September 1942 the Manhattan Project was created to construct the bomb under Oppenheimer's scientific direction, under the overall military command of General Leslie Groves, and under the civilian administrative authority of the University of California.[35] Owing to the new recognition of the contribution of theoretical physics for the military effort on both sides of the war, it is perhaps no coincidence that the scientific heads of the opposing German and Allied projects—Heisenberg and Oppenheimer—were both theorists, and they were both very much alike in more respects than they were different.

By the end of the war, nearly every American theorist of the current and future generation had studied or worked with Oppenheimer in some capacity. Oppenheimer himself was held in such high esteem that he was regarded by his colleagues and the general public as the dean of theoretical physics. And, because of the obvious military success of physics during the war, billions of dollars of military funding began flowing directly into physics research as the nation sunk into a competitive cold war for global supremacy with its former ally, the Soviet Union. Among the direct beneficiaries of such military largesse were Lawrence's Berkeley accelerators and MIT's electronics laboratory, and, with them, related theoretical research. By 1948 Lawrence's military-

34. The summer seminar and the Manhattan Project's physics are described by Serber and Crease (ref. 21), 68-72; Serber, *The Los Alamos primer: The first lectures on how to build an atomic bomb*, Richard Rhodes, ed. (Berkeley, 1992); and Lillian Hoddeson et al., *Critical assembly: A technical history of Los Alamos during the Oppenheimer years* (New York, 1993).

35. Histories of the Manhattan Project include Richard Rhodes, *The making of the atomic bomb* (New York, 1986); and David Hawkins, *Manhattan District history: Project Y, the Los Alamos project* [1961] (Los Angeles, 1983), vol. 1.

funded machines had finally achieved "high energy" with the production of mesons for the first time, and national accelerator laboratories were in the planning stages—all of which required more and better theorists.

At the same time, American theorists finally overcame the infinite hurdles in relativistic quantum mechanics still hindering the profession. This cast them onto the forefront of world research in fundamental physics and reversed the asymmetry between theory and experiment at high energies. Starting in 1947, the series of Shelter Island and Rochester conferences, again under Oppenheimer's overall leadership, were attended by many of Oppenheimer former students and wartime collaborators.[36] Their work brought forth the invention of renormalization, followed by a flurry of Nobel Prizes. Theory would now drive big-machine research in high-energy physics. But of the star performers—Schwinger, Dyson, Marshak, Tomonaga, and Feynman—only Schwinger was a direct product of the Oppenheimer school.[37] Although the school had helped establish the new discipline, it did not dominate the subsequent work of that discipline.

Even as the nation continued to pour millions into bomb and (later) reactor research during the 1950s, Lawrence and his colleagues convinced the authorities to direct large fractions of physics funding into high-energy particle research.[38] Although this policy continued to neglect other fields of practical interest, such as condensed matter and super- and semiconductors, theoretical research and theoretical physics rapidly expanded during the pre-Sputnik years of the 1950s, becoming the powerful force it is today.

As J. Robert Oppenheimer and others like him shifted the focus of their work during the war from the unraveling of arcane experimental results to practical military applications, American theoretical physics, and the nation itself, gained new levels of prestige and world significance. Then, fed by massive infusions of postwar American military money, talent, and ambition on the world stage, the ground prepared by the Oppenheimer school and the Manhattan Project blossomed during the early cold war into the distinctive American School of Theoretical Physics that has enjoyed unprecedented success in these and other areas of research down to the present.

36. Silvan S. Schweber, "Shelter Island, Pocono, and Oldstone: The emergence of American quantum electrodynamics after World War II," *Osiris, 2* (1986), 265-302; and Schweber, *QED and the men who made it* (Princeton, 1994).
37. This was also pointed out by Carson (ref. 23).
38. See AEC budget tables in Richard G. Hewlett and Jack M. Holl, *Atoms for peace and war, 1953-1961: Eisenhower and the Atomic Energy Commission* (Berkeley, 1989), 576-577.

Appendix. Oppenheimer's doctorates and their dissertation titles, as compiled by Raymond T. Birge. *Source*: BHDP, vol. 3, XIIIa-c.

1. Melba N. Phillips, May 6, 1933
   Problems in the Spectra of the Alkalies
   I. Photoionization Probabilities in Atomic Potassium
   II. Theoretical Considerations on the Inversion of Doublets in Alkali-like Spectra

2. Arnold T. Nordsieck, May 11, 1935
   The Scattering of Radiation by an Electric Field

3. Glen D. Camp, May 14, 1935
   Is a Relativistic Non-Conservative Theory of Mechanics Possible?

4. Willis E. Lamb, Jr., April 18, 1938
   A. On the Capture of Slow Neutrons in Hydrogenous Substances
   B. On the Electromagnetic Properties of Nuclear Systems

5. Samuel B. Batdorf, April 23, 1938
   An Investigation into the Penetrating Possibilities of Charged Particles of Arbitrary Magnetic Moment

6. Sidney M. Dancoff, Sept. 13, 1939
   Three Problems in Quantum Mechanics:
   A. Virtual State of He^5 and Meson Forces
   B. On Radiative Corrections for Electron Scattering
   C. The Calculation of Internal Conversion Coefficients

7. George M. Volkoff, Feb. 27, 1940
   I. The Equilibrium of Massive Neutron Cores
   II. The Oppenheimer-Phillips Process

8. Philip Morrison, April 23, 1940
   Three Problems in Atomic Electrodynamics
   A. Internal Conversion of Gamma Rays of Arbitrary Multiple Order
   B. Internal Scattering of Gamma-Rays
   C. Energy Fluctuations in the Electromagnetic Field

9. Hartland S. Snyder, April 25, 1940
   Five Problems: Cascade Theory; Quadratic Zeeman Effect; Gravitational Collapse; Mesotron Collisions; Energy Levels of Fields

10. Joseph M. Keller, Sept. 13, 1940
    Precise Determination of the Fine Structure Constant from X-Ray Spin Doublet Splitting

11. Robert F. Christy, April 22, 1941
    Cosmic-Ray Burst Production and the Spin of the Mesotron

12. Eugene P. Cooper, August 25, 1941
    Three Problems in Quantum Theory
    I. Internal Scattering of Gamma-Rays
    II. The Ground States of Be^{10} and C^{10}
    III. On the Separation of Nuclear Isomers

13. Shuichi Kusaka, Feb. 3, 1942
    Studies on the Spin of Elementary Particles
    A. Electric Quadrupole Moment of the Deuteron
    B. The Interaction of Gamma-Rays with Mesotrons
    C. Burst Production of Mesotrons
    D. Beta-Decay with Neutrino of Spin 3/2
    E. Quantization of the Wave Field for Particles with Higher Spin

14. Richard R. Dempster, March 10, 1942
    A. The Calculation of Transition Probabilities for Photoionization of Sodium from the 3P-State
    B. The Influence of Fluorescence upon the Central Intensities of the Solar D-lines

15. Roy Thomas, April 13, 1942
    Three Problems in Quantum Mechanics
    I. Internal Pair Production in Radium C′
    II. Burst Production by Mesotrons of Spin One-half and Zero Magnetic Moment
    III. High Energy Proton-Deuteron Scattering

16. Eldred C. Nelson, April 24, 1942
    Mesotrons and Nuclei
    I. Internal Conversion of Gamma-Radiation in the L Shell
    II. Isomeric Silver and the Weizsäcker Theory
    III. Note on the Neher-Stever Experiment
    IV. The Ground States of Be^{10} and C^{10}
    V. Pseudoscalar Mesotron Theory of Beta-Decay

17. Bernard Peters, May 8, 1942
    - I. Deuteron Disintegration by Electrons
    - II. Scattering of Mesotrons of Spin ½

18. Edward Gerjuoy, May, 1942
    Problems in Atomic and Nuclear Spectroscopy
    - A. On the Angular Distribution in the Reaction F^{19} (p, alpha)
    - B. Interference in the Zeeman Effect of Forbidden Lines
    - C. Tensor Spin-Orbit Forces in H^3 and He^4

19. Stanley P. Frankel, July 20, 1942
    Three Problems in Theoretical Physics (classified)

20. Chaim Richman, April 13, 1943
    A Problem in Theoretical Physics (classified)

21. Joseph W. Weinberg, June 4, 1943
    Studies in the Quantum Field Theory of Elementary Particles

22. David J. Bohm, June 5, 1943
    Theoretical Investigation of Scattering of Various Nuclear Particles (classified)

23. Leslie L. Foldy, June 15, 1948
    Four Studies in Theoretical Physics
    - I. The Theory of the Synchrotron
    - II. Theory of the Synchro-Cyclotron
    - III. On the Meson Theory of Nuclear Forces
    - IV. The Energy-Momentum Relation for Particles Interacting with Fields

24. Harold W. Lewis, June 16, 1948
    Three Problems in Theoretical Physics
    - I. The Multiple Production of Mesons
    - II. On the Reactive Terms in Quantum Electrodynamics
    - III. On the Analysis of Extensive Cosmic Ray Shower Data

25. Siegfried A. Wouthuysen, June 16, 1948
    Self-Energy and Covariance in Field Theory

CHAPTER TWO

J. Robert Oppenheimer's Path to Black Holes

Karl Hufbauer

J. Robert Oppenheimer's two papers with students Michael Volkoff and Hartland Snyder on the collapse of massive stars (into what have been called "black holes" since 1967[1]) are sometimes regarded as his most impressive piece of research. Jeremy Bernstein even characterizes that with Snyder as "one of the great papers in twentieth-century physics."[2] The present study provides a historical account of Oppenheimer's stellar theorizing before the Second World War. I first explore the backgrounds for Oppenheimer's interest in the physics of stars. Then I delineate his immediate path from a curiosity about the source of stellar energy to his and his students' engagement with the esoteric problem of stellar collapse. Finally I consider why their results engendered virtually no immediate follow-up.

Readers will soon discern that, as historical narratives go, the one I present has a high proportion of conjecture to fact. The primary reason is that this thread in Oppenheimer's scientific life, despite stretching from the mid-1920s to 1939, was extremely slender. Indeed, his interest in the physics of stars only added three papers—all of which appeared during 1938-1939—to his publication list. A supplemental

My thanks to David DeVorkin, Donald Osterbrock, and Bonnie Ludt for information regarding Oppenheimer and others mentioned in this study, to David Cassidy for discussion of Oppenheimer's reasons for getting involved in stellar theorizing, to Robert Smith and David Kaiser for alerting me to the prevalent lack of interest among physicists in general relativity in the late 1930s, and to Cathryn Carson for thoughtful editing.

The following abbreviations are used: *AMS*, *American men of science*; *ApJ*, *Astrophysical journal*; *CITWC*, California Institute of Technology, *Weekly calendar*, Archives, California Institute of Technology; *DSB*, *Dictionary of scientific biography*; HBP, Hans Bethe Papers, Special Collections, Cornell University Library; JRO, J. Robert Oppenheimer; *MNRAS*, Royal Astronomical Society, *Monthly notices*; *PASP*, Astronomical Society of the Pacific, *Publications*; *PR*, *Physical review*; RAC, Rockefeller Archive Center, Sleepy Hollow, NY; RCTP, Richard Chace Tolman Papers, Archives, California Institute of Technology; TBL, The Bancroft Library, University of California, Berkeley; *UCC*, University of California, *Calendar*, TBL.

1. For John Wheeler's coining of the term, see Kip Thorne, *Black holes and time warps: Einstein's outrageous legacy* (New York, 1994), 256-257.

2. Bernstein also speculates that had Oppenheimer "lived longer" these two papers might have been "honored...by a Nobel Prize"—see Bernstein, *Oppenheimer: Portrait of an enigma* (Chicago, 2004), viii, 48. As a caveat to Bernstein's view, I would note that those who selected the roughly one thousand seminal papers that appeared in *Physical review* during the journal's

reason is that Oppenheimer and his friends were disinclined as a matter of personal style and political prudence to preserve their letters and other manuscripts from the 1930s. Despite being rather conjectural, my narrative is more thoroughly grounded in interwar sources than the accounts of Oppenheimer's stellar theorizing recently offered by Jeremy Bernstein[3] and Kip Thorne.[4] It establishes that Oppenheimer's contacts with astronomy were increasingly substantial prior to his research on the problem of stellar collapse in 1938-1939. It also places these contacts and his stellar research in their astrophysical context. In making these contributions, it illuminates some of the broader patterns in Oppenheimer's physical theorizing.

Backgrounds for Oppenheimer's Stellar Theorizing

That Oppenheimer ever paid more than passing heed to the stars requires explanation. Relatively few of the many scientists who swarmed into theoretical physics during the interwar decades published on problems belonging to the nearby astronomical fields of theoretical astrophysics and cosmology.[5] Those who did so—we will encounter several in this story—had usually gotten acquainted with physics' value as an interpretive tool from specialists in these fields or from theoretical physicists who had earlier succeeded in bringing physics to bear on astronomical problems. Moreover, they had often built records as agile competitors who were on the prowl for problems with high importance-to-difficulty ratios where they could demonstrate their virtuosity. The backgrounds for Oppenheimer's involvement in stellar theorizing were similar.

In all likelihood, Oppenheimer got introduced to the idea that physics holds the key to the mysteries of the stars at Cambridge University during his 1925-1926 sojourn there as a postgraduate student. He seems—as will become evident—to have learned something then about Plumian Professor Arthur Stanley Eddington's recent, and cogent, introduction into stellar theory of the view that all ordinary stars, despite

---

first century (H. Henry Stokes, ed., *The* Physical review*: The first hundred years, A selection of seminal papers and commentaries* [Woodbury NY, 1995]) did not consider these papers, nor any of Oppenheimer's other five papers included in the overall CD collection, to be among the 200 most seminal papers republished in this work's paper version. See xvi, xix, 1288.

3. Ibid., 43-50. Cf. Bernstein, "The reluctant father of black holes," *Scientific American, 274:6* (1996), 80-85.

4. Thorne (ref. 1), 187-197, 212-220. Thorne's account of Oppenheimer's stellar theorizing gives a compelling portrayal of Oppenheimer's research style in attacking the problem of stellar collapse, which is based not only on Oppenheimer's three articles but also on Richard Tolman's relevant articles and archived papers and his interviews with Robert Serber, George Volkoff, and William Fowler.

5. Karl Hufbauer, "Dopo la vittoria: Le valutazioni degli addetti ai lavori sulla fisica teorica (1928-1939)," Giovanni Battimelli, Mario De Maria, and Arturo Rossi, eds., *La ristrutturazione delle scienze tra le due guerre mondiali* (2 vols., Rome, 1985), vol. 2, 81-87.

having a mean density greater than water, are essentially balls of gas.[6] Moreover, Oppenheimer's chief mentor was Ralph Howard Fowler,[7] a mathematical physicist who had begun to follow up on Eddington's breakthrough.[8] Oppenheimer's awareness of physics' value to astronomy would have deepened if, while working with Max Born at Göttingen University the following year, word reached him of Fowler's brilliant application of the new Fermi-Dirac statistics to interpret the very high densities of white dwarfs.[9] If this news did arrive, he might well have sought out and perused Eddington's recent monograph *The internal constitution of the stars* or its semi-popular sequel *Stars and atoms*.[10]

It is a few years later that we can see in Oppenheimer the first actual traces of a confidence in physics' usefulness to astronomy. In winter-spring 1929, having taken his Ph.D. at Göttingen, spent a postdoctoral year at Harvard and the California Institute of Technology, then returned to Europe as an International Education Board fellow, Oppenheimer was working with Wolfgang Pauli in Zurich. The one paper that he wrote there examined the characteristics of the radiation from a free electron that passed at relativistic velocity near an atom stripped of almost all its electrons.[11] He took up this problem because of its relevance to Pauli's current efforts with Werner Heisenberg to develop relativistic quantum mechanics. But, imbued with an "empiricist temper" by his earlier contacts at Harvard with Percy Bridgman, he wanted to give his results an evidentiary base. There being no pertinent laboratory data, he turned to what was known at the time about conditions in stellar interiors.[12] Oppenheimer's

6. Karl Hufbauer, "Eddington, the mass-luminosity relation, and the stellar-energy problem (1922-1926)," msc., 1995.

7. In his application (7 Mar 1928) for an International Education Board fellowship, Oppenheimer named Fowler as the supervisor of his Cambridge research in 1926 on quantum mechanics. See Oppenheimer's IEB folder, RAC, 51.3, box 64, 1095. For other evidence of their contacts in Cambridge, see Alice Kimball Smith and Charles Weiner, eds., *Robert Oppenheimer: Letters and recollections* (Cambridge, 1980), 96-97; and David C. Cassidy, *J. Robert Oppenheimer and the American century* (New York, 2004), 96, 102.

8. For Fowler's work as a theoretical physicist and contributions to theoretical astrophysics, see Subramanyan Chandrasekhar, "Ralph Howard Fowler, 1889-1944," *ApJ, 101* (1945), 1-5; and William McCrea, "Sir Ralph Howard Fowler, 1889-1944: A centenary lecture," Royal Society, London, *Notes and records, 47* (1993), 61-78.

9. Fowler, "Dense matter," *MNRAS, 87* (Dec 1926), 114-122. While in Göttingen, Oppenheimer wrote at least one letter to Fowler. See Smith and Weiner (ref. 7), 101.

10. To judge from the dates of the earliest reviews, Eddington's classic monograph appeared around November 1926 and the popularization around March 1927. Eddington's short appendix to the second book gives an account of white dwarfs, including Fowler's interpretation of them.

11. JRO, "Über die Strahlung der freien Elektronen im Coulombfeld" (submitted 6 May 1929), *Zeitschrift für Physik, 55* (5 Jul 1929), 725-737.

12. For Pauli's contemporary characterization of the paper, see Pauli to Arnold Sommerfeld, 16 May 1929, Armin Hermann, Karl v. Meyenn, and Victor F. Weisskopf, eds., *Wolfgang*

paper started by drawing attention to the bearing of his problem on Eddington's finding that the ability of the highly ionized atoms in stellar interiors to impede the outward flow of radiation—i.e., the stellar plasma's opacity—was ten times higher than the value derived from Hendrik Kramers's 1923 theory. Oppenheimer's up-to-date quantum-mechanical treatment of the problem indicated that free electrons in the stellar environment would radiate away energy much more rapidly than predicted by Kramers's semi-classical analysis. Accordingly, they would be captured more frequently, putting a larger fraction of the ionized atoms into states in which they could easily undergo photoelectric interactions and thereby increase the opacity. He concluded that his results were of "the right order of magnitude [to] account for the continuous absorption in the interior of a star."[13]

The following fall and winter, after returning to California for a split appointment at the University of California and Caltech, Oppenheimer got involved in helping two colleagues with astronomically-oriented projects. One such relationship was with Richard Chace Tolman, an eminent mathematical physicist at Caltech. Inspired by Edwin Hubble's discovery of the distance-redshift relation for galaxies, Tolman had embarked on a major study of the bearing of general relativity and thermodynamics on cosmology.[14] Oppenheimer, who had become friendly with Tolman during his postdoctoral months at Caltech in 1928,[15] read three longish manuscripts for him, providing helpful comments on sections in two of them.[16] It was probably somewhat later that Oppenheimer began giving assistance to Donald Menzel, a theoretical astrophysicist at

---

Pauli: Wissenschaftlicher Briefwechsel mit Bohr, Einstein, Heisenberg u. A. (New York, 1979), vol. 1, *1919-1929*, 500-503. For Oppenheimer's empirical orientation and this paper's place in his move to the forefront of theoretical research, see S.S. Schweber, "The empiricist temper regnant: Theoretical physics in the United States 1920-1950," *Historical studies in the physical and biological sciences, 17:1* (1988), 55-98, esp. 84-91; Schweber, *In the shadow of the bomb: Oppenheimer, Bethe, and the moral responsibility of the scientist* (Princeton, 2000), 65-66; and Cassidy (ref. 7), 109, 128-130.

13. JRO, "Über die Strahlung" (ref. 11), 725, 736-737, cited Eddington's *Internal constitution of the stars,* chapts. 9 and 10.

14. For a brief overview, see Judith R. Goodstein, "Tolman, Richard Chace," *DSB, 13* (1976), 429-450; and for his bibliography and a reporter's portrayal of his work in relativistic cosmology, see, resp., John G. Kirkwood *et al.*, "Richard Chace Tolman, 1881-1948," National Academy of Sciences, *Biographical memoirs, 26* (1952), 138-153; and Bernard Jaffe, *Outposts of science: A journey to the workshops of our leading men of research* (New York, 1935), 506-516.

15. Smith and Weiner (ref. 7), 134; and Judith R. Goodstein, *Millikan's School: A history of the California Institute of Technology* (New York, 1991), 140f. For other evidences of Oppenheimer's closeness to Tolman between 1929 and World War II, see Smith and Weiner (ref. 7), 137, 143, 182, 194f; Cassidy (ref. 7), 142, 187; and Robert Serber with Robert P. Crease, *Peace and war: Reminiscences of a life on the frontiers of science* (New York, 1998), 35.

16. Tolman thanked Oppenheimer for help in his papers "On the use of the energy-momentum principle in general relativity" (submitted 30 Dec 1929), *PR,* 35 (15 Apr 1930), 875-895, on 895; and "On the weight of heat and thermal equilibrium in general relativity," ibid., 904-924, on 919.

UC's Lick Observatory. Menzel was in the midst of using the observatory's sizeable collection of eclipse spectra to investigate physical conditions in the sun's chromosphere.[17] Oppenheimer not only showed him how to bring quantum mechanics to bear on the analysis of the data but also helped him assess the physical meaning of the results.[18]

Oppenheimer's references to Eddington's classic monograph and his interactions with Tolman and Menzel indicate that he was at least moderately attentive to theoretical physics' relevance to astronomy by 1930. He probably learned that year from reading *Nature* that Eddington's regnant stellar model was being vigorously challenged by the theoretical astrophysicist Edward Arthur Milne.[19] In June 1931, when his Cambridge mentor Fowler was visiting Pasadena, Oppenheimer surely learned more about this raucous dispute. Fowler sided with Milne, who was arguing that ordinary stars possess extremely hot cores with temperatures of billions of degrees. Taking the opposite view, Menzel's Princeton doctor-father Henry Norris Russell presented an argument that, though striving to strike a balance, ultimately stood behind Eddington's orthodox view that such stars are simply gas balls with central temperatures in the tens of millions of degrees.[20] Oppenheimer might well have picked out the basic physical differences between the two models from Fowler and Russell's reports. It is unlikely, however, that he anticipated at this time that he would ever get involved with stellar theory in more than incidental ways.

To understand why Oppenheimer would briefly, yet intensely take up such research in the later 1930s, we need to consider his broader scientific-intellectual trajectory. He was among the dozen or so most promising individuals in the large cohort that

17. Donald Osterbrock, "Young Don Menzel's amazing adventures at Lick Observatory," *Journal for the history of astronomy*, *33* (2002), 95-118.
18. Menzel acknowledged Oppenheimer's help in his papers "The deviation of stellar atmospheres from thermodynamic equilibrium," *PASP*, *42* (Apr 1930), 113-117, on 115; "Pressures at the base of the chromosphere: A critical study of Milne's theories," *MNRAS*, *91* (Apr 1931), 628-652 on 629, 652; and "On the broadening of lines in stellar spectra," *PASP*, *44* (Aug 1932), 260-261, on 261. During their discussions, the ebullient Menzel probably told Oppenheimer of his interest in the stellar-energy problem, since he referred to Oppenheimer's research in his paper "Annihilation of matter as the source of stellar energy," *PASP*, *43* (June 1931), 191-202, on 196.
19. Milne issued his challenge in "The masses, luminosities, and effective temperatures of the stars," *MNRAS*, *90* (Nov 1929), 17-54; and took it to the semi-popular level with his note "Stellar structure and the origin of stellar energy," *Nature*, *126* (16 Aug 1930), 238.
20. For a contemporary report on the June multi-society in Pasadena where the papers were given, see J.H. Moore, "The meeting of the Astronomical Society of the Pacific held in conjunction with the Pasadena meeting of the American Association for the Advancement of Science," *PASP*, *43* (Aug 1931), 278-281. Contrary to this report, Russell was unable to attend and so had his paper read by Theodore Dunham, Jr. For this correction and the background of the meeting, see David H. DeVorkin, *Henry Norris Russell: Dean of American astronomers* (Princeton, 2000), 240-243. A week before, Fowler spoke on "Milne's theory and the interior of stars," Mt. Wilson Observatory, 11 June 1931 (see *CITWC*).

entered theoretical physics during and just after the quantum mechanical revolution of the mid-1920s. Along with most others in this elite group, he worked both on the profoundly difficult question of uniting quantum mechanics and special relativity in their extension to field theory and on the comparatively easy task of using quantum mechanics to tackle a variety of more-or-less routine problems. Like the rest, he failed in the first endeavor and had numerous lesser successes in the second in his roughly twenty papers from 1930 through 1937.[21]

When choosing the specific problems upon which he published during these years, Oppenheimer apparently had various considerations in mind. He generally aspired to have at least one project going that would be of interest to his elite peer group. He also kept an eye on the novel results coming from the experimental studies of nuclear particles and cosmic rays at Berkeley and Caltech.[22] He needed as well to be on the lookout for problems that matched the proclivities and talents of those in his growing coterie of graduate students and postdoctoral fellows.[23]

Oppenheimer's research went well during these years. But its cumulative impact fell ever shorter of realizing the promise that others and he himself had seen in him in the 1920s.[24] Moreover, it was often bedeviled by minor yet irritating errors.[25] Oppenheimer's frustration at not living up to his own and others' expectations[26] contributed from about 1935 onward, I suspect, to the intensification of his interest in realms beyond physics proper—philosophy, literature, and especially left-wing politics. It may even have contributed to his dalliance with stellar theorizing in the late 1930s.[27]

Oppenheimer's Path *into* Stellar Theory, 1933-1938

It was in April 1933 that Oppenheimer gave the first evidence that he might someday take up stellar theorizing. Shortly after making his annual pilgrimage from Berkeley to Pasadena, he scheduled a talk on "Stars and nuclei" at the Mt. Wilson-Caltech Astronomy and Physics Club.[28] His title, a variant of Eddington's acclaimed *Stars and atoms* (1927), reflected his own and some of his Caltech colleagues' growing

21. Cassidy (ref. 7), 160-173. For a list of Oppenheimer's articles, see Smith and Weiner (ref. 7), 359-361.
22. Cassidy (ref. 7), 153. However, for evidence of Oppenheimer's unwillingness to limit himself and his students to serving as Ernest Lawrence's house theorists, see J.L. Heilbron and Robert W. Seidel, *Lawrence and his laboratory: A history of the Lawrence Berkeley Laboratory*, Vol. 1 (Berkeley, 1989), 256-258.
23. Schweber, "empiricist" (ref. 12); and Cassidy (ref. 7), 133-157.
24. Oppenheimer file (ref. 7).
25. Schweber, *In the shadow* (ref. 12), 68; and Cassidy (ref. 7), 118, 179.
26. Cassidy (ref. 7), 139, 171, 179.
27. Ibid., 181-210.
28. His talk was scheduled for 14 Apr 1933. See *CITWC*.

interest in nuclear phenomena.[29] It also indicated that he would focus on stellar interiors where, according to the prevailing view, energy is liberated by reactions between atomic nuclei stripped of their electrons by frequent collisions at the high ambient temperatures. If, as I suppose, his emphasis was on the stellar-energy problem,[30] he probably pointed out that adherents of Eddington's stellar model, which estimated stellar central temperatures in the tens of million degrees, had to identify one or more specific proton- or neutron-induced thermonuclear reactions that were capable of powering the sun and other stars at the low nuclear kinetic energies implied by such temperatures.[31] By contrast, those loyal to Milne's model—notably his acquaintance Theodore Eugene Sterne, another Fowler student—treated energy production in the stars as an incidental by-product of the myriad of thermonuclear reactions that could occur under the central temperatures in the billions of degrees envisaged by most proponents of that model.[32] Given Oppenheimer's ties to Fowler, it is conceivable that he presented the second of these approaches as the more promising one.

For the next four and a half years, we have no further evidence of Oppenheimer's incipient interest in the physics of the stars. His only encounters then that might have sustained this interest were with the émigré theoretical physicist George Gamow, who had dabbled from to time in stellar theory since 1928. Their first extended time together was in 1934, when both taught at the University of Michigan's annual summer Symposium on Theoretical Physics.[33] Oppenheimer just might have gotten a preview there of Gamow's forthcoming talk on "stellar nuclei and transformation of elements in stars," which was based on Lev Landau's suggestion that all stars have dense cores consisting entirely of neutrons (i.e., a variant of Milne's model).[34] In the summer of 1936, about

29. Shortly before Oppenheimer's talk, Charles Lauritsen's nuclear-physics group used Caltech's small accelerator to generate a neutron beam. See Goodstein (ref. 15), 233.
30. My conjecture about the focus of his talk is based on the fact, discussed below, that some four years later he set his student Volkoff to work on the stellar-energy problem.
31. For the research of Robert d'Escourt Atkinson, who was working within Eddington's framework, and its reception, see Karl Hufbauer, "A physicist's solution to the stellar-energy problem, 1928-1935," msc., 1997.
32. Sterne, "The equilibrium theory of the abundance of chemical elements" (submitted 15 Mar 1933), *PR, 43* (1 Apr 1933), 585-586; "The equilibrium theory of the abundance of the elements: A statistical investigation of assemblies in equilibrium in which transmutations occur" (submitted 15 Mar 1933), *MNRAS, 93* (Oct 1933), 736-767; and "A note on the liberation of energy by transmutations of nuclei in the stars" (submitted 30 Mar 1933), ibid., 767-769. For evidence of their acquaintance as sailing companions, see JRO to Sterne, 27 Mar 1936, Raymond T. Birge Papers, TBL, folder Oppenheimer. For Sterne's ties to Fowler, under whom he took a Ph.D. in 1931, see his NRC Fellowship folder, RAC.
33. Smith and Weiner (ref. 7), 181.
34. For the origins of Gamow's involvement in stellar theorizing, see Hufbauer (ref. 31), 7. His continuing interest in stellar theory before mid-1934 is indicated by an anonymous news report concerning a Leningrad talk he gave in support of Landau's neutron-core model of stars—see "Conservation of energy law declared unnecessary," *Science news letter, 22* (17 Dec

when Gamow was deciding to get more deeply involved in stellar theorizing, Oppenheimer again spent time with him first at the Seattle meeting of the American Physical Society and later at the family ranch in New Mexico.[35] This summer he doubtless heard more, or for the first time, about Gamow's attempts to develop Landau's model.[36]

When, in the course of the ensuing year, Oppenheimer learned of his new graduate student George Michael Volkoff's side-interest in astronomy,[37] he evidently saw an opportunity to follow up on his own side-interest in stellar theory. Volkoff was set to work on a review of the literature dealing with the stellar-energy problem. By early December 1937 he was ready to speak in the Berkeley Physics Seminar on "The source of stellar energy."[38] As Volkoff did not have a fresh solution to propose for this hoary problem, it is likely that he appraised the two main lines of attack. If so, he could have illustrated progress within the framework of Eddington's model by focusing on the recent, and promising, identification of several energy-generating nuclear reactions advanced by Heisenberg's protégé Carl Friedrich von Weizsäcker.[39] And to discuss developments within Milne's general framework, he could have reviewed the work of Sterne and Gamow.

Four or five months after Volkoff's seminar, Oppenheimer and his Pasadena colleagues William Fowler (Caltech) and Rudolph Minkowski (Mt. Wilson Observatory) arranged a session on "Nuclear transformations and their astrophysical significance" for the June meeting of the AAAS in San Diego. Oppenheimer chose "The physical

1932), 385-386—and Gamow and Landau, "Internal temperature of stars," *Nature, 132* (7 Oct 1933), 567. His lecture was the last of three that he gave at the University of London in October. See the flyer in his personnel file, George Washington University, Washington, D.C.

35. Paul Kirkpatrick, "Minutes of Seattle meeting, June 17-19," *PR, 50* (1936), 384-393; and Serber and Crease (ref. 15), 39-40.

36. Gamow's ongoing support for Landau's model is evident in "Nuclear transformations and the origin of the chemical elements," *Ohio journal of science, 25* (Sep 1935), 406-413; and the second edition of his monograph on nuclear physics, which was completed on 1 May 1936, *Structure of atomic nuclei and nuclear transformations* (Oxford, 1937), 231-238.

37. For Volkoff (1914-2000), see UC Berkeley Graduate Division, "Programme of the final examination for the degree of Doctor of Philosophy of George Michael Volkoff" (1940), in Volkoff thesis, UC Berkeley Library; *AMS*, 8th edn. (1949), 2586; and Erich Vogt, "George Michael Volkoff," *Physics today, 53:11* (2000), 76. Volkoff's attendance in 1934 at meetings of the Vancouver branch of the Royal Astronomical Society of Canada attests to an early interest in astronomy. See minutes of meetings for 13 Feb and 11 Dec 1934, in the *Journal of the Royal Astronomical Society of Canada, 28* (1934), 233-234 and *29* (1935), 38-39.

38. *UCC, 87:15* (29 Nov 1937), 1.

39. Weizsäcker, "Über Elementumwandlungen im Innern der Sterne. I" (submitted 22 Jan 1937), *Physikalische Zeitschrift, 38* (15 Mar 1937), 176-191. Among the many favorable reactions to this paper, the two most likely to have been seen by Volkoff were [William McCrea?], "Nuclear reactions in stars," *Nature, 139* (26 Jun 1937), 1116; and Henry Norris Russell, "The fuel of the stars," *Scientific American, 157* (Jul 1937), 12-13.

problem of stellar energy" as the title for his own talk.[40] He was probably motivated to help make these arrangements by signs that his elite peers in theoretical physics were zeroing in on it. During February through May 1938, new studies of the problem by Landau and Gamow, as well as a report on a major theoretical conference on "Nuclear processes as the source of stellar energy" that was organized by Gamow and Edward Teller and attended by, among others, Hans Bethe and John von Neumann, appeared in *Nature* and the *Physical review*.[41] Oppenheimer was most impressed—as we will soon see—by Landau's proposal that energy generation in stars results from the gravitational accretion of ordinary stellar matter onto their neutron cores.[42]

Shortly before the meeting in San Diego, Oppenheimer learned that Bethe was working on the stellar-energy problem. He quickly inquired about Bethe's progress, admitting that his own group's inability to come up with a plausible solution based on nuclear-reaction physics had led him to believe "that...one will be forced to [Landau-type] core models...in spite of many difficulties with them."[43] Bethe, who was just then winding up a preliminary paper for the *Physical review*, replied with an account of his carbon-cycle theory of energy generation within ordinary stars having the internal constitution predicted by Eddington's model.[44] According to an official account of the San Diego symposium, Oppenheimer in his talk not only[45]

> considered... [t]he details of the theory of the possible nuclear changes in the lighter elements and the possibility of their application to the interior of stars [but also] suggested a new model with a high central concentration of neutrons.

In his letter thanking Bethe for a quick response, Oppenheimer provided a somewhat fuller report. He granted that the relative abundances of carbon-12 and carbon-13

40. For the session's announcement, see "The San Diego meeting of the Society," *PASP, 50* (June 1938), 181-182, on 181.
41. News of elite interest in the stellar-energy was spread not just by the articles of Landau, "Origin of stellar energy," *Nature, 141* (19 Feb 1938), 333-334; Gamow, "Nuclear energy sources and stellar evolution," *PR, 53* (1 Apr 1938), 593-604; Gamow and Edward Teller, "The rate of selective thermonuclear reactions," ibid., 608-609; and the conference report of Chandrasekhar, Gamow, and Merle Tuve, "The problem of stellar energy," *Nature, 141* (28 May 1938), 982, but also by correspondence and word of mouth.
42. Oppenheimer's openness to Landau's paper is impressive in light of the fact that Landau had recently ventured into the pages of the *Physical review* to criticize his work. See Landau, "Stability of neon and carbon with respect to α-particle disintegration," *PR, 52* (15 Dec 1937), 1251; and JRO and Serber, "Note on boron plus proton reactions," *PR, 53* (15 Apr 1938), 636-638, on 637.
43. JRO to Bethe, 13 June [1938], HBP.
44. I infer the content of Bethe's letter, which has not survived, from Oppenheimer's response, JRO to Bethe, 24 June [1938], HBP. For Bethe's paper with Teller's student Charles Critchfield, see their article "The formation of deuterons by proton combination," *PR, 54* (15 Aug 1938), 248-254. They acknowledged Oppenheimer's comments regarding one point (on 250).
45. "The San Diego meeting of the Astronomical Society of the Pacific, June 22-23, 1938," *PASP, 50* (Aug 1938), 210-211, on 210.

supported Bethe's "regenerative cycle." Nonetheless, he doubted that "the astronomers would admit a high enough concentration of N in the sun, or a low enough concentration of H in Capella, to make your explanation tenable." He also thought there was "bad trouble" with Bethe's view that H could easily be captured by boron-10 to produce carbon-11. Turning from these misgivings about Bethe's theory, Oppenheimer admitted that "core models" also had serious problems. In particular, Landau had underestimated the "minimum stable core mass" by a factor "of about a thousand." Even for a star as massive as Capella, the core plus "the surrounding Fermi zone" of white-dwarf material would play havoc with the observational mass-luminosity-radius relation.[46]

Oppenheimer and Robert Serber, his postdoc since 1934 and close friend,[47] devoted some time that summer to looking further into the credibility of Bethe's putative solution of the stellar-energy problem and of Landau's claims for stellar neutron cores. Oppenheimer concluded both that Bethe had "scooped" him, as Serber put it,[48] and that Landau had indeed overlooked significant difficulties. In early September, Serber gave a mainly positive review of Bethe's theory in the Berkeley Physics Seminar,[49] and Oppenheimer and Serber mailed a one-page note "On the stability of stellar neutron cores" to the *Physical review*.[50] Their note readily acknowledged the success of Bethe's theory, remarking "that the formation of deuterons by proton collision, and at the least partially regenerative capture of protons by elements between carbon and oxygen [can] be made to account successfully for the main sequence stars." But, they opined, as the theory did not account for the energy output of very massive stars, serious consideration needed to be given to Landau's variant of Milne's alternative to the "standard Eddington model."

Landau had estimated that if a neutron core's self-gravity was to keep it from exploding the core must have a minimum mass somewhere between 5% and 0.1% of the sun's mass.[51] Using different assumptions about the nuclear force involved, Oppenheimer and Serber calculated a stable neutron core's minimum mass as being between 10 and "a few percent" of the sun's mass. However, the white-dwarf material that would surround a 10% solar-mass core would, in the case of the sun, nearly use up the remainder of the mass, confronting observers with a star far different than that so successfully represented by Eddington's model. Indeed, the mass of such a quantum-

46. JRO to Bethe (ref. 44). Oppenheimer seems to have been drawing here on the ongoing canvassing of the stellar-energy problem by his student Volkoff and postdoc Serber. See Serber and Crease (ref. 15), 49-50. For further evidence of Oppenheimer's initial skepticism about Bethe's theory, see JRO to Fowler, 9 Sep [1938], William Fowler Papers, Archives, California Institute of Technology.
47. For Serber, see Robert P. Crease's thoughtful "Introduction" as well as Serber's memoir, Serber and Crease (ref. 15).
48. Ibid., 49.
49. Serber, "Stellar energy," *UCC, 89:3* (5 Sep 1938), 1. I infer the character of his talk from Serber and Crease (ref. 15), 49-50.
50. *PR, 54* (1 Oct 1938), 540. They completed the note on 1 Sep 1938.
51. Landau (ref. 41).

mechanically degenerate envelope would not fall below that of the core unless—which they very much doubted—one of less than 3% solar mass could form. Their conclusion was "that even in the heaviest stars no core will be formed until practically all sources of nuclear energy have been, at least for the central material of the star, exhausted."[52]

Oppenheimer's Research on Stellar Collapse, 1938-1939

Having been scooped by Bethe on the question of energy generation, Oppenheimer might easily have set aside his stellar theorizing at this juncture. Instead, he turned in early fall 1938 to investigate what ultimately became of massive stars that had used up their thermonuclear reserves. Given his theoretical peers' unawareness of this issue and the astronomers' inability to provide any data whatsoever about such objects, Oppenheimer's decision was somewhat out of character. Two mutually reinforcing considerations probably led him in this direction. Oppenheimer surely knew that his student Volkoff, who had already been looking into the energy-generation problem for a year, would have found it disheartening to give up entirely on stellar theorizing. More important, in late-summer conversations with Tolman,[53] Oppenheimer learned that his friend had a new interest in bringing general relativity to bear on the stars. Tolman most likely went beyond reporting this interest to inform Oppenheimer that it had been stimulated by a request from their Caltech colleague Fritz Zwicky for theoretical assistance on the problem of stellar collapse.[54] If so, Oppenheimer would not have been happy to hear of the brilliant yet irascible Zwicky's interest in this question.[55] Still, he seems readily to have seen that taking up stellar collapse with the tools of general relativity offered the possibility of moving beyond his own and Serber's neutron-core results, which, like Landau's before them, had relied on Newtonian gravitational theory. He and Tolman soon agreed on the equations that needed solving and on the desirability of keeping one another apprised of their respective endeavors.

Presuming that a massive star that had spent itself would collapse into a neutron star,[56] Oppenheimer apparently began by inquiring whether the mass of such a star,

52. My reading of JRO and Serber's note draws heavily on Thorne (ref. 1), 190-191.
53. Oppenheimer usually stayed a few days with the Tolmans in Pasadena when driving back from the ranch in New Mexico to Berkeley. For oblique hints about the character of his talks with Tolman, see Tolman to JRO, 19 Oct 1938, RCTP, 3.20.
54. Tolman may also have reported that Zwicky had already submitted an article on the subject of stellar collapse. For Zwicky's anticipation of Tolman's assistance and of a future joint article, see Zwicky, "On collapsed neutron stars" (submitted 8 Aug 1938), *ApJ, 88* (Nov 1938), 522-525, on 523. Cf. Karl Hufbauer, "Zwicky, Fritz," *DSB, 18:supp. II* (1990), 1011-1013.
55. Oppenheimer's failure ever to acknowledge Zwicky's 1934 proposal with Walter Baade that supernovae engendered neutron stars was one indication that he did not take Zwicky seriously. Cf. Thorne (ref. 1), 178, 191-192, 196.
56. So far as I know, Oppenheimer always referred to what we call "neutron stars" as "neutron cores." His dislike of Zwicky probably accounts for his unwillingness to use Zwicky's terminology.

on the assumption that it was stable, had an upper limit and, if so, what it would be. A rough calculation using Newtonian gravity to counterbalance the outward pressure of a degenerate, almost liquid-like neutron gas indicated that neutron stars might be stable up to six solar masses, but no further. To refine this estimate, Oppenheimer needed to use general relativity to get a better estimate of the confining gravitational force within the non-Euclidean space near such a dense star's center. He and Volkoff first established the approximating equations needed for numerical work. Then Volkoff set about the laborious job of using a calculating machine to assess the stability of model neutron stars of different mass.

Tolman, who supposed that Oppenheimer was "trying to get approximate solutions for liquids of a prescribed physical character," was meanwhile on the hunt for "exact solutions for liquids of the physical character needed to get a mathematically simple result." In mid-October he wrote Oppenheimer about his success in coming up with such an ad hoc fluid. He had, he reported, found

> an exact solution for a sphere of compressible fluid, which exhibits no paradoxes. The solution is of course still a very special one, since the [fluid's] 'equation of state'...is not general, and the small 'compressibility'...is not independently adjustable.

Tolman hoped "to get a more general solution." Nonetheless, if this first solution survived Oppenheimer's scrutiny, he was considering sending it to the National Academy of Science's *Proceedings* for publication.[57]

Impressed by Tolman's solution, Oppenheimer discussed it with him in Pasadena sometime in the next week or so. Tolman reported that he had "found two more possible solutions for the gravitational field of a static sphere of perfect fluid." Not long afterwards, he sent Oppenheimer the details of the more interesting of these two new solutions so that they could discuss it during a brief visit he would soon be making to Berkeley. This solution had, in Tolman's view, two chief advantages. It was "not very complicated," as it relied on an equation of state for the fluid for which the density and pressure "decrease[d] continuously...out to the boundary of the sphere." And it had "the 'agreeable properties,' that a distant external observer would find [that] the gravitational mass [went] to infinity with the size of the sphere [and] that processes at the center [took] place at an infinitely slow rate."[58] Tolman's emphasis in this letter on how such a stable star would appear to a remote observer might well have led Oppenheimer, when he started thinking in the next month or so about non-static solutions to Tolman's equations, to begin considering how the process of relativistic collapse would appear from the distinct vantage points of distant and local comoving observers.

57. Tolman to JRO, 19 Oct 1938 (ref. 53).
58. Tolman to JRO, 9 Nov 1938, RCTP.

Sometime in early December, with the two efforts coming to a head, Oppenheimer, Volkoff, and Tolman met in Berkeley to talk over the results and publication. Their discussions apparently focused on Volkoff's surprising finding that the maximum mass of an idealized neutron star without nuclear forces was about 0.75 solar mass and on ways to bring Tolman's more general results to bear on actual stars.[59] They also seem to have agreed then to ask the *Physical review* to publish the mutually-dependent papers side by side, with Tolman's followed by Oppenheimer and Volkoff's.[60] In their paper Oppenheimer and Volkoff developed two potentially interesting lines of argument about the fate of massive stars that had consumed all "thermonuclear sources of energy." First, thanks to Volkoff's number crunching with the approximate equations and Tolman's analysis of the general equations, they were confident that, if neutron stars ever formed, their mass could not exceed roughly 0.75 solar mass. If so and if Landau was also correct in arguing that only stars above 1.5 solar mass could collapse beyond the white-dwarf stage, then "it seem[ed] unlikely that static neutron [stars] can play any great part in stellar evolution." In short, Oppenheimer and Volkoff concluded that, although old stars might occasionally end up as stable neutron stars, they very rarely did so. Second—and this result Oppenheimer seems to have regarded as their really interesting finding—they came away from the investigation believing that the usual fate of very massive stars was unceasing contraction. Since all stars possessing more than 1.5 solar mass would quickly exhaust their thermonuclear reserves, and since almost none, if any, would end up as stable neutron stars, most should "contract indefinitely, although more and more slowly, never reaching true equilibrium." Hence, in closing their paper, Oppenheimer and Volkoff announced that the "non-static solutions" to Tolman's relativistic equations which would provide an "understanding of the long time behavior of actual heavy stars...are now being investigated."[61]

In the first months of 1939, despite the excitement then among physicists about the discovery of nuclear fission, Oppenheimer undertook a closer study of the non-static solutions of Tolman's equations for massive stars. His collaborator was Hartland Snyder, a fourth-semester graduate student who seems to have gotten the nod because of exceptional mathematical ability.[62] In early February, Oppenheimer reported in a letter

59. The preceding account of Oppenheimer's path and collaborations during fall 1938 takes as its starting point Thorne's reconstruction of the story (ref. 1, 191-196). Also see Volkoff, "On the equilibrium of massive neutron cores" (given on 19 Dec 1938 at a meeting of the American Physical Society at UCLA), *PR, 55* (15 Feb 1939), 421-422.

60. Tolman, "Static solutions of Einstein's field equations for spheres of fluid," *PR, 55* (15 Feb 1939), 364-373; and JRO and Volkoff, "On massive neutron cores," *PR,* 55 (15 Feb 1939), 374-381. For a brief follow-up note, see Volkoff, "On the equilibrium of massive spheres" (submitted 28 Jan 1939), *PR,* 55 (15 Feb 1939), 413.

61. JRO and Volkoff (ref. 60), on 374, 380-381.

62. For Snyder (1913-1962), see UC Berkeley Graduate Division, "Programme of the final examination for the degree of Doctor of Philosophy of Hartland Sweet Snyder" (1940), in Snyder thesis, UC Berkeley Library; *AMS*, 8th edn. (1949), 2334; "Hartland S. Snyder," *Physics*

to his old friend George Uhlenbeck on the prior study with Volkoff and the ongoing study with Snyder, noting that "We have been working...on static and nonstatic solutions for very heavy masses that have exhausted their nuclear energy sources: old stars perhaps which collapse to neutron cores. The results have been very odd."[63] It seems that Oppenheimer suspected that the stellar material which entered into indefinite contraction was a massive neutron core. The project did not go swiftly, partly because of Oppenheimer's concurrent interests in fission, mesotrons, and left-wing politics, and partly because of the undertaking's intrinsic difficulty. Nonetheless, in early May 1939, Oppenheimer probably alluded to their study when speaking about "Stellar energies" at a Caltech Physics Research Conference.[64] Later that month, Snyder definitely reported on it in Caltech's Seminar on Theoretical Physics.[65] Around the same time, Oppenheimer may have discussed it with Tolman when commenting on a manuscript dealing with stellar stability and the origin of novae.[66]

Oppenheimer and Snyder brought their investigation to a close toward the end of June 1939. Introduced by Oppenheimer, Snyder gave a brief report on their work at a Stanford meeting of the American Physical Society.[67] Shortly thereafter, they submitted a succinct manuscript to the *Physical review*.[68] Their subject was the fates of stars above 0.7 solar mass that had "used up their nuclear sources of energy." If, during such a star's ensuing gravitational collapse, the star lost sufficient mass in one of three ways—ejection, transformation of mass into radiation, or rotational fission—then "the remaining matter could form a stable...white dwarf star." However, Oppenheimer and Snyder's focus lay elsewhere—on stars that did not lose enough mass to become white dwarfs.[69] In particular, simplifying the situation, they concentrated on the ideal case of

---

today, 15:7 (1962), 78-79; and Robert P. Crease, *Making physics: A biography of Brookhaven National Laboratory, 1946-1972* (Chicago, 1999), 124, 129, 144-145. For Oppenheimer's choice of Snyder, see JRO to Tolman [ca. Jan 1939], RCT. In this letter, which was written as Oppenheimer was beginning to look for a "general [non-static] solution [to] your equation," he reported that "I've showed it to Snyder and to Schwinger, who are both pretty expert." My guess is that Snyder ended up as Oppenheimer's collaborator because he was more willing than Schwinger to respond to the challenge. Cf. Thorne (ref. 1), 212; and Serber and Crease (ref. 15), 46.

63. Smith and Weiner (ref. 7), 208-209.

64. *CITWC*. Oppenheimer's use of "Energies" in his title suggests that he spoke about more than Bethe's highly acclaimed theory of energy generation in main-sequence stars.

65. Ibid.

66. Tolman acknowledged their discussion in his paper "On the stability of stellar models, with remarks on the origin of novae," *ApJ*, *90* (Nov 1939), 568-600, on 600.

67. Snyder, "Limiting solutions for collapsing masses" (talk given ca. 1 Jul 1939), *PR*, 56 (15 Oct 1939), 857.

68. JRO and Snyder, "On continued gravitational contraction" (received 10 Jul 1939), *PR*, *56* (1 Sep 1939), 455-459.

69. Ibid., 455. Note that they envisioned only two fates for stars above 0.7 solar mass, giving

a collapsing massive star without positive radial pressure. They interpreted the outcome of their analysis from two vantage points—that of "a local observer [who was] comoving with the stellar matter" and that of "a distant observer." To a local observer, "the total time of collapse" for a star of roughly the sun's mass would be "of the order of a day." By contrast, to "an external observer...the star [would] asymptotically shrink... to its gravitational radius [as it] close[d] itself off [from the rest of the universe except for its] gravitational field." These were indeed, as Oppenheimer had earlier confided to Uhlenbeck, odd results. In concluding, Oppenheimer and Snyder granted that "actual stars would collapse more slowly than [their idealized star] because of the effect of the pressure of matter, of radiation, and of rotation." Nonetheless, the basic agreement of their "qualitative" and "quantitative" arguments for the results gave them confidence that the predicted behavior "will be realized by all collapsing stars which cannot end in a stable stationary state."[70]

His paper with Snyder was Oppenheimer's last direct contribution to stellar theory. As he neared completion of his earlier papers with Serber and Volkoff, he had discerned obvious starting points for the next ones in line. Such was not the case with his paper "on continued gravitational contraction." So he dropped this line of research.

Epilogue, 1939-1940

The work by Oppenheimer and his disciples during 1938-1939 on the evolution of exhausted stars received little serious attention at the time. Oppenheimer's friend Tolman, who had provided crucial assistance, did write a follow-up paper, but he made no explicit reference to Oppenheimer's articles with Volkoff and Snyder.[71] Oppenheimer's difficult Caltech colleague Zwicky, who was also thinking about neutron stars, cited an abstract and note by Volkoff.[72] However, he passed over the articles for which Oppenheimer was a co-author, repaying Oppenheimer in kind for ignoring his priority. Harlow Shapley and his Harvard College Observatory colleagues, who by then included Menzel and Sterne, proposed Oppenheimer and Tolman for participation in a conference on stellar theory to be sponsored by the New York Academy of Sciences; but financial considerations seem to have ruled out invitations to them.[73] Further afield, William McCrea, a leading British expert on relativity, wrote

---

no hint here or later in the paper that they believed in the existence of stable neutron stars or cores. Also note that they avoided speculating about the fraction of massive stars that ended up on the path of indefinite contraction.

70. Ibid., 455-456, 459.

71. Tolman (ref. 66).

72. Zwicky, "On the theory and observation of highly collapsed stars," *PR, 55* (15 Apr 1939), 726-743, on 729, 742.

73. To judge from Shapley's files, it was in early spring 1939 that Oppenheimer and Tolman began to be considered as possible participants in a stellar-theory conference that would be

a fine abstract of Oppenheimer and Volkoff's work that took explicit notice of their brief remarks about non-static solutions.[74] Nonetheless, possibly because of the onset of war, he seems never to have followed up on this abstract or Oppenheimer and Snyder's ensuing study. One minor German physicist wrote an able abstract of Oppenheimer-Volkoff and another a perfunctory one of Oppenheimer-Snyder.[75] Unlike them, the astronomer Otto Heckmann of Göttingen was well-equipped to follow up on Oppenheimer's work; however, he also limited himself to a perfunctory abstract.[76] Finally and by contrast with all the preceding, Landau was reportedly so impressed with Oppenheimer and Snyder's study that he immediately added it to his "Golden List" of classic papers.[77]

Especially in comparison to the immediate reactions to Bethe's two principal papers on the stellar-energy problem, this was a decidedly meager response. Bethe's papers, which appeared in the *Physical review* on August 15, 1938, and March 1, 1939, had been widely hailed as a breakthrough by the first anniversary of the second paper.[78] By contrast, Oppenheimer's two papers on stellar collapse, which appeared in the same venue on February 15, 1939, and September 1, 1939, had—with the exception of Landau—been all but forgotten by the first anniversary of his second paper.

There were, I believe, five reasons for this early neglect of Oppenheimer's papers. First, unlike Bethe, who addressed a well-defined problem in a field with a dozen or so gate-keepers and a large following, Oppenheimer was tackling a problem that had only began to come to the fore as a result of Bethe's breakthrough. Second, unlike Bethe, who was able to draw on a fairly well-known and robust body of nuclear data and theory, Oppenheimer was invoking the sensational and well-tested but little-used theory of general relativity. Third, unlike Bethe, whose solution to the stellar-energy problem was fairly straightforward, Oppenheimer was offering a solution to the stellar-collapse problem that was completely counterintuitive. Fourth, unlike Bethe,

---

sponsored that fall by the New York Academy of Sciences. They were still being proposed as participants as late as early October, but only if travel expenses were available. See Harlow Shapley Papers 1930-40, Harvard University Archives, folder N.Y. Ac. Scis. According to David DeVorkin's reading of E.T. Miner (of the NYAS) to Russell, 5 Oct 1939, such monies turned out not to be at hand.

74. McCrea, abstract of JRO and Volkoff (ref. 60) and Volkoff (ref. 60), *Zentralblatt für Mathematik und ihre Grenzgebiete, 20:6* (3 Jul 1939), 284-285.

75. K. Burckhardt, *Physikalische Berichte, 20:15* (1 Aug 1939), 1658; and K. Bechert, ibid., *21:4* (15 Feb 1940), 443-444.

76. Heckmann, abstract of JRO and Snyder (ref. 68), *Zentralblatt für Mathematik und ihre Grenzgebiete, 22:6* (12 June 1940), 281-282.

77. Thorne (ref. 1, 219), who learned of this exceptional early interest in Oppenheimer-Snyder during an interview (1982) with Landau's associate Evgeny Mikhailovich Lifschitz, gives Landau's enthusiasm about the paper credit for engendering Russian theorizing about black holes in later decades.

78. Karl Hufbauer, "Bethe reception chron [1938-1942]," msc., 1998.

who reached out to potential audiences with citations, letters, talks, a successful prize application, and press interviews, Oppenheimer turned his attention to other matters and let his difficult papers fend for themselves. Last, unlike Bethe, who published his second paper a full half year before war began in Europe, Oppeheimer published his just as it began. But this seemingly plausible external explanation is nonetheless far less important than the others.[79] Ultimately, the reasons why Oppenheimer's work dropped like a stone lay with the state of play in the field and Oppenheimer's own lack of follow-through.

My account of this strand in Oppenheimer's research on stellar theory buttresses prior interpretations of his research career before World War II. It illustrates not only his virtuosity but also his good relations with his Caltech colleague Tolman and with his Berkeley coterie of graduate students and postdocs. In this instance and others, he clearly made the most of his joint appointment at Caltech and Berkeley. It also reveals that he began thinking about having a go at the stellar-energy problem at least as early as 1933, and it supports Serber's suggestion that he only took up the research that led him to black holes because Bethe scooped him on the stellar-energy problem. Finally, it gives a reading of Oppenheimer's three stellar-theory papers of 1938-1939 that, while drawing heavily on Thorne's insightful account, also signals Oppenheimer's increasingly strong doubts about the existence of stable neutron stars.

79. Those who are skeptical about this claim should see my account of the reception of Bengt Edlen's breakthrough on the coronal-line problem (published 1941-1942), Karl Hufbauer, "Breakthrough on the periphery: Bengt Edlén and the identification of the coronal lines, 1939-1945," Svante Lindqvist, ed., *Center on the periphery: Historical aspects of 20th-century Swedish physics* (Canton, MA, 1993), 199-237.

PART TWO

The Communist Question

Two big questions arise whenever anyone mentions Oppenheimer's relationship with Communism. First, was he ever a Communist, even in the very late 1930s and early 1940s? Second, does it matter, and if so, exactly how? The first question quickly leads to others, on which often turn the answers people give to the second question. When we ask if Oppenheimer was a Communist, do we mean actual membership in the Communist Party of the United States or something a bit different: a feeling of solidarity with the Party expressed through a willingness to work with some of its members?

A remarkable feature of the scholarly and popular discussion of Oppenheimer's relation with Communism is that the topic continues to generate enormous interest despite an almost total agreement that Oppenheimer was not a spy. There is no credible evidence that Oppenheimer ever transferred any technically sensitive information whatsoever to any person in contact with Soviet intelligence. This consensus has led some to conclude that whether or not Oppenheimer was Communist in *any* sense matters not at all. Among the historians who believe it does matter is Gregg Herken, who in one of the essays in this section explains why and how. Herken argues that Oppenheimer was almost certainly a secret member of the Communist Party.

Juxtaposed to Herken in this section are contributions by three other scholars who, like Herken, have performed exhaustive research on Oppenheimer's political activities and his personal relationship with those of his friends, relatives, and acquaintances who did join the Communist Party or strongly identified with it. Kai Bird and Martin Sherwin find inconclusive the evidence that persuades Herken of Oppenheimer's secret party affiliation. They detail a different picture, placing Oppenheimer close to the Party but not within it. Barton J. Bernstein wraps up the section with an overview of the controversy, woven together with methodological reflections on how his opinion has changed over time.

CHAPTER THREE

Was Robert Oppenheimer a "Closet Communist"? *The Debate and the Evidence*

Gregg Herken

Historians are always happy when their discoveries lead to a new and better understanding of their subject. Even more gratifying is when their work provokes new discoveries. After my book *Brotherhood of the bomb: The tangled lives and loyalties of Robert Oppenheimer, Ernest Lawrence, and Edward Teller* was published, in September 2002, my claim that Robert Oppenheimer and Haakon Chevalier had been members of a closed unit of the professional section of the Communist Party in Berkeley, during the years 1938-42, ignited some controversy.[1]

Although Chevalier later argued that their membership in this unit was something that he and "Oppie" had "least to be ashamed of," Oppenheimer steadfastly denied that he had ever "been a member of a Communist Party unit."[2] Because it seemed impossible at the time to prove who was right, in the first edition I described this group as a "rather naïve political coffee klatch," while noting that Chevalier considered it "something much more..."[3]

Since then, however, new evidence has appeared that, in my view at least, finally settles the matter—and proves that Chevalier, not Oppenheimer, was telling the truth.

In December 2002, the children of Gordon Griffiths, a former professor of history at the University of Washington who had been a graduate student at Berkeley in the late 1930s, donated their father's unpublished memoir—"Venturing outside the

The following abbreviations are used: JRO, J. Robert Oppenheimer; JROP, J. Robert Oppenheimer Papers, Manuscript Division, Library of Congress, Washington, D.C.

1. See, for example, "Book contends chief of A-bomb team was once a Communist," 8 Sep 2002, *New York Times*; "Oppenheimer bombshell," 24 Apr 2004, *San Francisco chronicle*; "The strange case of Robert Oppenheimer" and "The Oppenheimer case: An exchange with Gregg Herken," *New York review of books*, 4 Dec 2003 and 25 Mar 2004.

2. Chevalier to JRO, 23 Jul 1964, and JRO to Chevalier, 7 Aug 1964, JROP; also reproduced under "Documents" at www.brotherhoodofthebomb.com.

3. Gregg Herken, *Brotherhood of the bomb: The tangled lives and loyalties of Robert Oppenheimer, Ernest Lawrence, and Edward Teller* (New York, 2002), 31.

ivory tower: The political autobiography of a college professor"—to the Library of Congress. Griffiths died in January 2001, and had written the memoir near the end of his life as a kind of *apologia pro vita sua*, his children noted.[4]

Griffiths tells in the manuscript of how he had served, between fall 1940 and spring 1942, as the liaison contact between the Bay Area's Communist Party and "the Faculty group at the University of California." Griffiths confirms that Oppenheimer, Chevalier, and an English Department colleague, Professor Arthur Brodeur, were all members of the unit, and likewise that Oppie took "special pride" in his contributions to the unit's two "Reports to our colleagues," which were published in 1940 and signed "College Faculties Committee, Communist Party of California."[5]

Members of the Party's professional sections in closed units were not publicly identified with the Party—i.e., were not "card-carrying" Communists—since their affiliation was deliberately kept secret, not only to avoid compromising their usefulness to the movement, but also to protect the members themselves from reprisals.[6] While those belonging to the professional section were perhaps not subject to traditional Party discipline, in the same way as the Party's rank-and-file members, there can be little doubt that representatives of America's wartime counter-espionage apparatus, including the FBI and the U.S. Army's Counter-Intelligence Corps, regarded involvement in a secret Party unit as tantamount to Communist Party membership, and Robert Oppenheimer was surely aware of that fact.[7]

4. "Gregg Herken's book *Brotherhood of the bomb* has reopened the question of whether Oppenheimer was ever a member of the Communist Party. We are in possession of some material that bears on this issue, and—for whatever it is worth—thought we should make it available to responsible historians," their donation letter begins.

5. In an unpublished memoir, Chevalier claimed that Oppenheimer had had a major role in writing these "Reports," and likewise paid for their printing and distribution to colleges and universities along the West Coast. Chevalier confirmed Brodeur's membership in the closed unit in a 1973 letter to a researcher. The first "Report," portions of Griffiths's manuscript, and Chevalier's 1973 letter are reproduced under "Documents" at www.brotherhoodofthebomb.com.

6. "Nobody had cards in their own names. It was too dangerous." The existence and operations of the Communist Party's closed units was described in an interview with Edith Jenkins, who was the organizer for the Communist Party's professional sections in the Bay Area during the early to mid-1930s. Edith Jenkins, interview with Gregg Herken, 9 May 2002, National Security Archives, George Washington University, Washington, D.C. Jenkins, who chose the name of Mary Shelley's mother, Mary Wollstonecraft, for her "Party name," is also the author of a book describing her friendship with Robert Oppenheimer and Jean Tatlock. Edith Jenkins, *Against a field sinister: Memoirs and stories* (San Francisco, 1991).

7. As was Haakon Chevalier. In a 1959 *roman à clef* about his friendship with Robert Oppenheimer, Chevalier wrote of "Sebastian," the Oppenheimer character: "In any case, if he had regarded himself as a communist—which was rather in the perspective of history than in the context of a specific time and place—he had not been a communist in the commonly accepted sense, and technically had in fact not been a communist at all....So far as his actual involvement with the party was concerned, therefore, he could consider himself reasonably safe." Haakon Chevalier, *The man who would be God* (New York, 1959), 234. Chevalier had

Perhaps anticipating the argument that Oppenheimer was not a "real" Communist—since he was not a card-carrying member of the Party, nor did he pay dues, at least in the conventional manner—Griffiths wrote in his memoir: [8]

> Nobody carried a party card. If payment of dues was the only test of membership, I could not testify that Oppenheimer was a member, but I can say, without any qualification, that all three men considered themselves to be Communists.

The "Was Oppenheimer a Communist?" debate received fresh ammunition in spring 2005 with the publication of *American Prometheus: The triumph and tragedy of J. Robert Oppenheimer*, by Tufts University historian Martin Sherwin and biographer Kai Bird. In their book, Bird and Sherwin acknowledge the possibility—even the likelihood—that Oppenheimer had at one time been a "closet Communist," but they conclude that "we do not, and we cannot, know the extent of his commitment."[9]

While there is much that is new about Oppenheimer in Bird and Sherwin's admirable and extensive biography, their discounting of the importance of Oppie's prewar membership in a secret and closed Party unit—on the grounds that Oppenheimer was never actually a card-carrying Party member, and hence was not subject to the Party's traditionally strict discipline—misses the point that, to the FBI and the Army of that day, including General Groves, even a "closet" Communist represented a real security risk.[10] Consequently, Oppenheimer's need to hide his membership in the unit when he was chosen to head the bomb project at Los Alamos, and his later lying about it to security officials, gave him a secret he would—and did—keep unto death.

Ultimately, what is the historical significance of the fact that Oppenheimer was a Communist—"closet" or otherwise; *de facto* if not *de jure*—between the outbreak of the war in Europe and the Japanese attack on Pearl Harbor?

Former FBI Director J. Edgar Hoover spent much of his career attempting to convince Americans that communism was synonymous with treason. Regrettably, he

---

originally intended to write about his and Oppenheimer's membership in the closed unit in his 1965 memoir, *Oppenheimer: The story of a friendship*, until Oppenheimer discussed with his attorney the possibility of enjoining publication of the book. Chevalier to JRO, 23 Jul 1964, and JRO to Chevalier, 7 Aug 1964 (ref. 2).

8. Gordon Griffiths, "Venturing outside the ivory tower: The political autobiography of a college professor," on deposit with JROP.

9. Kai Bird and Martin J. Sherwin, *American Prometheus: The triumph and tragedy of J. Robert Oppenheimer* (New York, 2005), 152.

10. Among the discoveries in the book is new information about Paul Crouch, a former Communist Party member and subsequent FBI informant who identified Oppenheimer as the host of a secret Party meeting at Oppie's home in Berkeley during the summer of 1941. However, like the brief meeting a few months later between physicists Niels Bohr and Werner Heisenberg which inspired Michael Frayn's play *Copenhagen*, the so-called Kenilworth Court incident is likely to remain a demonstration of the fact that indeterminancy is a phenomenon in history as well as physics.

succeeded with all too many people—as Griffiths himself discovered, encountering "bitter hostility" from old friends to the revelations in his memoir concerning Oppenheimer. It was presumably this reaction which prompted Griffiths to write:

> [Oppenheimer's] defenders have always stoutly denied that he was ever actually a member of the Communist Party, and have sought to explain away his association with leftist causes with the assertion that he was politically naïve. A great deal of energy was spent by well-intentioned liberals who felt that this was the only way to defend his case. Perhaps at the time—at the height of the McCarthyite period—it was....But the time has come to set the record straight, and to put the question as it should have been put: not whether he had or had not been a member of the Communist Party, but whether such membership should, in itself, constitute an impediment to his service in a position of trust.

In this passage, written perhaps a decade ago, Griffiths has posed what, in my opinion, is the *real* question that should be asked about Oppenheimer and his past.

Since *Brotherhood of the bomb* first appeared, other authors, citing recently-discovered documents from former KGB archives, have contended that Oppenheimer was not only a Communist but a traitor—a Soviet agent, in essence, who willingly and knowingly abetted the U.S.S.R.'s wartime efforts to steal America's atomic secrets. In reality, such documents, I argue, tell us more about the KGB than about "Oppie."[11] There is no convincing evidence that Oppenheimer ever spied for the Soviets, or aided Russian espionage in any way.

Robert Oppenheimer, we now know, was an individual whom J. Edgar Hoover and his ilk would have had Americans believe was an impossibility: a member of a secret Communist Party unit who was also an American patriot. While the "well-intentioned liberals" about whom Griffiths wrote have spent a great deal of time and effort over the past half-century attempting to deny or explain away Robert Oppenheimer's affiliation with the Communist Party, they should now recognize—and rejoice—at this proof that J. Edgar Hoover was wrong.

The importance of Oppenheimer's protracted and previously unproven dalliance with communism lies, therefore, not in the fact of his having been a closet Communist, but in the explanation it offers of Oppenheimer's subsequent behavior. Absent an appreciation of a past that Oppie understandably felt he had to keep hidden, Oppenheimer's actions—particularly at the 1954 loyalty hearing which denied him a security clearance, and after—seem all but inexplicable. Why is it that Oppenheimer failed to defend himself at his loyalty hearing, which was, in effect, a political show trial? Why is it that Oppenheimer, unlike Andrei Sakharov, the Russian physicist-in-exile to whom he is often compared, failed to speak out passionately after that trial for the things he believed in, including an end to the nuclear arms race?

The story of Robert Oppenheimer, as I wrote in the Prologue to *Brotherhood of the bomb,* is, and remains, a "cautionary tale of arrogance, betrayal, and unforeseen

11. See "Author's Notes" and "Documents" at www.brotherhoodofthebomb.com.

consequences." Seen in the light of this new evidence about Oppenheimer's past, however, the outcome seems doubly ironic and tragic, in the classical sense of tragedy. Thus, the lie that allowed Oppenheimer to rise to prominence, as the director of wartime Los Alamos, would later, after the war, become the very weapon his enemies used to bring him down.

CHAPTER FOUR

Robert Oppenheimer and the Communist Party

Kai Bird and Martin J. Sherwin

Was Robert Oppenheimer a member of the Communist Party? Possibly, but the evidence is both scant and vague. Oppenheimer once described his friend Haakon Chevalier as "a parlor pink. He had very wide connections with all kinds of front organizations; he was interested in left-wing writers...he talked quite freely of his opinions."[1] The description might easily have been applied to Oppenheimer himself.

Without question, Robert was surrounded by relatives, friends, and colleagues who at some point were members of the Communist Party. As a left-wing New Dealer, he gave considerable sums of money to causes championed by the Party. But he always insisted that he was never a card-carrying member of the CP. Instead, he said, his associations with the Party were "very brief and very intense."[2] For a period in 1940-41, he participated in regular meetings in which dues-paying Communist Party members discussed current events. These meetings, encouraged and facilitated by the Communist Party, were specifically designed to involve independent intellectuals like Oppenheimer and blur the boundaries of Communist Party identity. But never having been a card-carrying Party member left Oppenheimer the option of deciding for himself how he wished to think of his standing with the Party. For a brief time,

This essay is adapted from *American Prometheus: The triumph and tragedy of J. Robert Oppenheimer*, published by Alfred A. Knopf in April 2005.

The following abbreviations are used: CP, Communist Party; FBI, Federal Bureau of Investigation, records at FBI Headquarters, Washington, D.C.; JRO, J. Robert Oppenheimer; *JRO Hearing*, United States Atomic Energy Commission, *In the matter of J. Robert Oppenheimer: Transcript of hearing before Personnel Security Board and texts of principal documents and letters* (Cambridge, 1971); JROP, J. Robert Oppenheimer Papers, Manuscript Division, Library of Congress, Washington, D.C.; TBL, The Bancroft Library, University of California, Berkeley. Interview transcripts can be found in Martin J. Sherwin's personal collection, soon to be donated to an appropriate archive.

1. Philip M. Stern with Harold P. Green, *The Oppenheimer case: Security on trial* (New York, 1969), 122; Hearings before the Committee on Un-American Activities, 81st Cong., 1st sess., 7 June 1949, National Archives and Records Administration, Records of the U.S. House of Representatives, Record Group 233, HUAC executive session transcripts, box 9, folder Robert Oppenheimer.

2. JRO, interview with Lt. Col. John Lansdale, 12 Sep 1943, *JRO Hearing*, 876.

he may well have thought of himself as an unaffiliated comrade. There is no doubt that in later years he minimized the extent of his associations with the Party. Quite bluntly, any attempt to label Robert Oppenheimer a Party member is an exercise in futility—as the FBI learned to its frustration over many years.

In fact, his associations with various Communists were a natural and socially seamless outgrowth of his place and station in life. As a professor at the University of California in the late 1930s, Oppenheimer lived in a politically active environment. Moving in such circles, he invariably left the impression with many of his friends who were formal members of the Party that he was with them. Robert, after all, wanted to be liked, and he believed in the social justice goals the Party espoused and worked for. His friends could think what they wanted to think. Not surprisingly, some in the Party thought he was a comrade. And naturally, when the FBI used wiretaps to monitor the conversations of these people talking about Oppenheimer, they occasionally heard bona fide Party members claiming him as one of their own. And yet again, other FBI wiretaps record Party members complaining about Oppenheimer's unreliability. Most importantly, there is no evidence that he ever submitted himself to Party discipline. He never had a Party card. He never censored his views to conform to the Party line, although, it must be said, like many Roosevelt Democrats, he agreed with many of the CP's social justice goals. Tellingly, he expressed qualms about the totalitarian nature of the Soviet regime. He openly admired Franklin Roosevelt and defended the New Deal. And while he was a member of various Popular Front organizations dominated by the Communist Party, he was also a staunch civil libertarian and a prominent member of the American Civil Liberties Union. In short, he was a classic fellow-traveling New Deal progressive who admired the Communist Party's opposition to fascism in Europe, and its championing of labor rights at home.

All this ambiguity is compounded by the fact that during these years of the Popular Front, the Communist Party's organizational structure, particularly in California, led to a confusion over the distinction between casual affiliation and actual membership. As Jessica Mitford wrote in her irreverent memoir of her experiences in the San Francisco branch of the Party, "In those days...the Party was a strange mixture of openness and secrecy." The conspiratorial-sounding "cell" of three to five members had been replaced by "branches" or "clubs"—"a nomenclature deemed more consistent with American political tradition." Hundreds of people might belong to these "clubs." Party business in these clubs was conducted in a fairly open and informal manner; everyone was welcome and people, often including FBI informers, attended weekly meetings in rented halls without too much attention to whether their party dues were up-to-date. On the other hand, Mitford reports that she and her husband "were at first assigned to the Southside Club, one of the few 'closed' or secret branches, reserved for government workers, doctors, lawyers, and others whose occupations could have been jeopardized by open affiliation with the Party."[3]

Many left-of-center, pro-union, anti-fascist intellectuals in the late 1930s never affiliated with the Communist Party. And yet, many who did join the Party chose to

3. Jessica Mitford, *A fine old conflict* (New York, 1977), 67.

hide their affiliation even if, like Oppenheimer, they were politically active on behalf of causes supported by the Party. So numerous were the Party's secret members that Communist Party chief Earl Browder griped in June 1936 that too many prominent figures in American society were hiding their Party identity. "How shall we dissipate the Red Scare from among the Reds?" he asked. "Some of these comrades hide as a shameful secret their Communist opinion and affiliations; they hysterically beg the Party to keep as far away from their work as possible."[4]

Years later, Haakon Chevalier insisted that Oppenheimer was one such secret party member. But when closely questioned about the "closed unit" of the Communist Party Robert allegedly belonged to, Chevalier described to Martin Sherwin in a lengthy recorded interview an innocuous gathering of friends more akin to the "discussion group" he reported in his 1965 memoir than an official "closed unit" described by Mitford. "We/he initiated it," recalled Chevalier, referring to Oppenheimer. "It was a closed unit and unofficial. There's no record of it....It was not known to anyone except one person. I don't know who he was, but [he was] in the top echelon of the party in San Francisco." This "unofficial" group known only to "one person" initially contained just six or seven members, though at one point as many as twelve were participating in its discussions. "We discussed things that were going on locally and in the state and in the country and in the world," recalled Chevalier.[5]

Chevalier's version of this *Rashomon* story is reflected in the FBI files. The FBI first opened a file on Oppenheimer in March 1941. His name had come to the Bureau's attention quite by accident the previous December. For almost a year the FBI had been wiretapping the conversations of William Schneiderman, the California Communist Party's state secretary, and Isaac "Pops" Folkoff, the party's treasurer.[6] The wiretaps were unauthorized by any court or by the Attorney General and were therefore illegal. But in December 1940, when one of the Bureau's agents in San Francisco overheard Folkoff referring to a 3:00 PM appointment at Chevalier's home as a meeting of "the big boys," an agent was sent to jot down license plate numbers. One of the cars found to be parked outside Chevalier's home was Oppenheimer's Chrysler roadster. By the spring of 1941 the FBI was identifying Oppenheimer as a professor "reported from other sources as having Communistic sympathies."[7] The FBI noted that he served

4. Harvey Klehr, *The heyday of American Communism: The depression decade* (New York, 1984), 413.

5. Haakon Chevalier, interview with Sherwin, 29 June 1982; see also Chevalier, *Oppenheimer: The story of a friendship* (New York, 1965), 19. Many years after her divorce, Barbara Chevalier noted in her unpublished memoir that Opje and Haakon had "joined a secret unit of the Communist Party. There must have been only six or eight members—a doctor, a wealthy businessman (maybe)...." Barbara noted that she had deliberately not wanted to remember the names of those involved. Barbara Chevalier manuscript, 8 Aug 1981, courtesy of Gregg Herken.

6. Born in Russia in 1905, Schneiderman came to the United States when he was three years old. In 1939, government prosecutors attempted to revoke his citizenship and deport him. The case was still under appeal at the time of his meeting with Oppenheimer; in 1943 the Supreme Court upheld Schneiderman's citizenship. Klehr (ref. 4), 484.

7. FBI report, 19 May 1941 (document 2), and FBI teletype, 16 Oct 1953, San Francisco bureau

on the Executive Committee of the American Civil Liberties Union—which the Bureau labeled "a Communist Party front group." Inevitably, an investigative file was opened on Oppenheimer that would eventually grow to some 7,000 pages. That same month, Oppenheimer's name was put on a list of "persons to be considered for custodial detention pending investigation in the event of a national emergency."[8]

Another FBI document, citing the investigative documents of "T-2, another Government agency," claimed that Oppenheimer was a member of a "professional section" of the Communist Party. One of these "T-2" documents found in Oppenheimer's FBI file included a two-page excerpt from a longer unidentified report listing the party membership of various branches of the Communist Party. Names and addresses are provided for the "Longshoreman's Branch," the "Seaman's Branch," and the "Professional Section." Nine members are listed for this "Professional Section": Helen Pell, Dr. Thomas Addis, J. Robert Oppenheimer, Haakon Chevalier, Alexander Kaun, Aubrey Grossman, Herbert Resner, George R. Andersen and I. Richard Gladstein. Oppenheimer clearly knew some of these individuals (Pell, Addis, Chevalier and Kaun) and it is equally clear that at least some of them were Communist Party members. But it is impossible to evaluate the credibility of this undated document.[9]

According to Chevalier, each member of this alleged "closed unit" paid dues to the Communist Party—except for Oppenheimer. "Oppenheimer paid his separately," Chevalier speculated, "because he probably paid a lot more than he was supposed to." Or, as Robert always insisted, he made contributions to causes, but never paid dues. "But the rest of us paid to one member who was also a known member, an open member [of the Party]," Chevalier continued. "I'm not supposed to say, but it was Philip Morrison." Otherwise, according to Chevalier, the group took no "orders" from the Party and functioned simply as a group of academics who met to share ideas

---

to FBI Director, FBI file of Haakon Chevalier, #100-18564, Part 1 of 2. The cable reports that when Schneiderman and Folkoff arrived, "there were observed parked in Chevalier's driveway cars registered to [blank] and J. Robert Oppenheimer."

8. N.J.L. Piper to Director, FBI, 28 Mar 1941, FBI file of JRO, obtained by Martin J. Sherwin from the FBI under the Freedom of Information Act, #100-17828, section 1, document 1.

9. FBI report, 18 June 1954 by Joe R. Craig, with attachment, "Excerpts from 97-1 (C-14)," FBI file of JRO (ref. 8). The attachment is undated but from the context of the excerpts it must have been written some time after August 1941 when Oppenheimer moved into his home at One Eagle Hill. Oppenheimer had met Helen Pell through their joint activities on behalf of the Committee to Aid Democratic Spain. (Pell was also a good friend of Steve Nelson. Nelson, interview with Sherwin, 17 June 1981.) Dr. Addis, of course, was Jean Tatlock's friend and the man who initially funneled Oppenheimer's donations on behalf of the Spanish Republic to the Communist Party. Alexander Kaun was a Berkeley professor who rented Oppenheimer his house for a time. In 1943 Oppenheimer told Lt. Col. Lansdale that he knew Kaun was a member of the American Soviet Council—but that he did not know whether he was a Party member. *JRO Hearing*, 877. George Andersen was identified as the "official Communist Party Attorney" in San Francisco. Aubrey Grossman and Richard Gladstein were attorneys for the union leader Harry Bridges.

about international affairs and politics.[10] Morrison, of course, has long acknowledged that he joined the Young Communist League in 1938 and the CP itself in 1939 or 1940.[11] When asked about Chevalier's memory, Morrison flatly denied that he had been in a Party unit with Oppenheimer. As a student, he pointed out, he never would have been assigned to a unit with faculty members.[12]

When asked by Sherwin in 1982, "What made you a member of the Communist Party as opposed to just a group of people who were Left?" Chevalier replied, "I don't know. We paid dues." When Sherwin pressed him again, "Did you receive any orders from the Party?" Chevalier said, "No. In a sense we weren't [regular Party members]." At the time, he explained, it was possible for men like himself and Oppenheimer to think of themselves as politically committed intellectuals who were nevertheless free from Party discipline. Members of this group contributed money to the Party's causes; they gave speeches at events sponsored by the Party and they drafted articles and pamphlets for Party publications. And yet, explained Chevalier, "*We both were and were not. Any way you want to look at it.*" Pressed further to explain this ambiguity, Chevalier said, "It had a kind of shadowy existence. It existed, but it wasn't identified, and that had some influence because we had our views about certain things that were happening which were transmitted to the center, and we were consulted about certain things....Apparently, the same thing happened in many other parts of the United States, closed units for professionals or people who didn't want to be identified in any way."[13]

The ambiguous nature of Oppenheimer's relationship to the CP, as described by Chevalier, is corroborated by Steve Nelson, a charismatic Communist Party leader in San Francisco and a friend of Oppenheimer's in the years 1940-43. Nelson saw Oppenheimer socially, but it was also his job to serve as one of the party's liaisons to the university community. "I met socially with this group," Nelson explained in a 1981 interview, "that included some Party members and some non-party people where they discussed freely what's ahead of us....This group was discussing questions of foreign policy. The general mood, which included Oppenheimer's mood, was that it would be tragic if the United States, England and France do not form some kind of alliance against Italy; it would be tragic. I don't remember now whether it was Chevalier or Bob [Oppenheimer] or any other member who expressed himself along these lines. But this was the tone of the meeting."[14]

Nelson reinforced Chevalier's ambiguous description of Oppenheimer's party membership. "I don't know that I could prove or disprove the point," Nelson said.

10. Chevalier, interview with Sherwin, 29 June 1982 (ref. 5).

11. Philip Morrison testimony, 7-8 May 1953, "Subversive influence in the educational process," U.S. Congress, Senate, Committee on the Judiciary, 83rd Cong., 1st sess., Part 9, 899-919.

12. Philip Morrison, interview with Sherwin, 21 June 2002.

13. Haakon Chevalier, interviews with Sherwin, 29 June 1982 (ref. 5), 15 Jul 1982.

14. Nelson, interview with Sherwin, 17 June 1981 (ref. 9).

"So I'll just leave it at that—that he was a close sympathizer. I know that to be a fact because we had a number of discussions of policies of the left....Now that doesn't mean that he was a member of the Party. I think he was a close friend of a number of Party members on campus."[15]

Nelson did not leave the Communist Party until 1957. In 1981 he published a memoir in which he briefly discussed his relationship to Oppenheimer. "My own estimate of Oppenheimer," Nelson remarked, "was that he had this association with the left. Whether one had a Party card or not didn't matter. He was associated with the causes of the left, and that was enough to murder him politically..."[16]

All the members of this allegedly closed Party unit are dead. But one member left behind an unpublished memoir. Gordon Griffiths (1915-2001) had joined the Communist Party in Berkeley in June 1936, just before leaving for Oxford. Upon his return in the summer of 1939, Griffiths quietly renewed his Party membership. But because his wife Mary had become disillusioned with the Party, Griffiths asked for a low-profile assignment. Eventually, he was assigned as "liaison with the Faculty group at the University of California." Griffiths took the assignment in the autumn of 1940 and left in the spring of 1942. In his memoir, he wrote that out of the several hundred faculty members in Berkeley only three were members of this "faculty Communist group"—Arthur Brodeur (an authority on Icelandic sagas and *Beowulf* in the English Department), Haakon Chevalier, and Robert Oppenheimer.[17]

Griffiths acknowledged that Oppenheimer denied that he had ever been a Communist Party member. Oppenheimer's defenders, Griffiths pointed out, have always explained Oppenheimer's fellow-traveling with the assertion that he was politically naïve. "A great deal of energy was spent by well-intentioned liberals who felt that this was the only way to defend his case. Perhaps at the time—at the height of the McCarthyite period—it was....But the time has come to set the record straight, and to put the question as it should have been put: not whether or not he had or had not been a member of the Communist Party, but whether such membership should, in itself, constitute an impediment to his service in a position of trust."[18]

Griffiths's memoir adds few details to Chevalier's description of what he called the 'closed unit.' Understandably, Griffiths clearly believed that Oppenheimer's attendance in these meetings qualified him as a Communist. He wrote that the group met regularly, twice a month, either in Chevalier's home or in Oppenheimer's home. Griffiths

15. Ibid.
16. Ibid. See also Steve Nelson, James R. Barrett, and Rob Ruck, *Steve Nelson: American radical* (Pittsburgh, PA, 1981).
17. Gordon Griffiths, "Venturing outside the ivory tower: The political autobiography of a college professor," unpublished manuscript, shorter version (courtesy of John Earl Haynes, Manuscript Reading Room, Library of Congress), 26. Griffiths produced two versions of this typed manuscript; the shorter, untitled manuscript names Oppenheimer as a member of the closed unit. Oppenheimer's name is not used in the longer manuscript; apparently, when Griffiths began to circulate the manuscript for possible publication, a friend persuaded him that he should not disclose Oppenheimer's name.
18. Ibid.

usually brought along recent Party literature to distribute, and he collected dues from Brodeur and Chevalier, *but not from Oppenheimer*. "I was given to understand that Oppenheimer, as a man of independent wealth, made his contribution through some special channel. Nobody carried a party card. If payment of dues was the only test of membership, I could not testify that Oppenheimer was a member, but I can say, without any qualification, that all three men considered themselves to be Communists."[19]

The faculty group, Griffiths recalled, didn't actually do very much "that could not have been done as a group of liberals or Democrats." They encouraged each other to devote their energies to such good causes as the Teacher's Union and the plight of refugees from the Spanish Civil War. "There was never any discussion of the exciting developments in theoretical physics, classified or otherwise, let alone any suggestion of passing information to the Russians. In short, there was nothing subversive or treasonable about our activity....The meetings were devoted mainly to the discussion of events on the world and national scene, and to their interpretation. In these discussions, Oppenheimer was always the one who gave the fullest and most profound explanations, in the light of his understanding of Marxist theory. To describe his attachment to left-wing causes as the result of political naiveté, as many have done, is absurd, and diminishes the intellectual stature of a man who saw the implications of what was happening in the political world more deeply than most."[20]

Kenneth O. May, the Berkeley CP functionary who had assigned Griffiths to this group, later told the FBI that Haakon Chevalier and other Berkeley professors attended this discussion group but that he "did not consider the people who participated in these gatherings to consist of a CP group."[21] Once a graduate student in Berkeley's math department, Ken May was a friend of Oppenheimer's. In 1936, May had joined the Communist Party and he visited Russia for five weeks in 1937 and again in 1939 for two weeks. He returned enamored with the Soviet model. During the 1940 local elections in Berkeley, May gave a speech before the school board defending the right of local Communist Party candidates to hold a meeting on the grounds of a public school. When the speech drew coverage in the local press, his father—a conservative U.C. Berkeley political scientist—publicly disinherited him and the university cancelled his teaching assistantship. The next year May campaigned as a Communist for a seat on the Berkeley city council.[22] Still a graduate student in the math department, May's affiliation with the Communist Party was no secret when he first met Oppenheimer. A friend of Jean Tatlock's, May was probably introduced to Oppenheimer at a meeting of the teacher's union sometime in 1939.

19. Ibid.
20. Ibid.
21. FBI report of interview with Kenneth May, 5 Mar 1954, FBI file of JRO (ref. 8).
22. Kenneth O. May to Dr. Lawrence M. Gould, President of Carleton College (confidential), 25 Sep 1950, Carleton College archives, Northfield, MN, courtesy of Carleton College archivist Eric Hilleman. May wrote an article in the *New masses* entitled "Why my father disinherited me." David Hawkins, interview with Sherwin, 5 June 1982.

Years later, after he had left the Party, May told the FBI he had visited Oppenheimer's home on several occasions to talk politics and recalled seeing him at "informal gatherings...which were held for the purpose of discussing theoretical questions concerning Socialism." He added that he did not consider Oppenheimer to be either a Party member or someone "under CP discipline." Oppenheimer was an independent intellectual, and as such, May explained to the FBI, "the CP tended to distrust intellectuals as a group in the management of CP affairs, but at the same time, the Party was anxious to influence the thinking of such people along CP lines and to gain the prestige and support of Communist objectives which they would lend to the Party. For this reason, May would keep in touch with the subject [Oppenheimer] and other professional people; he would discuss Communism with them and would provide them with CP literature."[23]

Oppenheimer, May explained to the FBI agents, was the kind of man who was quite willing to "agree with CP aims and objectives at any particular time if he had decided in his own mind that they had merit. He would not, however, condone those objectives with which he did not agree." May observed that "subject openly associated with whomever he pleased, Communists or not."[24]

The FBI would never resolve the question of whether or not Robert was a CP member—which is to say that there was scant evidence that he was. Much of the evidence in the FBI files on Oppenheimer's party membership is circumstantial and contradictory. If a few of the FBI's informants claimed that Oppenheimer was a Communist, most of the Bureau's informants merely painted a portrait of a fellow-traveler. And some emphatically denied that he was ever a Party member. Thus an informant who had been a Party member in Alameda County in 1938-40, and knew Ernest Lawrence, Oppenheimer, and Philip Morrison, told the FBI that it was "an 'open secret' that Morrison was a member of the Campus Professional Group" at the University of California. But T-1—identified as an FBI informant then living in Miami—insisted that Oppenheimer's name "was never mentioned 'even as a possible member' by any members of the Communist Party known to him."[25]

The Bureau had only its suspicions, and the conjecture of others. Only Oppenheimer himself knew—and he always insisted that he had never been a member of the Communist Party.

On August 24, 1939, the Soviet Union stunned the world by announcing that on the previous day it had signed a non-aggression pact with Nazi Germany. One week

23. FBI report of interview with Kenneth May, 5 Mar 1954 (ref. 21).
24. Ibid. May left the Party sometime during World War II. In 1946 he finally obtained his Ph.D in mathematics and that year he joined the mathematics department at Carleton College. John Dyer-Bennet, May's roommate at Berkeley, and Miriam May, May's third wife, interviews with Bird, 15 May 2001.
25. FBI memo to the Attorney General and Director, FBI, 29 Jul 1954, document 1913, section 47, FBI file of JRO (ref. 8).

later, World War II commenced when Germany and the Soviet Union simultaneously invaded Poland. Commenting on these momentous events, Oppenheimer wrote his fellow physicist, Willie Fowler, "I know Charlie [Lauritsen] will say a melancholy I told you so over the Nazi Soviet pact, but I am not paying any bets yet on any aspect of the hocuspocus except maybe that the Germans are pretty well into Poland. Ca stink."[26]

No issue was more vigorously debated within left-intellectual circles than the August 1939 Nazi-Soviet Non-Aggression Pact.[27] Many American Communists resigned from the Party. As Chevalier put it with marked understatement, the Soviet-German pact "confused and upset many people." But Chevalier remained loyal to the CP and defended the pact as a necessary strategic decision. In August 1939 he and four hundred others signed an open letter published in the September 1939 issue of *Soviet Russia today* which attacked the "fantastic falsehood that the U.S.S.R. and the totalitarian states are basically alike."[28] Oppenheimer's name did not appear on the letter.

According to Chevalier, it was in the fall of 1939 "that Opje proved himself to be such an impressive and effective analyst...Opje had a simple, lucid way of presenting facts and arguments that allayed misgivings and carried conviction."[29] At a time when Communists were suddenly extremely unpopular even among Californian intellectuals, he claims that Oppenheimer patiently explained that the Nazi-Soviet pact was not so much an alliance as a treaty of necessity motivated by the West's appeasement of Hitler at Munich.[30]

One evening after midnight, on his way home from a meeting of the League of American Writers, Chevalier dropped by Oppenheimer's home. Robert was still awake, working on a physics lecture. After Robert offered him a drink, Hoke (as he was known to his friends) explained that he needed his help in editing an anti-war

26. JRO to Fowler, 9 Sep 1939, Alice Kimball Smith and Charles Weiner, eds., *Robert Oppenheimer: Letters and recollections* (Stanford, 1995), 211.
27. Maurice Isserman, *Which side were you on? The American Communist Party during the Second World War* (Middletown, CT, 1982), 32-54.
28. *The Nation* reprinted this "Open letter." Stephen Schwartz, *From west to east: California and the making of the American mind* (New York, 1998), 290.
29. Chevalier, *Oppenheimer* (ref. 5), 31-32. In his novel *The man who would be God* (New York, 1959), on 21-22, Chevalier has his Oppenheimer character defend the Stalin-Hitler Pact with these words: "Even in the worst situation," he said in a low voice, "there is a right move, and there are many wrong ones. Since the Western powers violated their pledge to Czechoslovakia in Munich, Russia's situation has been dangerously exposed. This is surely the right move. Because it's the one move that foils the plot of a united attack on the Soviet Union by Germany and a coalition of Western nations—France and England, with American support....The pact is not an alliance with Germany. It is a quarantining of Germany against any combination with the West....This is going to be beastly to explain."
30. Numerous historians have lent credibility to this argument. See Alexander Werth, *Russia at war, 1941-1945* (New York, 1964), 3-39; Peter Calvocoressi and Guy Wint, *Total war: The story of World War II* (New York, 1972), 82.

pamphlet, sponsored by the League. Obliging his friend, Robert sat down and read the manuscript. When he had finished he stood up and said, "It's no good." He told Hoke to sit before his typewriter and then he proceeded to dictate new language. An hour later, Hoke left with "a completely new text."[31]

Robert was not himself a member of the League of American Writers, so what he had done by editing the pamphlet was a favor to his friend. As re-drafted, the pamphlet made an impassioned argument for keeping America out of the European war. Not all, but certainly many New Deal liberals were still anti-war during this period. Still, such arguments persuaded very few intellectuals. By the end of the year, Theodore Dreiser and Richard Wright were two of only a handful of prominent writers who retained their membership in the League of American Writers. Similar Popular Front organizations, like the American League for Peace and Democracy, lost thousands of members. A staunch liberal organization like the American Civil Liberties Union—of which Oppenheimer was a member—dismissed Elizabeth Gurley Flynn, a charter member, from its board of directors because of her membership in the Communist Party.[32] Those intellectuals who chose to remain on the Left thus felt increasingly isolated and marginalized.

Just as Robert had helped Chevalier edit a pamphlet sponsored by the League of American Writers, he may have helped to write or edit with two other pamphlets in February and April 1940. Entitled "Report to our colleagues," these pamphlets were signed "College Faculties Committee, Communist Party of California." Their purpose was to explain the consequences of the war in Europe. Over a thousand copies were mailed to individuals at various universities on the West Coast.

According to Chevalier, Oppenheimer not only drafted the reports, but he also paid for their printing and distribution. Not surprisingly, their discovery—combined with Chevalier's claim—has made them part of the debate as to whether or not Robert was a member of the CP.[33] Gordon Griffiths corroborates Chevalier's assertion of Oppenheimer's involvement with the production of these pamphlets. "They were printed on expensive bond, no doubt paid for by Oppie. He was not their sole author, but he took special pride in them....Free of jargon, these letters were stylistically elegant and intellectually cogent."[34]

31. Chevalier, *Oppenheimer* (ref. 5), 33.
32. Isserman (ref. 27), 38, 42. In 1941 the newly created "Fact-Finding Committee on Un-American Activities"—chaired by California State Senator Jack B. Tenney—held hearings to investigate allegations that the League of American Writers was in fact a Communist front. See Edward L. Barrett, Jr., *The Tenney Committee: Legislative investigation of subversive activities in California* (Ithaca, NY, 1951), 125.
33. Gregg Herken, *Brotherhood of the bomb: The tangled lives and loyalties of Robert Oppenheimer, Ernest Lawrence, and Edward Teller* (New York, 2002), 31; Chevalier, interview with Sherwin, 29 June 1982 (ref. 5); Chevalier, *Oppenheimer* (ref. 5), 35-36. Phil Morrison recalled helping Oppenheimer mail a pamphlet he had written analyzing the Soviet attack on Finland in the autumn of 1939. That pamphlet has not been found. Morrison, interview by Sherwin, 21 June 2002 (ref. 12).
34. Griffiths (ref. 17), 27-28.

"The outbreak of war in Europe," the pamphlet dated February 20, 1940, states, "has changed profoundly the course of our own political development. In the last month strange things have happened to the New Deal. We have seen it attacked, and *more and more surely* we have seen it abandoned. There is a growing discouragement of liberals with the movement for a democratic front and red-baiting has grown to a national sport. Reaction is mobilized."[35]

Chevalier insisted that the language here is distinctively Oppenheimer's. "You can recognize his style," he told Sherwin. "He has certain little mannerisms, using certain words. More and more surely. That's very characteristic of him. You wouldn't ordinarily find use of surely in such a context." Chevalier's claim is too thin a reed with which to identify Oppenheimer as *the* author of the pamphlet, but it does suggest that Robert might have had a hand in editing a draft of it. While "more and more surely" does sound like Oppenheimer, much else in the pamphlet decidedly does not.[36]

But what do these "Reports" propose? More than anything else they propose a defense of the New Deal and its domestic social programs. "The Communist Party is being attacked for its support of the Soviet policy. But the total extermination of the Party here cannot reverse that policy: it can only silence some of the voices, some of the clearest voices, that oppose a war between the United States and Russia. What the attack can do directly, what it is meant to do, is to disrupt the democratic forces, to destroy unions in general and CIO unions in particular, to make possible the cutting of relief, to force abandonment of the great program of peace, security and work that is the basis of the movement toward a democratic front."[37]

On April 6, 1940, the College Faculties Committee of the Communist Party of California issued another "Report to our colleagues." Like the first pamphlet this report carried no by-line. But again Chevalier insisted that Oppenheimer was among the pamphlet's anonymous authors. "The elementary test of a good society is its ability to keep its members alive. It must make it possible for them to feed themselves and it must protect their persons from violent death. Today unemployment and war constitutes so serious a threat to the well being and security of the members of our society that many are asking whether that society is capable of meeting its most essential obligations. Communists ask much more of society than this: they ask for

35. "Report to our colleagues: I," 20 Feb 1940, TBL, University Archives, Office of the President files, 1940, folder Communism. The pamphlets came to the attention of the university's president, Robert G. Sproul, who placed them in his presidential papers. During the course of an interview Chevalier brought out copies of the pamphlets and Sherwin read excerpts into a tape-recorder. Chevalier, interview with Sherwin, 15 July 1982 (ref. 13).
36. Thus, in the spring of 1941 Oppenheimer wrote his friend William Fowler, "as time goes on you'll have more and more a feeling of confidence...." JRO to Fowler, Saturday, Spring 1941, Smith and Weiner (ref. 26), 215. Similar language can be found in Oppenheimer's letter to his brother of 12 Mar 1932, in which he writes, "I know very well surely that physics has a beauty which no other science can match." JRO to Frank Oppenheimer, 12 Mar 1932, ibid., 155.
37. "Report to our colleagues: I" (ref. 35).

all men that opportunity, discipline, and freedom which have characterized the high cultures of the past. But we know that today, with the knowledge and power that are ours, no culture which ignores the elementary needs, no culture based on the denial of opportunity, on indifference to human want, can be either honest or fruitful."[38]

As in February, domestic issues are the focus of this report. The plight of the nation's millions of unemployed is examined and the decision of California and national Democrats to cut the budget for welfare relief is attacked. "The cutting of relief and the simultaneous increase in the budget for armaments are connected not only by arithmetic considerations. Roosevelt's abandonment of the program for social reform, the attack, where once there had been support, on the labor movement, and the preparation for war, these are related and parallel developments." From 1933-39, the pamphlet observes, the Roosevelt Administration had "followed a policy of social reform." But since August 1939, "not a single new measure of progressive purpose has been proposed...and the measures of the past have not even been defended against reactionary attack." Where once the Roosevelt Administration voiced its "disgust" for the antics of the Dies House Un-American Activities Committee, now it is "coddling" these reactionaries. Where once it defended organized labor, civil liberties, and the unemployed, now it is attacking labor leaders like John L. Lewis and pouring money into armaments.[39]

Roosevelt himself, a man whom the pamphleteers once thought "something of a progressive" has now become a "reactionary" and even a "war-monger." This transformation has happened because of the war in Europe. "It is a common thought, and a likely one, that when the war is over Europe will be socialist, and the British Empire gone. We think that Roosevelt is assuming the role of preserving the old order in Europe and that he plans, if need be, to use the wealth and the lives of this country to carry it out."[40]

If Oppenheimer had anything to do with this second pamphlet, his rational style had abandoned him. Is it possible that he really thought of Roosevelt as a "war-monger?" The one reference to the president in Oppenheimer's correspondence during this period suggests that he was disappointed with FDR, but hardly ready to denounce him.[41] If Oppenheimer had something to do with drafting these pamphlets,

38. "Report to our colleagues: II," 6 Apr 1940, TBL, University Archives, Office of the President files, 1940, folder Communism.
39. Ibid.
40. Ibid.
41. More than a year after the April 1940 pamphlet was published, he wrote his old friends Ed and Ruth Uehling: "My own views could, im Kleinem, hardly be gloomier, either for what will happen locally & nationally, or in the world. I think we'll go to war—that the Roosevelt faction will win over the Lindbergh. I don't think we'll get anywhere near the Nazis. Later I think the Hearst-Lindbergh side will kick the administration 'humanitarians' out. I see no good for a long time; & the only cheerful thing in these parts is the strength & toughness & political growth of organized labor." JRO to Edwin and Ruth Uehling, 17 May 1941, Smith and Weiner (ref. 26), 217.

his words portray someone primarily worried about the impact on domestic politics of a world teetering on the brink of a great disaster. If he ever thought of himself as a communist, this was the "very brief and very intense" period that he would soon describe in an interview with an army intelligence officer in 1943.[42]

By the late 1930s, Oppenheimer was a senior professor with a prominent public persona. He was giving speeches on political issues and signing public petitions. His name appeared on occasion in the local newspapers. San Francisco was a fiercely polarized city; the longshoremen strikes in particular had hardened the political extremes on both the left and right. And when the conservative backlash began, he was sensitive to the fact that his political activities somehow affected the reputation of the university. Indeed, in the spring of 1941 he confided to his Caltech colleague, Willie Fowler, "I may be out of a job...because UC is going to be investigated next week for radicalism and the story is that the committee members are no gentlemen and that they don't like me."[43]

"The University of California was an obvious target," observed a former graduate. "And Oppenheimer was very prominent because he was quite vocal and active. He would occasionally get somewhat alarmed about what was happening, and maybe he'd have to draw his horns in and he would become quiet. Then when something happened to provoke him...he became active. So he wasn't consistent."[44]

In contrast to Chevalier's assertions about Oppenheimer's communist sympathies in 1940, other friends saw Oppie's political views moving in a quite different direction—and for good reason. By 1938 American newspapers regularly reported on the wave of political terror orchestrated by Stalin against thousands of alleged traitors within the Soviet Communist Party. "I read about the purge trials, though not in full detail," Robert wrote in 1954, "and could never find a view of them which was not damning to the Soviet system." While his friend Chevalier gladly signed a statement in the April 28, 1938, *Daily worker* commending the Moscow trial verdicts

42. *JRO Hearing*, 876.
43. JRO to Fowler, Saturday, Spring 1941 (ref. 36), 216. We see no record that JRO was questioned by any investigative committee at this time, so perhaps he was not called. The University's president, Dr. Robert G. Sproul, did hear rumors about Oppenheimer's Communist associations. In May 1940 he was informed that an alumnus, a Miss Rassmussen, had complained that her cousin had joined the Young Communist League on campus. Her cousin "cited the presence of faculty members at their meetings as proof that their cause was just and serious. She [the cousin] constantly alluded to some member of the faculty as her authority for this statement or that, and as her guiding star. She called this faculty member 'Pop' or 'Pops' and Miss Rassmussen said she thought his real name was something like Oppenheimer but she would not want to say for sure." George A. Pettitt memo to Sproul, 29 May 1940, TBL, University Archives, Office of the President files, 1940, folder Communism security.
44. Martin D. Kamen, interview with Sherwin, 18 Jan 1979, 27.

against Trotskyite and Bukharinite "traitors," Oppenheimer never defended Stalin's deadly purges.[45]

In the summer of 1938 two physicists who had just returned from living in the Soviet Union—George Placzek and Victor Weisskopf—visited Oppie at his ranch in New Mexico. Over the next week, they had several long conversations about what was taking place in the U.S.S.R. Reining in his horse one day, Weisskopf told his friend, "It's worse than you can imagine. It's a morass." They told a "skeptical" Oppenheimer, "Russia is not what you think it is."[46] They talked about the case of Alex Weissberg, an Austrian engineer and communist who was suddenly arrested merely for associating with Placzek and Weisskopf. "It was an absolutely scary experience," Weisskopf said, "We called up our friends, and they said they didn't know us."[47] Oppie asked probing questions that showed how disturbed he was by Weisskopf's reports.

Sixteen years later Oppenheimer explained to his interrogators, "What they reported seemed to me so solid, so unfanatical, so true, that it made a great impression; and it presented Russia, even when seen from their limited experience, as a land of purge and terror, of ludicrously bad management and of a long-suffering people."[48]

There was no reason, however, why news of Stalin's abuses should cause him to renounce his sympathies with the American left. It was clear, as Weisskopf remembered, that Oppie "still believed to a great extent in communism."[49] Nevertheless, Weisskopf thought that this was the first time Oppenheimer was really shaken. "I know that these conversations had a very deep influence on Robert," Weisskopf said. "This was a very decisive week in his life, and he told me so....That weekend started Oppenheimer's turning away from the Communist Party." Weisskopf insists that Oppie "saw the Hitler danger very clearly....And in 1939, Oppenheimer was already very far from the Communist group."[50]

Oppie trusted Weisskopf. "He really had a deep attachment to me," Weisskopf recalled, "which I found very touching." Robert knew that Weisskopf, an Austrian social democrat, was not saying these things out of any antipathy for the left. "We were very much convinced—both sides—that socialism was the desirable development."[51]

45. *JRO Hearing*, 10; Chevalier, interview with Sherwin, 15 Jul 1982 (ref. 13); *The daily worker*, 28 Apr 1938. Chevalier was joined in this statement by nearly 150 prominent intellectuals, including Nelson Algren, Dashiell Hammett, Lillian Hellman, Dorothy Parker, and Malcolm Cowley.

46. Peter Michelmore, *The swift years: The Robert Oppenheimer story* (New York, 1969), 57-58.

47. Victor Weisskopf, interview with Sherwin, 23 Mar 1979. During World War II, Weissberg was eventually shipped to a Polish extermination camp. He jumped from a truck, however, and managed to escape into the woods where he became active in the Polish underground.

48. *JRO Hearing*, 10.

49. Victor Weisskopf, *The joy of insight: Passions of a physicist* (New York, 1991), 115.

50. Weisskopf, interview with Sherwin, 23 Mar 1979 (ref. 47).

51. Ibid.

Shortly after hearing from Weisskopf and Placzek, Oppenheimer expressed his concerns to Edith Arnstein, Jean Tatlock's old friend: "Opje said he came to me because he knew I would not be shaken in my political loyalties, and he needed to talk." He explained that he had heard from Weisskopf about the arrest of various Soviet physicists. He said he was reluctant to believe the report but neither could he dismiss it. "He was depressed and agitated," Arnstein later wrote, "and I suppose now I know how he was feeling, but then I was scornful of what I saw as his gullibility."[52]

That autumn some friends noticed that he was no longer as loquacious about his political views. "Opje is fine and sends you his greetings," wrote Felix Bloch to I.I. Rabi in November 1938. "Honestly, I don't think you wore him out but at least he does not praise Russia too loudly any more which is already some progress."[53]

Whatever his associations with Communist Party members, Oppenheimer had always been enamored with Franklin Roosevelt and the New Deal. His friends saw him as an ardent Roosevelt supporter. Ernest Lawrence recalled being lobbied vigorously by his friend in the days just prior to the 1940 presidential election. Oppie was incredulous that his old friend was undecided. That evening he presented such a passionate defense of Roosevelt's campaign for a third term that Lawrence finally promised he would cast another vote for FDR.[54]

Oppenheimer's political views were evolving, no doubt in reaction to the disastrous war news. In the late spring and early summer of 1940, he was clearly distressed by the collapse of France. That summer Hans Bethe encountered him at a conference of the American Physical Society in Seattle. Bethe had an inkling of Oppenheimer's political loyalties so he was struck one evening when he gave a "beautifully eloquent speech" about how the fall of Paris to the Nazis threatened all of Western civilization. "We have to defend Western values against the Nazis," Bethe recalled Oppenheimer

52. Edith Arnstein Jenkins, *Against a field sinister: Memoirs and stories* (San Francisco, 1991), 27. Edith chose as her Party alias the name of Mary Shelley's mother, Mary Wollstonecraft. She said that no one was a "card-carrying communist" in their own name: "It was too dangerous." In 1936-38, Arnstein was the official secretary and dues collector for a closed unit of the CP at Berkeley—but she left this position in 1938 when she quit law school. The professional section of the Communist Party at Berkeley, she said, was composed of several units with about eight individuals in each unit. She later said that Oppenheimer was certainly not a member of her closed unit, though she could not speak to this point for the years after 1938. Jenkins also remembered that Oppenheimer had once given her a small sum of money as a contribution to the Young Communist League (YCL). Jenkins, interview with Gregg Herken, 9 May 2002; Jenkins, interview with Bird, summer 2002.

53. Bloch to Rabi, 2 Nov 1938, Bloch Papers, Stanford University, box 1, General Correspondence, as cited in S.S. Schweber, *In the shadow of the bomb: Bethe, Oppenheimer, and the moral responsibility of the scientist* (Princeton, 2000), 108.

54. Herbert Childs, *An American genius: The life of Ernest Orlando Lawrence* (New York, 1968), 307.

saying. "And because of the Molotov-von Ribbentrop pact we can have no truck with the Communists."[55] Years later, Bethe told the physicist-historian Jeremy Bernstein, "He had sympathies to the far left, mostly, I believe, on humanitarian grounds. The Hitler-Stalin pact had confused most people with Communist sympathies into staying completely aloof from the war against Germany until the Nazis invaded Russia in 1941. But Oppenheimer was so deeply impressed by the fall of France [a year before the invasion of Russia] that this displaced everything else in his mind."[56]

On Sunday, June 22, 1941, the Chevaliers were driving to a picnic at the beach with Oppenheimer when they heard the news over the radio that the Nazis had invaded the Soviet Union. That evening everyone stayed up late listening to the latest news bulletins, trying to make sense of what had happened. Chevalier recalled Oppie saying that Hitler had committed a major blunder. By turning against the Soviet Union, Oppie argued, Hitler had "destroyed at one stroke the dangerous fiction, so prevalent in liberal and political circles, that fascism and communism were but two different versions of the same totalitarian philosophy."[57] Now communists everywhere would be welcomed as allies of the Western democracies. And that was a development both men thought was long overdue.

In the meantime, however, reports from the Russian front were bleak. A few days after the initial attack Oppenheimer and Chevalier were driving back to Berkeley from Pasadena when they stopped for a meal. In the restaurant they ran into Carey McWillams, the muckraking journalist who was then serving as California's Commissioner of Housing. McWilliams later remembered how pessimistic both men were about the ability of the Soviets to withstand the Nazi blitzkrieg. The world, they all thought, was headed for the dark ages.[58]

Soon after this, Oppenheimer attended a meeting of his Berkeley discussion group. Gordon Griffiths was there and later recalled, "Stalin had delivered a radio address calling upon the Soviet people to resist. [Stalin's speech was aired on July 3, 1941.] It was an eloquent speech and Oppie had brought the text to our meeting to read aloud. He was so moved that his eyes filled with tears. Not that it took Stalin's words to make us realize that the supreme test of the Soviet experiment had now come, but our confidence in the cause was strengthened by hearing rhetoric that did full justice to the awesome moment."[59]

55. Schweber (ref. 53), 108.
56. Jeremy Bernstein, *Hans Bethe: Prophet of energy* (New York, 1980), 65.
57. Chevalier, *Oppenheimer* (ref. 5), 42.
58. Ibid., 41-42; Stern with Green (ref. 1), 33.
59. Griffiths (ref. 17), 27.

After the Japanese attack on Pearl Harbor on December 7, 1941, the country was suddenly at war. "Our little group in Berkeley," Chevalier recalled, "inevitably reflected the country's changing mood." Chevalier said their small group "continued to meet irregularly"—though Oppenheimer himself rarely attended due to his busy travel schedule. "When we did meet," Chevalier wrote, "our business was largely confined to discussing the progress of the war and events on the home front."[60]

Chevalier always insisted that Oppenheimer, the man he considered his closest friend, shared his own leftist political views right up to the moment he left Berkeley in the spring of 1943: "we shared the ideal of a socialist society...there was never any wavering, any weakening of his position. He was firm as a rock." But Chevalier was clear that Oppenheimer was not an ideologue. "There was no blindness in him, no narrow partisanship, no automatic hewing to a line." Nevertheless, he also considered Robert a CP member, a charge that Oppenheimer always denied.[61]

Chevalier's description of Oppenheimer could easily portray a left-wing intellectual not under Party discipline. But over the years, as he turned to writing about his friendship with Oppie, Chevalier tried to suggest something else. In 1948, he outlined a novel in which the protagonist, a brilliant physicist working to build an atomic bomb, is also the de facto leader of a closed unit of the Communist Party. In 1950, Chevalier set aside the partially written manuscript after he was unable to find a publisher for it. But in 1954, after the Oppenheimer security hearing, he returned to the novel and in 1959 G.P. Putnam's Sons published it under the ponderous title *The man who would be God.*

In the novel, Oppenheimer's character, one Sebastian Bloch, decides to join the Communist Party—but to his surprise, the local CP leader refuses to let him formally join. "Sebastian would meet with the unit regularly, and in every way act as though he were a bona fide member, and the other members would so regard him; but he would pay no dues—he could make his own financial arrangements with the party, but outside the unit."[62] Later in the novel Chevalier describes the weekly meetings of this closed party unit as "informal seminars of the kind that were constantly being held on all sorts of subjects among professors and students on the campus." The members discussed "ideas and theory," current events, the "activity of this or that member of the Teachers Union," and "support to be given to a labor union campaign, a strike, an individual or group under attack on a civil liberties issue." After the Soviet Union invaded Finland in November 1939, Chevalier has the Oppenheimer character propose that the party unit should publish some essays that would explain the international situation "in language palatable to cultivated, critical minds." The Oppenheimer character pays for the printing and postage costs and does most of the

60. Chevalier, interview with Sherwin, 29 June 1982 (ref. 5); Chevalier, *Oppenheimer* (ref. 5), 46.
61. Chevalier, *Oppenheimer* (ref. 5), 187.
62. Chevalier, *The man who would be God* (ref. 29), 14-15.

writing himself. "It was his baby," writes the novelist. "A number of these 'Reports to the Faculty' appeared over the next few months."[63]

The thinly disguised novel did not sell well and Chevalier was unhappy with the reviews. *Time* magazine's reviewer, for instance, thought the "novel's underlying tone suggests an ex-worshipper stomping on a fallen idol."[64] But Chevalier could not let the matter drop. In the summer of 1964 he wrote Oppenheimer to say that he had nearly finished writing a memoir about their friendship. He explained, "I tried to tell the essential story in my novel. But readers in America were disturbed by the blend of truth and fiction, and it has become clear to me that for the record I have to tell the story straight…an important part of the story concerns your and my membership in the same unit of the CP from 1938 to 1942. I should like to deal with this in its proper perspective, telling the facts as I remember them. As this is one of the things in your life which, in my opinion, you have least to be ashamed of, and as your commitment, attested among other things by your 'Reports to our Colleagues,' which today make impressive reading, was a deep and genuine one, I consider that it would be a grave omission not to give it its due prominence." Chevalier then asked whether Oppenheimer would have any objections to the telling of this story.[65]

Two weeks later, Oppenheimer wrote back a terse note:[66]

> Your letter asks whether I would have any objections. Indeed I do. What you say of yourself I find surprising. Surely in one respect what you say of me is not true. I have never been a member of the Communist Party, and thus have never been a member of a Communist Party unit. I, of course, have always known this. I thought you did too. I have said so officially time and time again. I said so publicly in response to what Crouch said in 1950. I said so in the AEC hearings ten years ago.
>
> As ever,
>
> Robert Oppenheimer

Chevalier reasonably concluded that Oppenheimer's denial also meant to warn him that he could face a libel suit if he published a clear accusation that Oppenheimer had joined the Communist Party. So the following year he published *Oppenheimer: The story of a friendship* without the bald accusation. Instead, throughout the book, the alleged CP "closed unit" is described merely as a "discussion group."[67]

Chevalier told Oppenheimer that he had been compelled to write this book because "history, though coy, needs truth to be her handmaiden."[68] But in this case,

63. Ibid., 88-89.
64. *Time*, 2 Nov 1959, 94.
65. Chevalier to JRO, 23 Jul 1964, JROP, box 200, folder Chevalier, Haakon—Reference to Case.
66. JRO to Chevalier, 7 Aug 1964, JROP, box 200, folder Chevalier, Haakon—Reference to Case.
67. Chevalier, *Oppenheimer* (ref. 5), 19, 46.
68. Chevalier to JRO, 23 Jul 1964 (ref. 65).

the "truth" lies in each man's perception. Were all members of the Berkeley "discussion group" also members of the CP? Chevalier believed they were; Oppenheimer insisted that he was not a Party member. He would fund certain causes through the CP—the Spanish Republic, farm workers, civil rights, and consumer protection. He would attend meetings, offer his advice and even help the Party's intellectuals write position papers. But he did not have a Party card, he did not pay dues, he was totally free of Party discipline. His friends might have had reasons to think he was a comrade, but in his own mind he was clear that he was not.

He had never entered into that peculiar social contract expected of others who became Party members.[69] And so the ambiguity remains.

John Earl Haynes and Harvey Klehr—two historians of American Communism—have written that "to be a Communist was to be part of a rigid mental world tightly sealed from outside influences..."[70] This certainly does not describe Robert Oppenheimer at any time. He was reading Marx, but he was also reading the *Bhagavad-Gita*, Ernest Hemingway and Sigmund Freud—and the latter, a Communist heresy, was grounds for expulsion from the Party in those years.

Robert was probably closer to the party in the 1930s than he later admitted, or even remembered, but he was not nearly as close to it as his friend Haakon believed. This is neither surprising nor deceitful. The so-called "secret units" of the party—the sort of association that Oppenheimer is alleged to have had—were virtual organizations with no formal rosters, no established rules and little, if any, regimentation, as Chevalier explained to Sherwin. For obvious organizational reasons the party chose to see those individuals associated with "secret units" as having made substantial personal commitments. On the other hand, each "committed" member could set the limits of his commitment, and that commitment could change over time, very short periods of time, as it did, for instance, with Jean Tatlock.

69. As one of the FBI's informants put it, "although Oppenheimer may not have been actually brought into the Communist Party, the effort to bring him to acceptance of Communist Party philosophy and to secure his support for Communist aims was regarded by the Communists as successful." This FBI informant was Louis Gibarti, a Hungarian-born communist who spent the years 1923-1938 as a Cominterm agent. Gibarti, whose real name was Laszlo Dobos, left the Party in 1938 and then worked as a journalist. There is no evidence that Gibarti ever knew Oppenheimer. In 1950 he became an informant for the FBI. J. Edgar Hoover to Lewis Strauss, 25 June 1954, FBI file of JRO (ref. 8), Section 44, document 1800.

70. John Earl Haynes and Harvey Klehr, *In denial: Historians, communism, and espionage* (San Francisco, 2003), 39. John Haynes later wrote, "Oppenheimer, of course, would have been regarded by any party officer with any sense as a highly valuable ally. Further, he had no dependence on the party for organizational or other assistance. He was highly valuable to the party, but the party was not valuable to Oppenheimer except to the extent of his belief in its goals and objectives and whatever personal/fraternal ties he had developed with others in the movement. No skilled party leader would impose 'discipline' on someone like Oppenheimer; instead of giving orders, he would persuade, convince, cajole, ask politely and even plead if necessary." Haynes, email to Gregg Herken, 26 Apr 2004, courtesy of Herken.

Haakon seemed always to be pretty much committed, and in those days when he and Robert were close friends, it is not surprising that Hoke considered Oppie equally committed. Perhaps for a time he was, but we do not, and we cannot, know the extent of his commitment. But what we can say with confidence is that Robert's high level of commitment was short and did not last; in this he was more akin to Jean than to Haakon.

The bottom line is that Robert always was free to think for himself and to make his own political choices. In his actions, in his psyche, and in his politics he was no more under the control of the CPUSA than was any serious Democrat or Republican office-holder at the time. Commitments have to be put in perspective to be understood, and the failure to do that was the most damaging characteristic of the McCarthy period. The only truly important thing to know about Robert Oppenheimer was that in the 1930s he was devoted to working for social and economic justice in America and to resisting fascism in Europe. To achieve these goals he chose to stand with the Left.

CHAPTER FIVE

The Puzzles of Interpreting J. Robert Oppenheimer, His Politics, and the Issues of His Possible Communist Party Membership

Barton J. Bernstein

Lansdale	Now, have you yourself ever been a member of the Communist Party?
Oppenheimer	No.
Lansdale	You've probably belonged to every front organization on the coast.
Oppenheimer	Just about.
Lansdale	Would you in fact have considered yourself at one time a fellow traveler?
Oppenheimer	I think so. My association with these things was very brief and very intense. [Of Haakon Chevalier,] I wouldn't be surprised if he were a member [of the Party], he is quite a Red.[1]

Positions involving responsibility for, or access to, matters which should remain secret, need to be filled by men who can be trusted whether they are scientists or not. In assessing trustworthiness, considerations of loyalty and conviction are obviously pertinent; but considerations of character, experience and temperament are more frequently involved and are equally important. The measures involved in such assessment are not, as the experience of the last years has shown, easily perfected. Neither the secret investigations of the employee's record, nor the act of judgment itself, which must resolve doubt not in favor of the individual but in favor of security, can be subject to our traditional juridical safeguards.[2]

Introduction: The Puzzles

Probably no major physicist of the twentieth century, and undoubtedly no American-born physicist of that century, who was deeply involved in government activity

The following abbreviations are used: AEC, U.S. Atomic Energy Commission; AEC Records, RG 326, NARA; *BAS, Bulletin of the atomic scientists*; CP, Communist Party; FBI, Federal Bureau of Investigation; *HSPS, Historical studies in the physical and biological sciences*; JRO, J. Robert Oppenheimer; *JRO Hearing*, U.S. Atomic Energy Commission, *In the matter of J. Robert Oppenheimer: Transcript of hearing before Personnel Security Board and texts of principal documents and letters* (Cambridge, 1971); JROP, J. Robert Oppenheimer Papers, Manuscript Division, Library of Congress, Washington, D.C.; MED Records, Records of Manhattan Engineer District, RG 77, NARA; NARA, National Archives and Records Administration, College Park, MD; NAS, National Academy of Sciences; RG, Record Group; TBL, The Bancroft Library, University of California, Berkeley.

1. JRO, interview with Lt. Col. John Lansdale, 12 Sep 1943, *JRO Hearing*, 876-877.

2. NAS to President Harry S. Truman, 3 Feb 1949. See NAS, "A statement adopted by the

and in nuclear weapons work, has created so many interpretive problems—about his politics, his related beliefs, and his personality, for his own contemporaries, for later historians, and for others in his own lifetime and in the aftermath—as the often charismatic J. Robert Oppenheimer. The very complexities and uncertainties in interpreting Oppenheimer, as well as the fact of his power and his involvement in significant public events, add to both the interest in him and the difficulties of explaining him.

To state all that about Oppenheimer is not to minimize the troubling interpretive problems raised about understanding some others—including two notable European-born physicists connected with important weapons work: the brilliant Werner Heisenberg, about whom it has long been asked whether or not he truly sought a wartime German A-bomb; and the very different, and less able Edward Teller, Heisenberg's former student and by the late 1940s, if not somewhat earlier, Oppenheimer's enemy, who became a Washington insider, an ardent advocate of H-bomb development, and a proponent of many weapons systems.

But the Oppenheimer puzzles—the problems in interpretation on so many matters—seem to eclipse those of Heisenberg, Teller, and others. The Nobel Prize-winning physicist Hans Bethe, a longtime friend and supporter of Oppenheimer, uneasily acknowledged much of the puzzle.[3] Another Nobel physicist, I.I. Rabi, noting that he himself was not the easiest person to understand, stressed (in some dismay, not admiration) how much more complicated Oppenheimer was and seemed. Rabi, like some other contemporaries, felt that Robert Oppenheimer did not have a firm sense of identity, was conflicted about whether or not he was Jewish, and dwelled often in the penumbra of ambivalence and ambiguity about many matters.[4]

Understanding Oppenheimer politically and personally is very difficult. He was intentionally convoluted, and by temperament probably could not have been straightforward. He was often overly dramatic, frequently operating as if he were on a stage, and more interested than many others in the public realm in the impression he made on associates and others. Self-critical, he was also remarkably self-deceptive. He delighted in parading his knowledge of subjects distant from science, and offered himself at times as a kind of modern-day renaissance man. That impressed many, was simply accepted by some, and annoyed others like the future Nobel Prize-winning

---

Council of the National Academy of Sciences authorized for transmission to the President of the United States," 3 Nov 1949, with Alfred Richards, president of NAS, to Members of NAS, 14 Feb 1949, JROP, box 134, and NAS Papers, Washington, D.C. The statement was written by Oppenheimer, Harvard president James Conant, and Bell Laboratories head Oliver Buckley and adopted unanimously by the NAS Council. Both Buckley and Conant testified for Oppenheimer in the 1954 loyalty-security hearing, in which both the majority of the Gray hearing board and the majority of the AEC used very similar standards to those in the NAS statement in finding against Oppenheimer.

3. Bethe, interviews and discussions with Barton J. Bernstein, mostly 1989-92.

4. Rabi, interview with Bernstein, 1979. Literary analyst Francis Fergusson, a lifetime friend of Oppenheimer, made a similar point in 1978 interviews. Also see Rabi in Jeremy Bernstein, *Experiencing science: Profiles in discovery* (New York, 1976), 68-69.

physicist Felix Bloch, who regarded Oppenheimer as excessively arrogant and certainly both pretentious and pompous.[5]

Oppenheimer's involvement in the significant nuclear-weaponry events of his time—most notably, the development and use of the atomic bomb, and the disputes about the development of the H-bomb—add richly to the interest in Oppenheimer and thus to the concern with understanding him. The painful fact of the 1953-54 loyalty-security case against him, in which the government declared him a security risk and permanently stripped him of his security clearance, further attracts analytical attention to him, his career, and his enemies. Many interpreters view him as, ultimately, a tragic figure—struck down by others. Some also add that he possessed fatal flaws contributing powerfully to his destruction.

Oppenheimer, by most judgments, never did the very significant theoretical physics which he had once seemed destined to accomplish. That alone has justifiably raised questions. Was he overrated by contemporaries, or have historians, unduly influenced by later testimonials, exaggerated the estimates seemingly made in the late 1920s and into the mid- or late '30s? If not significantly overrated in that early period on the basis of the evidence then available to contemporaries, what went askew for Oppenheimer? Or is this question possibly unfair, and is too much—by his contemporaries then, and by historians later—being expected of Oppenheimer in physics?[6]

Beyond that, of course, is his relationship to the crucial events of the first half of the wartime 1940s—the development and use of the atomic bomb. Was he unalloyed in his hopes and in his wartime bomb-building activities, or somewhat uneasy in his effort at Los Alamos to develop the bomb? Did he feel before or shortly after the Alamogordo test of mid-July 1945 any desire not to have the weapon used on Japan, and especially on its cities? Why is it that he never raised any issues, in the 1945 target committee or elsewhere in the pre-Hiroshima months, about using the bomb on Japanese cities and heavily on noncombatants there?[7]

5. Bloch, discussions with Barton J. Bernstein, 1965-69, and Marvin Chodorow (confirming Bloch's opinion), discussions with Bernstein, 1983-85.

6. In interviews in 1980-83, Luis Alvarez contended that Oppenheimer's black hole work in the 1930s merited a Nobel prize, but that is a generally minority judgment. Emilio Segre, for example, strongly disagreed in interviews in the early 1980s, as did a cluster of other notables whom I queried. An exception (though not a notable one in physics) is Jeremy Bernstein, *Oppenheimer: Portrait of an enigma* (Chicago, 2004), viii, 47-50. See generally Kip Thorne, *Black holes and time warps: Einstein's outrageous legacy* (New York, 1994), 187-220, and Karl Hufbauer's essay in this volume.

7. Major F.F. Derry and Norman Ramsey to Major General L.R. Groves, "Summary of Target Committee meeting on 10 and 11 May 1945," 12 May 1945, MED Records, Groves files. Also see Interim Committee minutes, 31 May 1945, and Scientific Advisory Panel report (written by JRO), 16 June 1945, MED Records, Harrison-Bundy files; Groves-Oppenheimer transcript, 6 Aug 1945, in MED Records, Groves files. The conclusion about Oppenheimer is based heavily on many archival records including Los Alamos materials, supplemented slightly by JRO, interview with Alice K. Smith, 11 Nov 1957, courtesy of Smith.

How much can, and should, historians trust Oppenheimer's postwar claim that at the mid-July Alamogordo test, as he watched the world's first atomic explosion, he thought of the fateful, evocative words from the *Bhagavad-Gita*, "I am become death, the destroyer of worlds"? Or should historians instead have relied on meteorologist Jack Hubbard's diary entry that Oppenheimer, observing the atomic blast, remarked that it renewed his faith in the capacity of the human mind?[8] Avoiding an either/or polarity, could it be, perhaps, that Oppenheimer had one set of thoughts from the *Bhagavad-Gita*, but uttered very different words to Hubbard, and was in fact deeply conflicted without openly acknowledging it at the time?

And after Hiroshima and Nagasaki, did Oppenheimer feel doubt, possibly even regret, and perhaps guilt? Many have interpreted his alleged October 1945 comments to President Harry S. Truman about Oppenheimer having blood on his hands as evidence of guilt, not simply regret. But there is good reason for serious doubt about whether Oppenheimer, in dealing in the White House with authority and political power in the person of Truman, would have said anything like that. Having moved toward political power, Oppenheimer was not generally inclined to challenged authority in such a way—and certainly not with the president who had authorized the atomic bombings.[9]

So often, with Oppenheimer, more than most of his contemporaries in science who were involved in weaponry and political advising, there are various myths or near-myths of what he said, felt, or did. He contributed to such beliefs; and contemporaries—sometimes in admiration and sometimes in dislike—also contributed to the

8. Most often, as in Jon Else's documentary *The day after Trinity*, the quotation uses the noun "destroyer," but sometimes "shatterer" instead occurs. See "Education," *Time* (8 Nov 1948), 77. Jack Hubbard diary, 16 Jul 1945, courtesy of Ferenc Szasz. The diary was typed years after the events, and there is evidence in some places in the extended diary of later additions and revisions. Also see Barton J. Bernstein, "Four physicists and the bomb: The early years," *HSPS*, *18:2* (1988), 231-263. Probably unintentionally but carelessly further reaching beyond the mid-July 1945 evidence, writers sometimes contend that Oppenheimer claimed he had *uttered* the *Bhagavad-Gita* words at the Alamogordo test.

9. On May 7, 1946, after receiving a turn-down by Oppenheimer to participate in Bikini nuclear tests, Truman claimed to remember the content of a meeting "some five or six months ago"—it was probably on October 25, 1945, and also included Secretary of War Robert Patterson—in which Oppenheimer allegedly "spent most of his time [w]ringing his hands and telling me they had blood on them because of the discovery of atomic energy." Years later, Dean Acheson, who according to the President's appointment book had not been present at the Oval Office meeting, also reported that Oppenheimer said similar words to Truman, and pungently reported alleged words by Truman that reached beyond those in Truman's memorandum. Acheson was probably loosely remembering Truman's note of May 7, 1945, to Acheson, a copy of which is in President's Secretary's Files (PSF) 201, Harry S. Truman Library, Independence, MO. The Appointment Book for 25 Oct lists a 10:30 meeting, which probably continued to 11 a.m., whereupon Truman met with the postmaster from Joplin, MO, probably a longtime acquaintance. See, for example, Philip Stern with Harold Green, *The Oppenheimer case: Security on trial* (New York, 1969), 90.

legends. Teller, for example, claimed to recall (no doubt incorrectly) that Oppenheimer soon after Hiroshima and Nagasaki proposed closing the Los Alamos weapons laboratory and not continuing atomic-weapons development.[10]

In 1947, and in later years, what did Oppenheimer's eminently quotable 1947 remark about the physicists knowing sin mean to him?[11] Was he, as some former A-bomb physicists believed in 1947 and afterward, taking moral responsibility and even apologizing for the 1945 atomic bombing of Japan, or the development of the bomb, or both? Or was his meaning—as so often with Oppenheimer—evasively elusive and possibly so intended?

How might one square Oppenheimer's postwar words with his advice in 1943 to the Nobel laureate Enrico Fermi about the prospects of devising a secret plan to develop a radiological-warfare capacity for poisoning enemy food with a standard of possibly killing about a half-million people?[12] Was such thinking, in 1943, by Oppenheimer an anomaly? Was this perhaps part of what he might have been thinking about years later when he spoke of physicists knowing sin? Such questions can help to focus the dialogue about understanding Oppenheimer and his politics, especially if politics is defined in the broadest sense.

How and why in World War II, and then a few years later, could he give severely damaging testimony to security investigators and at least one congressional investigating committee—sometimes involving serious exaggeration—about former students? He knowingly "did them in," and yet never even warned them after virtually betraying them. Even some of Oppenheimer's friends rebuked him for such action, but none parted company with him on such matters.[13]

10. Teller, interviews with Bernstein, 1968-88.
11. Oppenheimer's words that "the physicists have known sin" presumably first gained wide attention in early 1948, and also appeared in "Education," *Time* (8 Nov 1948), 77. The 1947 talk in which the statement first occurred is reprinted in JRO, *The open mind* (New York, 1955), 88; I have not checked for small revisions in the 1955 book. For annoyance and critical responses by some A-bomb alumni, see Freeman Dyson, *Disturbing the universe* (New York, 1979), 52-53, who further elaborated on this in an interview.
12. JRO to Fermi, 25 May 1943, JROP, box 33. This letter was later published, with brief commentary, in Barton J. Bernstein, "Oppenheimer and the radiological poison plan," *Technology review* (May/June 1985), 14-17, and the historical issues of radiological warfare were further pursued in "Radiological warfare: The path not taken," *BAS, 41* (Aug 1985), 44-49. Ultimately nearly two dozen Manhattan Project scientists were queried on this subject. Former Oppenheimer student Philip Morrison was unique in strongly opposing any publication mentioning the 1943 Oppenheimer letter, which, significantly, Alice K. Smith, an Oppenheimer friend, had chosen to exclude from the generally impressive volume she co-edited: Alice Kimball Smith and Charles Weiner, eds., *Robert Oppenheimer: Letters and recollections* (Cambridge, 1980). Smith and, in April 2004, Weiner, interviews with Bernstein. The response of Smith and Morrison, while in the minority, suggests the great difficulty of resting analyses heavily on pro-Oppenheimer recollections by Oppenheimer stalwarts.
13. See, for example, "Testimony of Dr. J. Robert Oppenheimer" before U.S. House of Representatives, Committee on Un-American Activities, 7 June 1949, declassified JRO files (made

There is also the related set of important questions about the so-called Haakon Chevalier - George Eltenton - Robert Oppenheimer affair, involving possible Soviet efforts at espionage against the A-bomb project in winter 1942-43. What is known is that Oppenheimer's friend, Chevalier, after talking with Eltenton, a Bay Area Communist, approached Oppenheimer in late 1942 or early 1943 about Communist interest in the top-secret project in which Oppenheimer was engaged. Whether Chevalier was warning Oppenheimer of that Communist interest, or seeking information, remains uncertain. Exactly what Chevalier said to Oppenheimer is similarly unclear, as is what Oppenheimer said to Chevalier. But Oppenheimer told various stories—crucially contradictory on some matters—to Manhattan Project officials, implicating and helping injure Chevalier but never choosing to inform him of what was said to security-concerned officials.[14]

In view of Oppenheimer's own loyalty-security difficulties during World War II on the top-secret Manhattan Project, and in 1946-47 involving his membership on the Atomic Energy Commission's General Advisory Committee, why did he in 1949 help craft and then endorse the National Academy of Sciences statement recommending that in loyalty-security cases, because of the importance of national security, the state's judgment about individuals when the evidence was unclear had to be resolved in favor of the American state, not the individual?[15] Did Oppenheimer in 1949 sincerely believe in such a standard, or was he reluctant to push for more protection for the individual and risk provoking opposition? Did he not understand that he had been treated differently, and even that he might again need similar benefits in the future?

Adding to the issues in the puzzles, why did he betray his Caltech friend and sometime colleague, the physicist Richard Tolman, by having an affair with his wife, the psychologist Ruth Tolman, during the early part of Oppenheimer's own marriage to Katherine (Kitty) Puening Oppenheimer? According to at least two of Oppenheimer's enemies, that affair, when Richard Tolman learned about it, led to his stroke and death in the late 1940s. Did the affair also add great strains on Kitty, already rather high-strung, nervous, and a heavy drinker?[16]

---

available ca. 1992), AEC Records. Also see Bethe to JRO, 26 June 1949, and Victor Weisskopf to JRO, 23 June 1949, JROP; and Edward Condon to JRO, 27 June 1949, Condon Papers, American Philosophical Society, Philadelphia, PA. Years later, in interviews in the 1990s, former Oppenheimer students Rossi Lomanitz and Joseph Weinberg were cool about Oppenheimer but unwilling to condemn him for the injury he had done them.

14. This has been dealt with in depth in a number of places, but also see Haakon Chevalier statement, 26 June 1946, [author unidentified], San Francisco, on "Haakon Chevalier," 27 Mar 1947, FBI file of Chevalier, #100-18654-[no. unreadable], part 1, secured under FOIA from FBI; and JRO to Chevalier, 24 Feb 1950, JROP. On Chevalier in 1954 after the hearings, Chevalier to JRO, 20 Sep 1954 (in English translation), Legat (Paris) to Director, FBI, 9 Dec 1954, FBI file of Chevalier, #100-18654-[no. unreadable], part 1.

15. NAS statement, 3 Nov 1949 (ref. 1).

16. Virginia Walker (Admiral Lewis Strauss's longtime secretary), interview with Bernstein,

And given the fact of Kitty's alcoholism, why did he help contribute to her problems by pushing strong liquor, while uneasily and clumsily making excuses for her problems, and in effect clearly not helping her but adding to her drinking problems? To ask such questions is not to impose the standards of the post-Oppenheimer years about alcoholism and co-dependency (not a word used in this context in the 1940s and '50s), for these are issues that his associates and friends wondered about in the two decades before his own death in 1967.[17]

There are also the issues of his very name and his relationship to being Jewish. He normally insisted, falsely, that the initial "J." in "J. Robert Oppenheimer" stood for nothing, but actually it was for his first name, Julius.[18] Why did he choose to deny that? And, as Rabi among others noted, Robert seemed uncertain about whether he was Jewish or not, and what that meant. To enemies like Atomic Energy Commission commissioner Lewis Strauss, a religious Jew, this was a powerful cause for doubt about Oppenheimer's character, helping to provoke Strauss's suspicion and enmity. To Teller, in contrast, this was a meaningless issue.[19]

Why did Oppenheimer sometimes support the quest for the H-bomb during World War II, oppose it briefly in postwar 1945, endorse it in 1947-49, but strongly—on moral and strategic grounds—oppose it in late 1949, after the Soviets tested their first atomic bomb? Afterward, when Truman in January 1950 ordered a more vigorous effort to develop the H-bomb, did Oppenheimer accede to that order or did he seek in various ways to delay the bomb's development? Even after the Teller-Ulam scientific breakthrough of early 1951, was Oppenheimer still inclined against the H-bomb? Or, as he later stated, did it then seem a "sweet" thing, something to be endorsed and pursued?[20]

1979; Molly Lawrence (widow of Ernest O. Lawrence), interview with Bernstein, 1979. See the Ruth Tolman letters in JROP, box 72. Jeremy Bernstein, *Oppenheimer* (ref. 6), 140-142, expresses grave reservations that there was such an affair, but the evidence in the letters from various contemporaneous sources seems overwhelming. Also see Lewis Strauss to files, 9 Dec 1957, Strauss Papers, Herbert Hoover Library, West Branch, IA, box 1 (initially used in the Washington, D.C., office of Lewis H. Strauss).

17. Kitty's problems were discussed with, among others, Frank Oppenheimer in 1980-84; much later with Judy Oppenheimer (Frank's daughter and Kitty's niece), and with Robert Bacher, in the 1990s. Efforts to discuss this issue usefully with Peter Oppenheimer (Kitty's and Robert's son) were not successful.

18. Robert P. Crease double-checked and secured a copy of the birth certificate, with the name Julius, as Crease confirmed to me.

19. Virginia Walker (Strauss's secretary), interview with Bernstein, 1979; this theme was also suggested by Richard Pfau, discussions with Bernstein, late 1970s. Teller, various interviews, dismissed this issue on a number of occasions.

20. Bernstein (ref. 8) and Peter Galison and Barton Bernstein, "In any light: Scientists and the decision to build the Superbomb, 1942-1954," *HSPS, 19:2* (1989), 267-347, place Oppenheimer and the H-bomb issues in a broad context. (Because of a typographical error, the latter essay's subtitle appears in published form as 1952, but it is usually corrected without

Trying to understand Oppenheimer is a task of seeking to make sense of his inconsistencies, unraveling his deceits, probing the nature of his charismatic appeal, explaining his success at Los Alamos, poking behind his dramatic self-portrayals, explaining his many cruelties to others, critically assessing the accusations and innuendoes as well as the purposes by his enemies (especially the physicist Ernest Lawrence, AEC commissioner and chairman Strauss, and Teller) and interpreting Oppenheimer's frequently elliptical, evasive prose. He chose on many important subjects—including, notably, the use of the A-bomb, his relationship with some who had befriended him, his attitudes toward the H-bomb, and his own loyalty-security case—to speak in ways that make understanding him difficult.

Most people, in living in the world and dealing with different situations and people, project somewhat different senses of self, depending upon the audience and the purpose. Few if any of the significant physicists of his generation were as straightforward as the estimable Hans Bethe. Somewhat like Oppenheimer, many more had peculiar and often self-serving relationships with women as did Albert Einstein, a pacifist and socialist, who championed the values of decency, peace, and generosity while in his private life sometimes operating very differently in dealing with women. And few could be so openly quirky and impish, but without meanness, and so delighting in enthusiasm and in political crusades, as the mercurial Leo Szilard, who could extol the values of being somewhat childlike. Yet, these individuals—definitely in political behavior and beliefs, and sometimes in their personal lives—usually seem considerably easier in many ways to understand than Oppenheimer.

Adding to the interpretive problems for historians, Oppenheimer cultivated that rare but sometimes self-destructive talent of implying what he did not utter, and of suggesting what he may not have meant. Such wordings sometimes allowed him to evade responsibility, though he might also seem to be taking responsibility for acts and decisions that were not truly his. He could and did use language deftly: to conceal and to evade, to evoke and to provoke. His phrasing could delight some and annoy others, sometimes seeming pretentious and sometimes profound.

Take, as a brief example of a segment of the Oppenheimer puzzles, a not-untypical letter he wrote in 1964 to the theoretical physicist Max Born, his aged German-Jewish mentor. In the postwar period, Born was a strong opponent of nuclear weapons and of the 1945 atomic bombing of Japan. He also had regretted, and resented, that Oppenheimer as head of the prestigious Institute for Advanced Study at Princeton

---

comment to 1942, as intended.) Often, interpreters of Oppenheimer, in focusing on his comment of "technically so sweet," fail to recognize that this 1954 phrasing—presented by him as accurately representing his 1951 thinking—was made in his loyalty-security hearing, where he was basically on trial partly because of his opposition to the H-bomb. See *JRO Hearing*, 251, 250, 229. Oppenheimer in that hearing had a vested interest in appearing as much for the H-bomb as possible for the period after Truman's pro-H-bomb public announcement of January 31, 1950. Generally, interpreters of Oppenheimer minimize this context and uncritically accept pro-H-bomb evidence for the post-Truman decision period.

had never even invited him to journey across the Atlantic to speak and stay for a time at that well-endowed institute.[21]

Significantly aiding Oppenheimer in an earlier time when the young Oppenheimer was struggling to find his way in physics, Born had helped rescue him in the mid-1920s from the pain of unsuccessful experimental work at the Cavendish Laboratory in Cambridge, invited him to Göttingen in physics, and helped set him on the path to achievement and recognition in theoretical physics. In 1964, a decade after Oppenheimer's loyalty-security case, and on the occasion of celebrating Oppenheimer's sixtieth birthday, the *Reviews of modern physics* had invited Born, among other notables, to publish testimonials about Oppenheimer in the respected journal. Despite some unexpressed ambivalence, Born had done so, in a statement that concealed any hurt he felt about personal mistreatment and Oppenheimer's earlier years of work in weaponry.

In 1964, acknowledging Born's kind words that had appeared in the journal, Oppenheimer replied privately to Born, thanking him in a letter that in part bears reading and pondering. Oppenheimer wrote: "Over the years, I have felt a certain disapproval on your part for much that I have done. This has always seemed to me quite natural, for it is a sentiment that I share; but for this reason the words you wrote were doubly welcome."[22]

What was Oppenheimer disapproving of in his own past? Developing the A-bomb, which Born deplored? Promoting its use on Japan, and especially on its cities in 1945, which Born condemned? Continuing to advise on developing nuclear weaponry, and how best to expand the American nuclear arsenal—all of which Born disapproved of—until Oppenheimer lost his clearance in 1953-54? Mistreating and slighting Born? All this? Some part? Something else?

For the historian, seeking to understand Oppenheimer's thinking in the post-1954 period about his earlier nuclear-weaponry past, this letter nicely indicates part of the nature of the Oppenheimer puzzles. What did Oppenheimer mean? Did he know? Was he enticed by an evocative phrase and perhaps not committed to any clear meaning on the subject? Was he purposely evading specificity, while intentionally acknowledging their great differences up to 1954 about nuclear weapons, their development, and use, and the responsibility of scientists?

The quest to understand Oppenheimer can illuminate more than Oppenheimer, his politics and personality, and the problems of his time. The very process of interpretation—the assessing of evidence, the weaving together of a narrative, the rejection

21. Born to JRO, 12 May 1964, JROP, box 22; and Born, "What is left to hope for?", *BAS* (Apr 1964), 2-5.

22. JRO to Born, 16 Apr 1964. This also followed Born's letter of 11 Dec 1953, in which Born complained to Oppenheimer that Oppenheimer's Reith Lectures for the BBC, dealing in part with the 1920s revolution in physics, had entirely omitted any mention of Born's pioneering work. Oppenheimer replied on 16 Dec 1953. On Born's resentment, stated in guarded ways, see Born to JRO, 12 May 1964, all JROP, box 22.

of some interpretive frameworks, the recognition and then the questioning of earlier assumptions in the Oppenheimer literature—offers a valuable opportunity, especially for someone like myself, who has wrestled with parts of the Oppenheimer puzzles for about three decades, to think critically about the logic and the psychology, in practice, of historical interpretation.

At what point do older interpretative frameworks get scrutinized, and how and why are some slowly rejected and others retained? Why in the writing about Oppenheimer, especially in the past dozen years of the word processor when rewriting is easier, are particular interpretations cast on paper and then overturned a few days or months later? How much more difficult is it to give up on an interpretation, and be truly open to new information and to rethinking a problem, if the historian (as in my own case) has already published a particular interpretation and thus staked out a position in public? Revising oneself is difficult, and there is usually considerable but rather unrecognized reluctance to give up a published interpretation, and even virtually to controvert it.

And what, for the historian, is the subtle intersection of the psychology and the logic of the historian's process of interpretation and then reinterpretation? Moreover, in the assessment of key evidence, and the decision of whether it proves a point or not, are not the judgments frequently but normally, implicitly, probabilistic? For example, do not many judgments rely—usually without clear statement—on various standards like "beyond a reasonable doubt," or "far more likely than not," or "somewhat likely," or "based upon the preponderance of evidence"?

In looking at a web of evidence on a troubling subject, should one aggressively pick each piece apart like a litigator and if finding a weakness, discard that piece? If so, is there not a danger of erecting unreasonably high standards for reliability of evidence? If there is a loose web of accumulating evidence, should not the whole amalgam of that evidence be assessed? Or, most reasonably, should not the historian synergistically range between these two strategies of assessment, and seek to be aware that most difficult interpretive matters are probabilistic, not definite?

But shouldn't higher standards for the reliability of evidence be imposed on evidence that makes a crucial interpretive difference on a major issue, such as whether someone was a Communist, and usually lower standards on whether, say, an unimportant conversation occurred on a Monday or a Tuesday? Or, as some historians may contend, should the standard for evidence be the same regardless of the importance of the particular issue? If so, can a sustained history be researched and all the relevant evidence assessed by using such consistent standards?

Oppenheimer, The Loyalty-Security Case, and Longtime Assumptions in Understanding His Left-Wing Politics

It is in this extended, yet briefly discussed context of the general Oppenheimer puzzles, and the related questions about the nature of evidence and interpretation, that the problem of Oppenheimer's political beliefs and activities can be analyzed. Ideally, were this paper many times longer, a host of questions could be considered. But for various reasons, this paper is more narrowly focused on Oppenheimer himself, his leftist past, its meanings, how that past can be interpreted, the problems of interpretation, and why and how it matters interpretively whether or not he had been a member of the Communist Party or just a fellow traveler.

With only a few exceptions, until the past handful of years, the standard interpretation by historians and most Oppenheimer defenders took Oppenheimer's 1940s words to security investigators and his affidavit and answers in his 1954 loyalty-security hearing at face value, or nearly so, regarding his left-wing involvements of the 1930s and very early 1940s. By his own admission, he had been—he repeatedly stated, in various words—a fellow traveler, a member of many front organizations, briefly and intensely so, but definitely not a member of the Communist Party. Thus, partly for that reason, to most defenders and later historians, the 1953-54 Oppenheimer case had been very unfair, the "prosecution" unwarranted, and Oppenheimer himself in fundamental ways innocent and deserving to retain his security clearance.

My own studies of the 1980s and early '90s of Oppenheimer and his loyalty-security case, while seeking to introduce more complexity into the analysis, ended up in basically the same interpretive place and with rather familiar assumptions: Oppenheimer had not been a Communist Party member, and implications and suspicions to the contrary were incorrect. What my work had sought was to provide a fuller appreciation of the political context of 1947-54, and thus to stress that the wrong questions had often been asked by historians and others about the origins and nature of the 1953-54 loyalty-security case. In view of the overwhelming evidence that no one else in the federal government with a high-level security clearance in 1947-54 could have had Oppenheimer's political past and associations, a key question about the origins of his case is not only why and how did that case occur in 1953-54, but, as importantly, why didn't it happen earlier? Why did it occur so *late*? How had Oppenheimer escaped earlier charges triggering an earlier case?[23]

Literally, no American physicist with Oppenheimer's self-admitted past of fellow traveling in the late 1930s and early '40s, or any other American non-scientist with a similar record, would have retained any security clearance, let alone a high-level security clearance into 1947, if that long. Added to that were the facts that Robert's brother and sister-in-law had been Communist Party members, that Robert's longtime girlfriend and brief fiancée (Jean Tatlock) had been a Communist, and that Robert himself was on the record, in 1943 and 1946, giving various conflicting stories to investigators about the 1942-43 Chevalier episode. Virtually even one of these items, if taken alone, would have made it impossible for anyone else to hold a security clearance, let alone the highest clearance. That highest clearance, in effect, gave him access to what were

23. Barton J. Bernstein, "In the matter of J. Robert Oppenheimer," *HSPS, 12:1* (1982), 195-252; Bernstein, "The Oppenheimer loyalty-security case reconsidered," *Stanford law review, 42* (Jul 1990), 1383-1484; and Bernstein, "The Oppenheimer conspiracy," *Discover* (Mar 1985), 22-32. The 1985 essay was written for a popular audience and often is not as carefully phrased.

regarded as the nation's most precious secrets—how to make nuclear weapons, what improvements and new programs were being devised and implemented, and the size of the nuclear stockpile.

In the security-conscious—one might say, not unfairly, security-crazed—postwar decade, Oppenheimer was, literally, unique in having his kind of past and associations and his kind of clearance. So, in part, the analytical problem to address is why and how Oppenheimer had received unique, truly exceptional treatment. Related to that was the problem of determining why the exceptionalism lapsed, why such damning formal accusations were made in 1953-54, why they were taken seriously by President Dwight D. Eisenhower and by the AEC, why Oppenheimer exercised his right to have a formal hearing, whether (as seems likely) he expected to triumph in the hearing, whether that expectation was reasonable even if wrong, and why and how he ultimately lost in the 1954 loyalty-security hearing.

Guided by such questions, that entire analytical process—with its awareness of contingencies and its effort to separate the highly contingent from the virtually inevitable—comfortably assumed that Oppenheimer, as he had contended, had not been a member of the Communist Party. There was no need to assert, much less to defend, that assumption in studies of the case and the man. It seemed self-evident and reasonable to most historians, including myself.

Not only had Oppenheimer repeatedly testified that he had never been a CP member, but the top officials in the FBI then and afterward, and postwar AEC investigators, had usually concluded that there was no good evidence that he had been a member. Lurking in the FBI and AEC files were pieces of evidence to the contrary from wiretaps, but they were fragmentary, often elliptical, suggestive at best, and certainly not conclusive, despite occasional contrary judgments in FBI reports.[24] If FBI Director J. Edgar Hoover, who reviled and mistrusted Oppenheimer, could not spur his agency to find good evidence that Oppenheimer had been a CP member,

24. See, for varying examples of assessments, "Julius Robert Oppenheimer," 5 Aug 1947, FBI file of JRO, #100-17828-16, sect. 7; [probably FBI director Hoover] to Tolson, Ladd, Belmont, and Nichols, 19 May 1953, #100-17828-[no. unreadable], sect. 14; A.H. Belmont to D.L. Ladd, 19 Nov 1953, #100-17828-549; and A.H. Belmont to L.V. Boardman, 13 Apr 1954, #100-17828-1135[?], sect. June mail. In the late 1970s, I received the Oppenheimer FBI files under FOIA, and in 1990 requested a new review, resulting in newly reviewed files (with one section reportedly lost) released in 2001-02. The process took *more than a decade*. Unfortunately, the FBI's reviewers rigidly and naïvely apply privacy restrictions to exempt materials, and thus on privacy grounds delete names of people who actually died *many* years before. The reviewers often have no knowledge of the field, including which of Oppenheimer's associates were dead or still alive in 2000-01. Futhermore, the FBI will not provide a general list of names mentioned in documents under review, so it seems impossible to inform the FBI of who is dead and then to provide the required evidence (usually an obituary). The 2001-02 release of files did not indicate in any document (among about 7,000 pages) what new lines or words had been restored since the 1970s review, so these could be disclosed only by a document-by-document comparison.

that failure seemed to establish easily for historians that Oppenheimer was truthful and had not been a member.

The issue of evidence, based on the substantially declassified Oppenheimer FBI files, was not simply one of the availability of legally admissible evidence. The issue was really the absence of anything that seemed like reasonably good evidence. The anti-Oppenheimer contentions by Paul and Sylvia Crouch, two former CP members, about an alleged 1941 Communist meeting at Oppenheimer's Berkeley home was a subject that the FBI vigorously pursued but could never really pin down. Far more was needed than the testimony by two ex-Communists, and the necessary firm evidence could not be found, though Hoover, deeply suspicious of Oppenheimer and frustrated by the absence of solid evidence, seemed to keep hoping that his agents could uncover some.[25]

Even in 1954, with Hoover and AEC "prosecutor" Roger Robb virtually "cherry-picking" for evidence that Robert had been a CP member, they produced nothing in the loyalty-security hearing that was truly damning, or even nearly persuasive, that he might well have been a CP member in earlier years. Thus, the FBI files could not make a powerful case on this issue against Oppenheimer; they could only suggest slivers of unsettling contrary evidence.[26] That was not enough to make open-minded historians uncomfortable, and certainly insufficient to persuade pro-Oppenheimer historians that he had been or even *might* have been a CP member years before. Any historian suggesting the contrary in print up through the mid-1990s or slightly later would have seemed unsound, and might even have been charged with red-baiting.

Chevalier: A Sometimes Dissenting Voice on Oppenheimer's Past

Up to about the mid-1990s, the most unsettling evidence on the subject, claiming that Oppenheimer had been a Communist Party member, was Haakon Chevalier's privately stated contention on a few occasions. In July 1964, as Chevalier was preparing to finish a non-fiction book on his relationship with Oppenheimer, he informed Oppenheimer that the forthcoming volume, allegedly partly written at the urging of Niels Bohr, was going to tell the whole story of the Oppenheimer-Chevalier relation-

25. On Paul and Sylvia Crouch and their contention of a secret CP meeting in July 1941 at Oppenheimer's house, see, for example, Paul Crouch, "Broken chains" (unpub. msc.), chapt. 29 ("Atomic espionage"), Paul Crouch Papers, Hoover Institution, Stanford University, box 12. Also see Paul Crouch to *New York herald tribune*, 15 Apr 1954, Crouch Papers, box 7. There are literally dozens of FBI reports and directives to seek to confirm the Crouchs' claims of this CP meeting in Oppenheimer's home. See, for example, Director, FBI, to Assistant Attorney General Warren Olney III, 7 May 1954, FBI file of JRO (ref. 24), #100-17828-1410, sect. 33; and Joe Craig, 24 May 1954, #100-17828-1584, sect. 37. On Oppenheimer's denial, JRO, 9 May 1950, Robert Bacher Papers, Archives, California Institute of Technology, Pasadena, CA, box 5.
26. See, for example, Ladd to Belmont, 19 Nov 1953, FBI file of JRO (ref. 24), #100-17828-549.

ship. That included, Chevalier stressed, their 1938-42 membership "in the same unit of the CP."[27]

Tucked in Oppenheimer's papers at the Library of Congress, Chevalier's 1964 letter, couched in the superficial form of decency and honesty, read (it seemed to me and others) actually like a kind of emotional blackmail ten years after the loyalty-security case. In 1954, Chevalier had slowly broken with Oppenheimer, when, upon reading the published transcript of the hearings, Chevalier had learned for the first time that Oppenheimer had implicated him in events of 1942-43 that sounded suspiciously like an indirect espionage attempt by Chevalier for the Soviets. After 1954, Chevalier had tended to believe, unreasonably, that all his postwar difficulties in getting jobs were because of Oppenheimer's secret contentions involving their meeting.

Chevalier's letter to Oppenheimer of July 23, 1964, merits brief quotation, drawing only on the two crucial paragraphs. Addressed familiarly to "Dear Opje," as Oppenheimer was often called by friends, Haakon Chevalier, no longer truly a friend, said: [28]

> The reason for my writing you is that an important part of the story concerns your and my membership in the same unit of the CP from 1938 to 1942. I should like to deal with this in the proper perspective, telling the facts as I remember them [in my book]. As this is one of the things in your life which, in my opinion, you have least to be ashamed of, and as your commitment, attested among other things by your "Reports to Colleagues," which I consider today make impressive reading, was a deep and genuine one, I consider that it would be a grave omission not to give it due prominence. Before going ahead, however, I think it only fair to give you an opportunity to say whether you have any objections. For reasons that are only to a minor degree personal, I shall do my best to respect your wishes in the matter.

Read in context, this letter seemed nasty, designed to punish, and conceived to make Oppenheimer, already crushed by the 1953-54 loyalty-security case, suffer again. Chevalier, based partly on his earlier mawkish novel, a lame *roman à clef* dealing somewhat with Oppenheimer, was, it seemed, heavy-handed, embittered, and eager for revenge. Chevalier was not a man whose judgment—in the view of most historians who read this July 1964 letter—should be trusted.

The evidence for such mistrust seemed further established by Oppenheimer's firmly clear reply on August 7, 1964. Unlike Oppenheimer's frequent use of multiple negatives and elliptical phrasing, he was direct and politely forceful in answering Chevalier's letter: [29]

> Your letter asks whether I should have any objections. Indeed I do. What you say of yourself I find surprising. Surely in one respect what you say of me is not true. I have

27. Chevalier to JRO, 23 Jul 1964, JROP, box 200.
28. Ibid.
29. JRO to Chevalier, 7 Aug 1964, JROP, box 200.

> never been a member of the Communist Party, and thus have never been a member of a Communist Party unit. I, of course, have always known this. I thought you did too. I have said so officially time and again. I said so publicly in response to what [Sylvia and Paul] Crouch said in 1950. I said so in the AEC hearings ten years ago.

Chevalier's rather mean-spirited and often distasteful book, *Oppenheimer: The story of friendship*, published in 1965, made no claims of Communist Party membership for Oppenheimer.[30] That seemed, further, to substantiate Oppenheimer's much earlier statements to the Manhattan Project, the AEC, and elsewhere, as well as his August 1964 reply to Chevalier, that Oppenheimer had never been a Communist Party member.

In the late 1970s, when Chevalier was interviewed on camera by the filmmaker Jon Else for his Oppenheimer documentary *The day after Trinity*, Chevalier never made any comments suggesting that he or Oppenheimer had been Communist Party members. By tone and emphasis in his on-camera comments and in off-camera discussions, Chevalier actually seemed offended by any such implications.[31] But Chevalier in the late 1970s, a decade after Oppenheimer's 1967 death and a quarter-century after the two men had privately broken, did come across on camera and off as a hopeless romantic and virtually a jilted lover. In his overly dramatic way, he seemed pathetic, somewhat like a suppressed hysteric, and given to great exaggeration—basically, an unsound person of unstable judgment.[32]

All that seemed further to confirm that his July 1964 charges should be ignored. But within a few years, after Jon Else's film had been produced and shown, Chevalier—to Else's shock—stated in private conversation with Else that both he, Chevalier, and Oppenheimer had indeed been secret Communist Party members in the late 1930s and early 1940s. Else, by his later reports, did not know what to think and tended not to believe Chevalier's seemingly peculiar accusations about Oppenheimer.[33] Else's judgment on this matter seemed eminently reasonable. After all, to repeat, Chevalier was still bearing a grudge and deeply resentful. There was also the obvious question of why Chevalier, if his accusation about Oppenheimer from the early 1980s was true, had not made the charge in the late 1970s to Else at the time of the filming of the Oppenheimer documentary.

In two sets of interviews with Chevalier in the early 1980s, the thoughtful Oppenheimer biographer Martin J. Sherwin, seeking to understand Oppenheimer and Chevalier's renewed claim that the physicist had been a secret Party member, sought to press Chevalier on just what Party membership meant and how it was even determined.

30. Haakon Chevalier, *Oppenheimer: The story of a friendship* (New York, 1965).
31. Jon Else, discussions with Bernstein, 1979-81 and 2000-05.
32. *The day after Trinity*. In consulting on this documentary film, which went through a number of stages as it changed from a rather long version of a few hours to the present version of nearly ninety minutes, I probably saw lengthier "takes" of Chevalier than are in the final version.
33. Jon Else, discussions with Bernstein, 2001-05. My compressed summary undoubtedly does not do justice to Else's evolving thinking.

Why, for example, asked Sherwin, did Chevalier conclude that he, himself, had been a member of the Communist Party, not just someone on the Left in a group of similar-minded people? Chevalier answered: "I don't know. We paid dues." What about being subject to Party discipline? asked Sherwin. Chevalier explained, in response to such queries by Sherwin, that they did not receive orders from the Party. "No," Chevalier said. "In a sense we weren't [regular Party members]."[34]

Chevalier went on to talk about the "kind of shadowy existence" in Berkeley of not being under Party orders, of not being open with others about Party membership, of being intellectuals who felt committed to Communism, paid dues to the Party, gave speeches, and wrote pamphlets. "Apparently," Chevalier further explained, "the same thing happened in many other parts of the United States, closed units for professionals or people who didn't want to be identified in any way."[35]

All this could certainly suggest that Party membership was a rather slippery matter, to be determined at the time, and even years later, by the person. By that conception, some participants in the closed group might think of themselves as members, and others not. How the Party itself conceived of these people was not clear, nor whether the Party's judgment should be regarded as binding on an historian in an effort, years later, to determine who was a CP member or not and what that person had reasonably believed at the time.

More Evidence on the Subject: Reportedly from the NKGB/KGB Archives

But in 1999, more evidence on the subject of Oppenheimer and CP membership—somewhat unsettling, if taken by itself—was presented in a book, *The haunted wood: Soviet espionage in America in the Stalin era.* The evidence was a quoted document from February 1944 in KGB files, sent to Moscow to Vsevoled Merkulov, the NKGB head, reporting that Oppenheimer had been a secret member of the Communist Party.

That 1944 document, not more precisely dated in the book and with specific authorship not provided, read in its key part:[36]

> "Chester"—Oppenheimer, Robert, born in 1906, American Jew, secret member of the compatriot organization [the American Communist Party], professor at the University of California, works on "Enormoz" in the field of rapid neutrons, directed the construction of the California cyclotron. In connection with the special significance

34. Chevalier, interviews with Sherwin, 1982, briefly summarized and with some quotations, Kai Bird and Martin Sherwin, "Talking points on J. Robert Oppenheimer and the Communist Party," presented at the J. Robert Oppenheimer Centennial at Berkeley, 22-24 Apr 2004.
35. Ibid.
36. Allen Weinstein and Alexander Vassiliev, *The haunted wood: Soviet espionage in America—The Stalin era* (New York, 1999), 184. The two authors, for emphasis, put part of the second paragraph in italics.

> and importance of his work, [he] is allegedly under special guard in consequence of which the compatriot section [the following is in brackets in the book as an explanation: the Communist cell to which Oppenheimer had belonged prior to beginning his work on the project] received from its center an instruction to cease relations with "Chester" to avoid compromising him.
>
> [Oppenheimer] represents a very large interest for us. His membership in the compatriot organization and friendly attitude toward our country provide grounds for counting on a positive result of his cultivation.

The two authors of *Haunted wood* further stated, "The assertion in the memo to Merkulov *and in other NKGB documents* [my emphasis] that Oppenheimer was a 'secret member' of the American Communist Party cannot be independently corroborated." [37]

The problems of assessing this February 1944 document, and the contention about the content of "other NKGB documents" with similar claims, were, and are, formidable. The book's endnote for the February 1944 document—to File 82702, page 143—was not really helpful, because those files were closed to others.[38] Adding to the problems, the book's claims about the content of "other NKGB documents" was entirely undocumented. But the problems were even greater, because neither the book's authors nor the nature of their project commanded trust in some academic quarters, especially for those who wished to doubt the document's authenticity and the accuracy of the claim about "other NKGB documents" regarding Oppenheimer's CP membership.

Haunted wood was and is a strange volume. It is the result of a peculiar commercial venture, one involving payments by Random House to the KGB's retired agents group, so that a few designated scholars and journalists could receive privileged access to previously closed NKGB/KGB files. Among the people benefiting from this arrangement were the book's co-authors—the American historian Allen Weinstein and a former Soviet journalist, Alexander Vassiliev, who had worked for Soviet intelligence as a KGB agent.[39] In the 1970s, Weinstein had written a dubious book on Alger Hiss and the Hiss perjury/spy case.[40] After publishing his study, and amid a controversy about his sources and interpretation, Weinstein had promised to make his files available to at least one unfriendly critic (Victor Navasky of *The nation*). But Weinstein continued to renege on that promise up to 1999 and apparently into 2004. Weinstein had also

37. Ibid., 184.
38. Ibid., 362, n. 54.
39. Ibid., xi-xii, and Weinstein, "Introduction," ibid., xv-xix.
40. Allen Weinstein, *Perjury: The Hiss-Chambers case* (New York, 1978). Weinstein was often praised and also not infrequently the subject of criticism in journals and elsewhere, including Victor Navasky, "Weinstein, Hiss, and the transformation of historical ambiguity into Cold War verity," Athan Theoharis, ed., *Beyond the Hiss case: The FBI, Congress, and the Cold War* (Philadephia, 1982), 215-245.

reportedly pledged to give his files to a major research library and presumably to open them, but had not done so into early 2004.[41] None of that inspired confidence for many in Weinstein, especially regarding his reports on the contents of otherwise closed Soviet intelligence documents in otherwise closed files. It was also hard to have confidence in a former Soviet intelligence agent who had worked for the KGB. Adding to the problems, inquiries to Weinstein about gaining a copy of the original Soviet document were unanswered. It was even unclear who had translated the document, if indeed the document existed, and how accurate the translation was.[42]

In addition, the reported 1944 document itself erred on some simple facts. Oppenheimer was born in 1904, not 1906. He really did not oversee the construction of Berkeley's cyclotron. He gave some advice, but physicist Ernest O. Lawrence had overseen the construction. If the document erred on such easily determined factual matters, how reliable was it in its basic claim about Oppenheimer's CP membership? Might a Communist agent in the U.S., seeking to make a good record to impress his Kremlin bosses, have exaggerated and reported to Moscow an achievement—Oppenheimer as a secret CP member—when he was not? Short of seeing more Soviet records from about the time on Oppenheimer and others, this was certainly a possible explanation of the troubling document, if it existed.

These many problems, overwhelming when amalgamated, left so many major questions that there seemed no good reason firmly to trust the book's report and quotation, but there was good reason to wonder about earlier assumptions about Oppenheimer and his claims of not having been a CP member. Those assumptions were coming to seem somewhat vulnerable, and possibly meriting close reexamination. What was once a comfortable assumption was being converted, slowly, into a question.

More Evidence, Further Unsettled Assumptions

The publication in Russia in 2000 of a collection of documents on the Soviet atomic project contained some similar information about Oppenheimer: that he had been a member of the American Communist Party. The volume's title (in translation) was *Atomic project of the U.S.S.R.: Documents and materials*, with L.D. Rybav as general editor; the relevant segment was Volume II, *The atomic bomb, 1945-1954*, Book 2. The key item was an appendix to a document, apparently prepared for Lavrenty Beria, Soviet commissar for security, by spymaster P.A. Sudoplatov. It was entitled, "The state of development of the problem of using atomic energy in the capitalist countries,"

41. In 2004, about the time Weinstein was nominated as archivist of the United States, he began making serious arrangements to send his papers to the Hoover Institution, Stanford University, and they may be opened—depending upon the speed of Hoover archivists finishing a preliminary inventory, and then Weinstein's granting final approval—in early 2006.

42. On translation matters, see *Haunted wood*, xii (mentioning three translators) and xvi (Vassiliev's role is less than fully clear).

and was apparently from no later than early January 1946. It stated, in discussing "Oppenheimer, Robert, 1906," that he was a "secret [literally non-open] member of the American Communist Party."[43]

Perhaps this early postwar document was simply a rehash of the February 1944 document quoted in *Haunted wood.* This new document erred identically on Oppenheimer's birth date, asserting 1906. Short of seeing this newly released Sudoplatov document in the original, and ideally in context in the actual archival file where it had been found, it was hard to assess its accuracy and its possible multiple purposes in 1946. It certainly seemed likely that the document was authentic, because there was no reason about ten years after the collapse of the Soviet Union for Russia to fabricate such a document for publication. Such disinformation on this issue did not seem likely by 2000. Furthermore, other information in this document—for example, that physicist Frédéric Joliot-Curie was a member of the French Communist Party—was accurate. But the information in the document on the German-born physicist Klaus Fuchs did not mention his role as a Communist spy or his membership in the German Communist Party. How could one interpret that omission? Was that subject too sensitive for inclusion on paper in this document for Beria, and, if so, why?

Dismissing this postwar Soviet document's statement about Oppenheimer was difficult. Trusting the document seemed somewhat risky. Cautious agnosticism—constituting a retreat from earlier assumptions that Oppenheimer had definitely not been a CP member—seemed the safest conclusion. Such a retreat, however, did not lead to the rather different conclusion: that Oppenheimer had been a member. Put otherwise: on the basis of that document, especially if taken alone, or even with Chevalier's 1964 letter and his early 1980s claims, and the 1944 document in *Haunted wood,* don't build any substantial analysis on Oppenheimer's membership or non-membership, keep an open mind on the subject, worry about the problem, and discuss it with other scholars. Above all, recognize that this was now an important question in understanding some aspects of Oppenheimer.

More Evidence—and/or Charges

In 2002, in a notably careless, highly contentious, and very dubious book, *Sacred secrets: How Soviet intelligence operations changed American history*, two American journalists with privileged access to some former Soviet spy files published another Soviet document contending that Oppenheimer had been a secret CP member. The two American authors, a husband-and-wife team, were Jerrold and Leona Schecter. Their book was often a peculiar hodgepodge of confusion, strident assertion, and very strained inference on espionage matters, including mostly Soviet espionage in the United States.

43. The translation and a copy of the title page and the relevant document on 371 were provided by David Holloway.

The Schecters' book printed an October 1944 document, in both the Soviet original and in translation, asserting that Oppenheimer had in 1942 been a secret member of the American Communist Party. Dated October 2, 1944, and apparently from or for file 1107M, the document was from Merkulov to Laverenty Beria, the Soviet commissar of state security. The document stated (with the parenthesis in the original and underlining by the translator for words handwritten in the original) that "in 1942 one of the leaders of scientific work on *Uranium in the USA,* professor Oppenheimer (handwritten) (unlisted [nglastny] member of the *apparat* of Comrade Browder) informed us about the beginning of work."[44]

Based partly on other paragraphs of the extended text beyond that quoted here, the Schecters sought to stretch the meaning of this document to conclude that Oppenheimer had been a Soviet spy.[45] That very dubious interpretation, pushing far beyond the limited and somewhat ambiguous evidence, was not essential to the more limited significance of the document: another statement, from long-classified Soviet files, that Oppenheimer had been a Party member, and a secret one.

This claim of secret CP membership, when added to the material in *Haunted wood* and in the Russian volume of 2000, and taken with Chevalier's letter of 1964 and his statement of the early 1980s to Jon Else, was very hard to dismiss. Taken alone, this October 1944 document could seem very suspicious. Given the fact that other writers could not gain access to the relevant Soviet/Russian file, and assess the context and authenticity of the document, there was room for some, maybe considerable, doubt.[46] Moreover, it was possible—though not provable—that this October 1944 document was simply a restatement of the February 1944 document that had been quoted in *Haunted wood* and that might have been the basis for the postwar report by the spymaster Sudoplatov.

44. Jerrold and Leona Schecter, *Sacred secrets: How Soviet intelligence operations changed American history* (Washington, 2002), 315 (Russian version) and 316-317 (translated document).
45. Ibid., 46-56, including their claim that in 1942, local agents following Moscow's orders persuaded Oppenheimer to break contacts with the Party. The Schecters' charge that Oppenheimer was a spy was the subject of a 2002 session at the Wilson Center in Washington, with Gregg Herken arguing strongly that the evidence did not establish that claim and Jerrold Schecter (it was unclear in reports whether Leona joined him in the presentation) seeking to sustain the accusation of Communist spying.
46. The document is not in the Jerrold Schecter Papers, a single box with many materials used in their book, deposited at the Hoover Institution, Stanford University, in 2002 and fully opened that year or early in 2003. *Sacred secrets* (ref. 44), xxvii, states that all the documents used in that book had been sent to the Hoover Library (meaning the Hoover Institution archives) and would be "available in ten years." I wrote to the Schecters to ask for a copy of the document, and did not receive a reply, but—partly because of the other two Soviet intelligence documents—saw no reason to question the existence of this document or to write a second time. David Holloway checked the translation from the Russian of the key segment of the document, using the published Russian version in *Sacred secrets.*

In view of so many contingencies, no firm conclusion on Oppenheimer's membership in the CP seemed strongly reasonable. But that also meant that the long-held assumption that he clearly had not been a member was also fully in question. It was difficult not to worry considerably that long-held assumptions had been wrong. It started to seem that *quite possibly*—no stronger words were acceptable—Oppenheimer had been a member of the American Communist Party. If so, it was important to think more deeply, and critically, about what this changed in the larger framework of Oppenheimer interpretation and even about the criteria for evidence on an issue like whether or not Oppenheimer had been a secret CP member.

Uneasy Recollections: Reconsidering Older Evidence—Discussions with Frank Oppenheimer

All that also meant reassessing earlier evidence, even very soft evidence—from discussions with Frank Oppenheimer, Robert's brother. In the late 1970s and into about the mid-1980s, I had probably about seven or eight phone conversations, as well as a few face-to-face meetings, with Frank Oppenheimer, sometimes in connection with our speaking at showings of Jon Else's documentary *The day after Trinity*. A number of times, in privately discussing his brother and aspects of Robert's 1953-54 loyalty-security case, and Frank's and his wife's (Jackie's) late 1930s and early '40s membership in the Communist Party, as well as FBI statements about Frank's paying dues to the Party, we moved uneasily onto issues of Robert's left-wing politics and the 1942-43 Chevalier-Robert Oppenheimer meeting.[47]

On such subjects, Frank was normally very edgy, sometimes uncomfortably asking me why I was inquiring, and more than occasionally squirmed. My own efforts were loosely—even naïvely—conceived; I wanted to know more, and I was somewhat puzzled about some events. For example, I knew that General Leslie Groves, the head of the wartime Manhattan Project, had always thought that Frank had been approached by Eltenton or Chevalier or someone else on A-bomb matters, and that Robert, in his shifting testimony about the Chevalier-Oppenheimer meeting, was at least partly seeking to protect Frank.

Occasionally, I asked Frank why he and his wife, as well as Jean Tatlock, had joined the Communist Party and why Robert had not. That line of questioning by me was desultory and occasional. It always assumed that Robert's many reports of not having been a Party member but only a fellow traveler were truthful. I was seeking to understand Robert's politics, not to challenge his reports of his politics.

47. Frank Oppenheimer was interviewed on camera (probably by Else) for *The day after Trinity*, and I was on two panels with him after the documentary appeared. I also sent him copies of my 1982 Oppenheimer case essay in manuscript and in published version, and we discussed it, as we did a copy I sent him of Robert Oppenheimer's May 1943 radiological warfare letter.

Yet, Frank's answers why Robert had not joined never seemed satisfactory, and Frank usually seemed very uneasy about these issues. Only years later, after the publication of some of the Soviet documents on Robert's membership, did Frank's behavior, viewed through the prism of this new information, leave me very suspicious. Perhaps Frank had been protecting Robert, knew that Robert had been a CP member, and was struggling, uneasily, to explain what Frank knew was untrue: that Robert had not been a member.

There was obvious danger in my engaging in a critical rethinking of those rather distant conversations with Frank. My notes on those earlier discussions were occasional and skimpy, nothing was rigorously recorded, and possibly I was recalling more uneasiness and squirming by Frank than had occurred. Or, as likely, a discussion of those themes involving his dead brother was upsetting to Frank for other eminently plausible reasons, and not because Robert had been a Communist Party member.

Thus, it seemed intellectually dangerous, and unfair to the evidence and the issues, to reach any comfortable conclusion from those remembered distant conversations with Frank about Robert and the question of Party membership. At most, by about 2002, it seemed warranted to conclude that there was good reason to worry about the rectitude of Robert's many denials of having been a CP member. Reinterpreting those earlier discussions with Frank Oppenheimer seemed somewhat to tilt the balance further *against* accepting Robert Oppenheimer's earlier denials of Party membership. Reinterpreting the conversations with Frank thus slightly added to the emerging conclusion that Robert *quite possibly* had been a secret Party member.

Conversations, Interviews, and Added Evidence: Two Who Had Known Robert in 1938-42

But there were other sources, people who had known Robert in 1938-42 who might be helpful on this subject. In the years after 2000, in connection with my research and teaching on Oppenheimer-connected subjects, I had occasionally spoken with Haakon Chevalier's ex-wife, Barbara Chevalier (she and Haakon had separated in about 1950), about her husband, his 1930s and '40s politics, the Berkeley left-wing world, some of Robert Oppenheimer's former students, Robert Oppenheimer himself, his personality and politics, her responses to the atomic bombing of Japan and the 1953-54 loyalty-security case, and related matters.

When we spoke about her ex-husband Haakon, whom she obviously did not like, she usually acknowledged that he had been in a secret unit of the Communist Party. But normally—with varying levels of less than conviction in her voice—she reported that she did not know about whether Robert Oppenheimer, too, had been a member. By mid-2002, as I came slowly to the conclusion that Robert quite possibly had been a member, I felt that I should call her to inform her of my emerging conclusion, because I was writing two essays involving Oppenheimer and I did not want her to be surprised by learning, for the first time, of my conclusion when it appeared in print.

In planning to discuss the subject with her, I felt very divided and even sketched some notes about her likely answers, what I preferred, and how her answers would influence my emerging interpretation. On the one hand, I rather hoped that she might cite and provide good evidence that I was wrong in my emerging conclusion, and thus I could return to my old and still-preferred assumptions: that Robert had not been a Party member. On the other hand, because the evidence to the contrary was accumulating, I hoped that she would strongly confirm it and thus reduce my doubts or even erase them, allowing me to write that Robert very probably had been a member. Her affirmative testimony on this subject, I decided in advance, would push me beyond a "quite possibly" judgment.

In our mid-2002 phone conversation, when I sketched the accumulating evidence (the Soviet documents) and explained my emerging conclusion about Robert having quite possibly been a Party member along with Haakon, she interrupted, stating abruptly, "Yes, they were both in the cell. But they weren't spies." To her, the secret unit was not the interesting issue—it was distant history and nearly irrelevant. She speedily went on to talk about what really mattered to her in mid-2002: 9/11, the possibility of political purges in the United States, another red scare, etc. The very speed of her reply that Oppenheimer had been in a secret unit, her clarity, and her emphatic voice were very different from her earlier comments, in the preceding period, about Robert and the Communist Party.[48]

It seemed clear that in earlier conversations she had been intentionally evasive, even knowingly deceptive, on the subject of Robert Oppenheimer and Communist Party membership. In this mid-2002 conversation, when I moved, however, to ask whether she had been a member, she uneasily replied, seeming variously to say first yes and then no. That subject was not central to my interest, and it seemed unfair, and indelicate, to push on the issue.

In a few later conversations, and most notably in early 2003, I asked her on a few separate times whether it was possible, and perhaps likely, that she and Haakon could have considered the group a "secret Communist Party cell" but that Robert might have thought of it as, simply, a left-wing discussion group. Her answer was that "Robert had to know that it was a secret Party cell." She repeated variations of this a few times, each time emphasizing that Robert definitely knew. She was indignant about the question, because it seemed, in her way of thinking, that Robert's intelligence or political

48. After interviews and conversations in 2000-01, the key interviews and conversations—in which Barbara Chevalier said that Oppenheimer had been in a secret Communist Party cell—were on 22 June, 25 June, and 19 Jul 2002, and then on 3 Jan 2003. The quotations are from the 22 June conversation, but she used similar words later. In two conversations, she also mentioned a journal/memoir on some of these issues and apparently read some lines from it to me on the phone. Gregg Herken read parts of that journal/memoir and placed excerpts on the web. For Herken's website see his essay in this volume; Herken also sent me some of these materials.

sophistication, or both, were being challenged.[49] When queried about the names of other cell members besides Haakon and Robert, she demurred. She replied, variously, that she could not remember, that she did not want to provide other names, and that she felt she was being asked to "name names" and she did not want to do that.

The interview in mid-2002 on Robert's membership, especially after her earlier denials and evasions, greatly tipped the evidential balance for me on the question of whether Robert had been a CP member. She had obviously been reluctant earlier to admit Robert's Party membership, and there was no reason thus to distrust her in 2002 when she spoke otherwise. She seemed to be speaking against interest, and even then perhaps somewhat reluctantly. She was not eager to talk in any depth about Robert and his CP membership. I soon learned that she had provided the historian Gregg Herken with similar conclusions on Robert's membership in a secret unit with Haakon.

Thus, the conclusion that Robert had been a CP member was looking rather strong. At about the same time some additional, but considerably less important information from another interview source seemed further to confirm matters. Between 2000 and 2002, in pursuing my interest in Robert Oppenheimer, I had a few conversations with Edith A. Jenkins, who in the 1930s and at least into the mid-'50s had been a Bay Area Communist and even, briefly in the '30s, a minor Party functionary. Most of my questions in discussions with Jenkins were about Robert Oppenheimer's personality, his relationship with women, and about Jean Tatlock (a friend of Edith's). Occasionally, the questions of Robert's politics and whether he had been a Communist Party member came up.

In our early conversations, Jenkins variously stated that Haakon Chevalier had been a Party member or might have been a member, and she said she did not know about Robert, while seeming to imply that he had not been a CP member. But by mid-2002, in a conversation, when I mentioned that substantial evidence was uneasily accumulating that Robert had been a Party member, apparently in a secret unit, she shifted and said, "Yes, I'd heard that." She was unclear whether she heard that in the 1938-42 period or perhaps somewhat later, and she said she could not recall the source. But she seemed to treat the conclusion as reliable—not as loose hearsay.[50] At one point, I asked her whether she could put a probability on the likelihood that Robert had been a member. She did not like the concept of probabilities; she thought of herself as a poet, not a scientist, she stated politely. But, finally, when I continued to suggest likely probabilities, she said that "something like fifty-five, sixty percent or maybe somewhat higher" seemed reasonable.[51]

49. Barbara Chevalier, interview with Bernstein, 3 Jan 2003. Barbara Chevalier was born in 1907, but her mental faculties and hearing seemed fine in 2002-03.

50. Edith Jenkins, interview with Bernstein, 25 June 2002.

51. She was born in October 1913, her nephew later reported, and her hearing in 2002 seemed much worse than in 2000. By the time of the composition of this essay, in 2005, when I sought to tell her of the recently released Gordon Griffiths memoir (see ref. 53) and to check on whether she had more to say about Oppenheimer and the CP issue, her hearing was reportedly so bad

Such evidence, if taken alone, was skimpy and rather weak: an aged woman, though apparently with full mental faculties, but hard of hearing, recalling issues from about sixty years before involving someone else. Admittedly, Robert Oppenheimer was not simply "someone else." And my suggesting and pushing for a probability had the liability that it may have prodded Jenkins beyond her autonomous judgment on the matter.

I queried her about what defined CP membership, and secret cells. She acknowledged that the standard of dues-paying normally defined CP membership, but she admitted that the issue could not be so simply resolved. On secret units, she mostly indicated—sometimes saying more than that, and often less—that she knew very little about such cells. Like Barbara Chevalier, Jenkins believed that Oppenheimer, if he had been involved in such a unit (and she thought it was very likely that he had been) could not have misunderstood what it was. He could not have believed that it was simply a secret discussion group, she said. That would have made no sense in Robert's case, she asserted.

Edith Jenkins's report added slightly to the balance on the Oppenheimer/CP issue. The considerably more important evidence was Barbara Chevalier's report, and the fact that Barbara's report was very similar to what she had told Herken, constituted very strong evidence *because of the Soviet documents*. All that supplemented and gave significant credibility to Haakon Chevalier's 1964 letter and his early 1980s comments to Jon Else. The uneasiness of Frank Oppenheimer, reinterpreted through the prism of this other evidence, seemed to add a bit more weight—not a great deal—to the accumulating evidence that Robert Oppenheimer had been a secret Communist Party member.

New Evidence: The Memoir of a Former Communist—Gordon Griffiths and Oppenheimer

Gregg Herken's *Brotherhood of the bomb: The tangled lives and loyalties of Robert Oppenheimer, Ernest Lawrence, and Edward Teller,*[52] with its claim that Oppenheimer

---

that she could not use the phone. Her nephew took my questions, seemed to believe that my queries were a kind of red-baiting, but promised to ask my questions of her. One day later, he informed me that she believed Oppenheimer had not been a member of the Party. Her nephew did not know what she had stated in 2002 to me. It was reasonably clear from this phone conversation with her nephew that, despite my earlier conversations with her, she did not remember who I was and may too have suspected that my query was part of right-wing red-baiting, which painfully reminded her of the postwar decade of McCarthyism. Thus, I am reluctant to base any judgment on her recent testimony (as reported by her nephew), but want to stress that her earlier testimony plays a marginal role in my final judgment and that the absence of that testimony—or even counter-testimony directly from her at this juncture—would not alter my conclusions in 2005.

52. Gregg Herken, *Brotherhood of the bomb: The tangled lives of Robert Oppenheimer, Ernest*

had been a member of a secret Communist cell, provoked the release of some added information on this subject: a useful memoir by a former secret Communist Party member involved in Berkeley in the early 1940s with Oppenheimer. The longtime University of Washington history professor Gordon Griffiths had written in his later years a still-unpublished memoir, in which he had described a secret Berkeley faculty group supporting the Communist Party.[53]

Griffiths had reportedly remained through his later life a man of the Left, but in apparently about the mid-1940s he left the Party. Many years later, he had written the memoir partly for his children and grandchildren, so that they would understand how and why he, a decent, humane man, had been in the Party, even though in later years Stalinism and the Party itself seemed so distasteful.[54] His memoir, in manuscript form, was provided with an introduction by "the children of Gordon Griffiths," dated December 2002. Gordon Griffiths had been a CP member, his adult children stated, from the mid-1930s to the mid-1940s. His memoir made clear that he had dropped out of the visible Party in fall 1940, partly because of conflicts with his wife, a British alien and recent British Communist Party member who feared that his membership in America would get them into trouble and hurt them. Unwilling either to leave the Party entirely or to be a visible open member, he still continued in the Party. According to the early draft of his memoir, he chose a comparatively safe, secret role as the liaison, from autumn 1940 to about mid-1942, to the Berkeley group. He named three members: "Arthur Brodeur, an authority on Icelandic sagas and *Beowulf* in the English Department, Haakon Chevalier of the French Department and Robert Oppenheimer of Physics."[55]

Griffiths's own phrasing asserted that this was a group of Communists who believed themselves to be Communists. He wrote of collecting "dues [he meant Party dues] from Brodeur and Chevalier [but] I was given to understand that Oppenheimer, as a man of independent wealth, made his contribution through some special channel." Griffiths went on, however, to state: [56]

Lawrence, and Edward Teller (New York, 2002). For my own stated judgment in mid-2003, see Bernstein, "Reconsidering the 'atomic general': Leslie R. Groves," *Journal of military history, 67* (Jul 2003), 897-898, and note 22.

53. Perhaps Herken and possibly Bird, or both, in late 2003 or early 2004 provided me with pages of the first (pp. 25-28) and second (pp. 15-16) versions of Gordon Griffiths, "Venturing outside the ivory tower: The political autobiography of a college professor." Griffiths was born in 1914 and died in 2001, though some publications incorrectly claim a 1915 birthdate. His son confirms 1914, as does the Library of Congress.

54. "The children of Gordon Griffiths," 30 Dec 2002, attached to parts of the memoir, and provided to me by David Griffiths. David Griffiths, interviews with Bernstein, Apr 2004 and Sep-Oct 2005. David Griffiths provided me with the entire memoir, in both versions.

55. Griffiths (ref. 53), draft one, 26; see his note, phrased in such a way that Griffiths strongly sides with Chevalier's claim that Oppenheimer *was* a CP member.

56. Ibid., 26-27.

> Nobody carried a Party card. If payment of dues was the only test of membership, I could not testify that Oppenheimer was a member, but I can say, without any qualification, that all three considered themselves to be Communists.

But that phrasing, by sliding from implying Party membership to a statement about those who believed they were Communists, could lead someone, in possibly somewhat straining the evidence, to conclude that some members were Communist Party members and some were actually not, and that Oppenheimer may well have considered himself a Communist but not a Party member. Griffiths did seem to lean heavily toward the conclusion that Oppenheimer was a Communist Party member, but only implied strongly—without clearly stating—that Oppenheimer had been a Party member.

Griffiths's less-than-clear phrasing could be used by those like Gregg Herken who were sure that Oppenheimer had been a Party member, and by those like Oppenheimer biographers Martin J. Sherwin and Kai Bird who seemed reasonably sure of the opposite: that Oppenheimer had not been a Party member, but only a fellow traveler. In my own judgment, as well as in the 2004 judgment offered by his son (Reed College physics professor David Griffiths), Griffiths seemed to make a reasonably strong case, but less than a definitive statement, for Oppenheimer having been a Communist Party member. That would further explain why an associate of Griffiths, in reading his memoir, strongly reacted to Griffiths's naming names and why Griffiths, in a later draft, removed these names from his manuscript.[57]

The fact that Griffiths served as a secret liaison agent for the Communist Party to this group added strongly to the evidence—but did not fully prove—that the group was actually a secret unit. But why else a secret liaison agent? After all, there were many in the open Party who could have performed the necessary tasks—distributing some Party literature, and collecting some dues, if this was only a discussion group. The fact of a secret liaison agent implied far more.

After Gordon Griffiths died, and after physicist Philip Morrison, an Oppenheimer student who had been a Communist Party member, also died, David Griffiths stated that his father had confided that Morrison had been the preceding liaison agent to the Chevalier-Oppenheimer group. As his son explained, Gordon Griffiths did not want to put this in his manuscript, even though Morrison had long ago publicly admitted his CP membership. Such an effort at discretion by Gordon Griffiths, who did not want to risk injuring people, was not surprising.[58]

But, significantly, Morrison apparently had not disclosed his formal liaison activity to Sherwin, when Sherwin had asked Morrison whether he had been a member of the Chevalier-Oppenheimer group. The crucial question was actually quite different, and Sherwin had no reason to know the right question to ask of Morrison. In turn, Morrison—apparently guilefully not forthcoming on something that might injure

57. "The children of Gordon Griffiths" (ref. 54).
58. David Griffiths, interview with Bernstein, 30 Sep 2005.

Oppenheimer more than a third of a century after the latter's death—carefully omitted this important fact.[59]

Would knowing that make a difference in the analysis of the Oppenheimer/CP membership issues? Perhaps. Should recognizing Morrison's selectively narrow answer make one wary of relying for other important, sensitive matters on his replies in an interview, even one conducted more than a half-century after the 1938-40 period? Probably—unless his other answers did not seem unduly protective of Oppenheimer and, ideally, could be checked in some key ways against reliable archival materials.

What for historians are important questions—in trying to make sense of Oppenheimer's past, and trying not to engage in red-baiting regarding matters now more than a half-century in the past—are questions about events that retained great salience for Oppenheimer's friends, even when they looked back after five or six decades. Understandably, they were often protective of him and of his past. The problem for historians, in using that interview material, is to seek to determine what should probably be trusted, what should probably be discarded, and how to define the process of judgment. All this raises very troubling problems about how to assess evidence, and how wary to be of some pro-Oppenheimer sources, especially on sensitive issues. For years, in contrast, most historians in dealing with interpreting Oppenheimer have been very wary, justifiably, about relying on Edward Teller or Lewis Strauss on many matters.

Assessing the Evidence

Sherwin's interview with Chevalier in the early 1980s, Griffiths's manuscript comments, and other materials help to emphasize that there is an important analytical problem in defining what was Communist Party membership, how one knew about one's being a member or non-member, and about others being members or not, and whether everyone in or near the Party agreed at the time, or years later, on how to determine membership. There is substantial evidence—even from interviews with others not involved in the Oppenheimer problem, and not knowing him—that membership was a somewhat slippery matter, that criteria shifted and were sometimes unclear.

Most Communist Party members, but apparently not all, had cards. So having a card did definitely mean membership. But not having a card did not definitely mean non-membership. The issue of Party discipline was also slippery. To many in the Party, apparently, there was Party discipline. To some, however, judging from their much later reports, they were not under discipline.[60] All of this undoubtedly was more complicated in the case of secret cells. Could someone be in a secret cell, not believe it was a cell,

59. See Kai Bird and Martin J. Sherwin's essay in this volume, and Morrison in U.S. Senate, Committee on the Judiciary, "Subversive influence in the educational process," 83rd Cong., 1st sess., Part 9, 899-919. My own interviews and discussions with Morrison in the late 1970s and early 1980s about Oppenheimer never got into this area.

60. The looseness of criteria in defining membership was underlined in interviews with Carl

think of it as only a Marxist discussion group, and thus feel that he or she was not a member of the Party? Possibly. Was this very likely? Apparently not.

Was it more or less possible that Oppenheimer, in view of his intelligence, but also in view of his capacity for evasiveness and equivocation, could have misunderstood the nature of the group? Or could others instead have misunderstood and transformed in the case of this group a study/discussion group into a Party cell, even a secret one? Barbara Chevalier, apparently not herself a member, believed the answer was no in the case of Robert. Might Frank Oppenheimer, however, have answered differently? If so, would he have been honest?

In the early 1950s, when questioned, the mathematician Kenneth May, who knew of the group in about 1938-42 and had recommended that Griffiths serve as the secret liaison agent starting about autumn 1940, had described it as not being a Party group. But May, having broken with the Party and still on the Left, was giving answers to representatives of the FBI, whom he obviously did not like but did not want to deeply offend. By his politics of the early 1950s, he was presumably not a man who was going to implicate others, and not Oppenheimer.[61] Should May's testimony to the FBI be trusted and used to determine the nature of that 1938-42 group and Oppenheimer's Party membership? Very probably not.

There are few easy questions and no easy answers in the case of Robert Oppenheimer and Party membership. Too much time has elapsed since 1938-42, all the alleged participants are now dead, and even their testimony, for various reasons, may well seem suspect. How would one assess candor? If their testimony was against interest, it should seem candid. But if consistent with interest? The effort at answers has to be woven into the framework that deals with the other related evidence, including, notably, but not exclusively, the three Soviet documents from 1944 to 1946.

One analytical strategy, which Oppenheimer's defenders on the issue of Communist Party membership are likely to adopt, is to work through each piece of evidence separately, set very high standards for acceptance, and then reject each piece, because some holes—or reasons for doubt—can be presented. In many ways, Sherwin and Bird,

---

Marzani, 1970-71, and Frank Oppenheimer, 1980-84, but there may be some analytical danger in using Jessica Mitford (see Bird and Sherwin's essay in this volume, n. 3) on the crucial 1938-42 period because she did not join the Party until 1944 (according to her): Mitford, *A fine old conflict* (New York, 1977), 62-64. Similarly, using CP leader Earl Browder's 1936 complaint (Bird and Sherwin, n. 4) as meaningful for 1938-42 is questionable, because 1936 was a very different time.

61. Kenneth May, interview, 4 Mar 1954, in Keith G. Teeter, 5 Mar 1954, FBI file of JRO (ref. 24), #100-17828-891. Comparing Griffiths's brief comments on May in his memoir with this FBI interview strongly indicates how much May was not forthcoming with the FBI. That is hardly surprising. But it should lead to some skepticism for historians analyzing what May told the FBI and treating it as candid and useful on Oppenheimer's 1938-42 political activities and involvements. For further evidence of May's elusiveness, see May to L M. Gould, President of Carleton College, 25 Sep 1950, Carleton College archives, Northfield, MN, courtesy of Carleton College archivist Eric Hillemann.

in their presentations at the 2004 Oppenheimer centennial in Berkeley, followed this thoughtful, honest strategy. Another strategy, pursued in the impressive Bird-Sherwin biography and in their intelligent essay in this volume, is to note some of the evidence, not refute some important parts (primarily the Soviet documents), stress the complexity of defining Party membership, emphasize that Oppenheimer obviously did not follow Party discipline, and conclude that he was basically honest and accurate in his later statements: he was not a Party member.[62] To make that case seem stronger, Bird and Sherwin thoughtfully weave pieces of evidence into a counter-narrative to seek to establish an alternative statement: he was a fellow traveler, but close to the Party, even perhaps closer than he admitted. Much of their counter-narrative, at crucial junctures, relies heavily upon later interview evidence—including, notably, from the physicists Hans Bethe and Victor Weisskopf, who were Oppenheimer loyalists. But there is no effort in the Bird-Sherwin essay to consider such problems of possible intentional or unintentional bias by Bethe and Weisskopf precisely because of their loyalty to Oppenheimer.

Bird and Sherwin, while starting out in their essay by saying that Oppenheimer was "[p]ossibly" a CP member, retreat speedily to assert that any effort at labeling Oppenheimer a member is "an exercise in futility," and ultimately—in both their narrative and their assertions—seem to conclude: he definitely was not a Communist Party member. They greatly privilege their counter-narrative by declaring—but *without any* trustworthy evidence in this key case—that the group's meetings were designed to "blur the boundaries of Communist Party identity." Adding to their interpretive problems, the Bird-Sherwin counter-narrative, unfortunately, sometimes places the crucial twenty-two-month period from the mid-1939 Nazi-Soviet pact to the mid-1941 German invasion of the Soviet Union as a time of the Popular Front in the United States, when it was not. For many anti-fascists on the American left, the August 1939 pact was a betrayal of values and meant a deep split with Communists.[63] Further

62. Kai Bird and Martin J. Sherwin, *American Prometheus: The triumph and tragedy of J. Robert Oppenheimer* (New York, 2005), 135-152. On 190-191, in a lengthy footnote, they discuss the two 1944 Soviet documents (not the 1946 one), but in the context of espionage charges—though perhaps the biographers intended their criticisms of inconsistencies between the two 1944 documents to impugn the reliability of those two documents regarding their statement that Oppenheimer was a secret CP member. If so, the discussion in the footnote is inadequate and requires far more substantial analysis. My criticism of this aspect of their distinguished book, and even the fact that I disagree with some other parts of the book, does not detract from my general admiration. Theirs is a very impressive biography of Oppenheimer, by far the best ever published. Oppenheimer is not the kind of subject, however, for whom there will ever be a "definitive" biography—too much of his life will remain subject to interpretive dispute. Yet all future writing on Oppenheimer will have to deal with *American Prometheus* as the landmark volume on the subject.

63. On the Popular Front and left-wing intellectuals, as well as the CP itself, see Maurice Isserman, *Which side were you on? The American Communist Party during the Second World War* (Middletown, CT, 1982), 32-110; and Harvey Klehr, *The heyday of American Communism* (New York, 1984), 389-409.

adding to interpretive problems, Bird and Sherwin seek—but unsuccessfully—to establish that Oppenheimer in November 1940 was a strong FDR supporter. Their single cited source, a biography of Ernest Lawrence, does not state, contrary to their assertions, that Oppenheimer *strongly* endorsed Roosevelt. Indeed, that biography is not even clear on whether Oppenheimer supported FDR in the November 1940 election—and that may be an additional key weakness in the Bird-Sherwin argument about Oppenheimer's politics in 1940.[64]

A competing strategy on the issue of Oppenheimer and the CP question, generally followed by those who contend that Oppenheimer definitely had been a CP member, is to gather each piece of evidence, use rather loose standards for assessing it, and reach a conclusion of "definite" Party membership. Herken's vigorous presentation in his thoughtful, honest April 2004 conference presentation comes close to this approach, as does the essay based upon it that he contributed to this volume.[65]

A middle way is to look first at each piece of evidence, then to look at the entire set of evidence, and to seek to offer a probabilistic conclusion. On such an analysis, there would be a clear rejection of the strategies used by the Sherwin-Bird advocates and the Herken advocates. It seems most reasonable to acknowledge that the evidence leans heavily in the direction that Oppenheimer was a CP member, but that any contention that he was definitely a member is excessive and any conclusion that he definitely was not is also excessive.

This middle way—even if some practitioners conclude that Oppenheimer probably was not a CP member—has the great advantage of avoiding the stark alternative of "definitely a member" versus "definitely not a member." Pushing the evidence from the available Russian/Soviet and American sources to either of those extreme positions seems dubious.

If the available evidence on the subject of Oppenheimer and CP membership were presented in a criminal case, in which "beyond a reasonable doubt" is the prevailing standard, Oppenheimer would be found "not guilty"—i.e., not a Party member. But if the evidence was weighed in the way that evidence is often assessed in civil cases, by the standard of "preponderance," where even fifty-one percent is adequate, then it seems likely that most analysts would conclude that Oppenheimer had been a Communist Party member. That means making very clear that the conclusion is something

64. Herbert Childs, *An American genius: The life of Ernest O. Lawrence* (New York, 1968), 307—though cited by Bird and Sherwin in their book and essay—offers *no direct, explicit* evidence on Oppenheimer's attitude toward FDR, and it is possible that Oppenheimer was mostly arguing against Wendell Wilkie, not for FDR. Childs's sourcing, 544-545, is not very helpful, but it might be possible to pin this matter down by working through his interview materials in TBL. Unfortunately, years ago, when I used them, the question of Oppenheimer and his 1940 politics did not seem especially important.

65. See, for example, Gregg Herken and Daniel J. Kevles, "The Oppenheimer case: An exchange," *New York review of books* (25 Mar 2004), 46-49. Also see Herken's essay in this volume. Herken's *Brotherhood of the bomb* (ref. 52) presented both Chevalier's and Oppenheimer's versions, but strongly leaned (on 31-32 and in our 2002 conversations) to Chevalier's position.

like "more probably [a member] than not," or perhaps even "far more probably than not [a member]," but certainly not "definitely [a member]."

Reaching beyond the important issues involving Oppenheimer, the problems raised in interpreting the evidence of his possible membership in the CP emphasize the need for historians, in analyzing and assessing difficult evidence, to move self-consciously and explicitly to using, even loosely, probabilistic standards—and then, in communicating with readers (or listeners) to make clear the probabilistic nature of the assessment. It is one thing to say that the likelihood on an important matter is about ninety-five percent or better. It is a very different matter to conclude that it is fifty-one or sixty percent or so. In the second situation, the evidence clearly leans near the margins, and a contrary judgment—that the evidence slides to the other side of the margins—may seem equally plausible or nearly so.

Unless historians on such subjects make some effort to communicate probabilities, they can easily talk past one another. They can also leave readers (and listeners) unaware of the standards being employed for assessing crucial evidence. The use of explicit probabilistic statements, even at the risk of suggesting undue specificity in judgment, may both improve the dialogue between historians on competing sides of issues, and educate readers (and listeners). Perhaps such awareness of probabilistic bases for judgment can, at least sometimes, even help historians resolve interpretive differences.

Dealing with the question of Oppenheimer's possible Communist Party membership creates other problems. For the historian who has already staked out a public position, in writings or lectures or both, in the years before the unsettling evidence began accumulating, it is very difficult to reassess the problems, freshly reconsider the evidence, and revise one's interpretation. It is far easier, and more comfortable, to reject the new evidence, to create very high standards for evidence, and stick with one's old position.

When the stakes are high for the interpreter, when professional reputation may be an issue, there is in most cases an understandable tendency to create, often without heightened awareness, standards for evidence that are so demanding that the old interpretation is maintained. In turn, those who desire to overturn older interpretations—and often there are professional advantages to such efforts, because the profession rewards newness—may frequently use, without so recognizing it, lower standards for assessing evidence. Put bluntly, but perhaps too simply, the criteria for evidence employed by a historian may frequently—though certainly not always—be in line with professional interest.

To me, looking back on my own struggles with the new evidence as it emerged, I think that I had a vested professional and psychological interest in not giving up the older framework: that Oppenheimer had been truthful, and had not been a Communist Party member. That may have led me, for a time, to erect unusually high standards for assessing evidence. In turn, possibly because I was not under public attack for my earlier operating assumptions, and in some years taught about the Oppenheimer interpretive problems to new classes (who did not know what I had written), it may have been easier, than it would have been otherwise, to move slowly between 1999 and 2002 to revise my older assumptions. Had my earlier work on Oppenheimer been very central to my scholarship in the past quarter-century, rather than an interesting problem dealt with periodically, I think I would have been even more resistant to change.

Had I greatly liked Oppenheimer, and not felt often ambivalent about him on other issues, undoubtedly my resistance would have been greater. Briefly describing my own process allows me to indicate on paper to others some of the psychological problems of resistance, and to suggest the nature and difficulty of revising conclusions.

If Oppenheimer Was a Party Member, What is the Significance?

What changes analytically, for understanding and interpreting Oppenheimer, if indeed, as seems likely, he had been a Party member and in a secret unit around 1939-42? Basically, such a conclusion does not significantly alter the understanding of his political beliefs in left-wing reform at home in the United States. His own statements over the years emphasized that he had been, in effect, a fellow traveler. He stressed his concern about injustice, about a very deep sense of right and wrong, in becoming involved in various left-wing and Party-supported organizations and causes.

But if he was a Communist Party member in the post-August 1939 period, *after* the Soviet-Nazi pact of that year, that does open some serious questions about his attitudes toward the Soviet Union in the critical period up to the German attack on the Soviet Union in June 1941. Pushing further inquiry into that twenty-two-month period might help untangle matters and thus enrich the understanding of Oppenheimer's attitudes toward the Soviet Union, and the process and dating of his moving away from a pro-Soviet perspective. At present, the 1938-42 period, in terms of his political beliefs and activities, remains in some important ways rather shadowy. If he sharply criticized the Soviet Union in 1940, and very much wanted the United States with the fall of France in 1940 to enter the war in Europe, there would be one conception of Oppenheimer in this crucial period. If, on the other hand, the fall of France upset him, but he opposed American armed intervention, the conception of Oppenheimer would be rather different.[66]

66. Herken relies heavily on two Communist Party pamphlets from 1940, one of February 20 and one of April 6. For the pamphlets, see College Faculties Committee, Communist Party of California, "Report to our colleagues: I" (20 Feb 1940), and ibid., "II" (6 Apr 1940), TBL, University Archives, Office of the President files, 1940, folder Communism. Herken concludes that the pamphlets help establish that Oppenheimer was a CP member, because they include his phrasings in places, because they use poetry that he liked, and, further, because Chevalier stated that Oppenheimer was the author. Bird and Sherwin, in their essay and in their book, struggle to whittle away such claims leading to Herken's conclusion. They treat the first pamphlet as partly by Oppenheimer, but are rather unsure about the second, using some "if" phrasings. They contend that the pamphlet's assertion of FDR's being a "war monger" is at odds with other (alleged) evidence on Oppenheimer in 1940. But they fail to emphasize that such evidence is, first, largely from *much later* interview and memoir statements and, second, a strained interpretation

Probably as important in interpreting Oppenheimer is the fact that, if he was actually a CP member, he lied about the matter, repeatedly and sometimes under oath. In view of other lies in which he was caught—whether those of the heart or groin involving Jean Tatlock, Kitty, Ruth Tolman and Richard Tolman, or on the Chevalier-Oppenheimer matter—there would be nothing greatly shocking to learn that he lied on one more matter: CP membership. It is not as though Oppenheimer is normally viewed as staunchly and painfully honest, a man who almost never stooped to duplicity on personal or political matters. Most biographers, if only because of the Chevalier-Oppenheimer matter, have a more alloyed and thus more nuanced view of Oppenheimer, a very complicated man. After all, he could not even be candid on the comparatively small issue of whether the "J." in his name was an abbreviation or not.

What would make his lies about CP membership strikingly distinctive, and perhaps unique for him in his life, is that these lies sometimes involved perjury, with great legal penalties. It would also mean, having earlier lied repeatedly about former CP membership, that in 1953-54, as he moved toward a loyalty-security hearing and then participated in it, he was greatly vulnerable to some firm evidence that might emerge from the FBI inquiry to establish that he had earlier committed perjury.

On the eve of the 1954 loyalty-security hearing, one top FBI official was thinking along these lines: that Oppenheimer would again, under oath, deny Communist Party membership, and that the FBI evidence might lead to a "possible perjury proceeding." Shortly after the hearing ended, J. Edgar Hoover explained to a lower-level FBI official: "Since his [Oppenheimer's] testimony was under oath [that he was never a Communist Party member], investigations may [he meant should] now be directed with a view toward possible prosecution providing legally admissible evidence can be obtained of

---

of a few sentences in the biography of Ernest Lawrence. Bird and Sherwin cite for support of their position a May 1941 letter by Oppenheimer to Edwin and Ruth Uehling, in Smith and Weiner (ref. 12), 217, but that letter in my judgment is very ambiguous on the key matters.

I have wrestled with the problems of the pamphlets, and the issue of Oppenheimer's authorship, especially for the second pamphlet, and with what conclusions can be drawn if (at most) he only wrote pieces of the second. There are so many contingencies, and the evidence is of such a nature, that it is difficult for me to come to a useful conclusion—whereas Herken seems sure that the evidence decisively establishes his position, and Bird and Sherwin strongly conclude it does not, generally pushing the pamphlet evidence to support their conclusion that he was a fellow traveler but very close to the Party. After two or three years of looking at the evidence and mulling it over, I remain torn about what conclusions from the pamphlet evidence are *most likely* to be reasonable. I think that definite judgments on the pamphlets in determining the question of Oppenheimer and CP membership push too far.

The virtual absence in Oppenheimer's own papers of useful political correspondence for 1938-42 seems peculiar. Robert Serber destroyed all of their correspondence, fearing that it might be used in later loyalty-security matters. Frank Oppenheimer's papers, in TBL, are not helpful. Nor are other collections, including Richard Tolman's at Caltech and Felix Bloch's at Stanford. I have not double-checked Hans Bethe's (Cornell) and I.I. Rabi's (Library of Congress), nor examined Victor Weisskopf's (MIT).

actual Communist Party membership."[67] Hoover and top associates were eager to find the evidence to indict and convict Oppenheimer. Other anti-Oppenheimer partisans like Strauss would have been delighted.

Most scholars who have studied the Oppenheimer loyalty-security transcripts have watched him crumble under the clever assaults by "prosecutor" Robb. The brilliant, normally impressive Oppenheimer—known for his verbal agility, for his skillful equivocation, for his arrogant assertiveness, and for his intimidating presence—withered in the inquiry room. The normal explanation has been, reasonably, that such an inquiry, like a criminal trial, is a venue where even the stout-hearted and the skillful, because their own being is at risk, prove vulnerable, lose a sense of presence, and often fall apart. All that long seemed to make good sense of what happened to Oppenheimer in the nearly four-week hearing in early 1954.

But *if* it turns out that he had also earlier committed perjury on the issue of his CP membership, and also did so at the hearing, as seems likely, then he knew what his attorneys at the hearing could not: that he might be plunged into a situation in which it was not simply the loss of his security clearance and the question of his loyalty that was at stake. The lurking, greater threat was an indictment and conviction for perjury, with painful criminal penalties.

Some readers of this essay may conclude that this implication—the danger of a perjury indictment and conviction—is pushing "if" matters too far. But if Oppenheimer had been an earlier CP member, placing that possibility of a perjury charge in the context of the 1954 hearing does not seem unreasonable. Viewing him as unusually vulnerable, because of his earlier, literally criminal lies, makes reasonable sense and seems analytically appropriate.

All this indicates the importance of doing more research and more thinking about Oppenheimer and the question of Communist Party membership. The case is not closed, research remains to be done, and there is need for continuing thoughtful dialogue by scholars seeking to understand Oppenheimer, his politics, and his travails. Such dialogue, ideally, should involve, as it has for key participants, the sharing of information and archival materials in the honest—and sometimes spirited—wrestling with these problems.[68]

67. A.H. Belmont to L.V. Boardman, 13 Apr 1954 (ref. 24); and Director, FBI, to SAC, Newark, 12 May 1954, FBI file of JRO (ref. 24), #100-17828-1444, sect. 33. There are other FBI documents from this period and in the summer, seeking to pursue the perjury issue.

68. This interpretation emerges, in addition to the 2004 Oppenheimer Centennial in Berkeley, from many discussions with David Holloway, especially in 1999-2005, and in a number with other longtime friends, Martin J. Sherwin, Kai Bird, and Gregg Herken in recent years. Parts of this interpretation have been developed in teaching at Stanford and in various programs' seminars and lectures at Stanford and elsewhere. The differences with Herken on the one side and Sherwin and Bird on the other in no way alter my basic respect for their general work, and these are simply differences—sometimes strongly stated, and sometimes not—between friends who have been privately discussing (and sometimes disagreeing about) Oppenheimer for years and often sharing various archival and related materials to help one another.

None of this, however, should be conflated with the unproven and apparently very exaggerated charges by the Schecters, Sudoplatov, and a few others that Oppenheimer was an atomic spy. Sherwin and Bird, and Herken, as well as myself—while disagreeing on the important issue of whether Oppenheimer was actually a Communist Party member for a period—fully agree that there is no reasonable evidence that J. Robert Oppenheimer was a spy. Those who claim otherwise have stretched skimpy, questionable, and often ambiguous sources to conclusions that cannot be supported on the basis of the available evidence and that therefore should not be accepted by responsible interpreters of Oppenheimer or of modern American and Cold War history.[69]

69. Significantly, the historian Harvey Klehr, who recently co-authored with John Earl Haynes a very spirited book condemning specified historians for not stressing, or sometimes not finding, communist espionage (*In denial: Historians, Communism, and espionage* [San Francisco, 2003]), did not contend in his recent review that Oppenheimer was a CP member, though Klehr noted Herken's conclusions. Klehr, "Father of the A-bomb," *Weekly standard* (6 June 2005), 31-33, called to my attention by Charles Palm. *In denial* does not discuss the question of Oppenheimer's possible CP membership. Bird and Sherwin's essay in this volume quotes from an e-mail message from Haynes to Herken (n. 69), but the quoted segment does not deal directly with the key issue of Party membership. It seems to deal mostly with incentives and Party discipline, and no historian arguing that Oppenheimer was or may have been a CP member has asserted, to my knowledge, that Oppenheimer operated under Party discipline. For a thoughtful essay raising critical issues about FBI records and espionage conclusions by historians, see Shawn Mullet, "Intelligence files and the unlearning of atomic espionage: Joseph Weinberg and Giovanni Rossi Lomanitz" (unpublished msc.).

PART THREE

The Soviet Comparison
New Insights

From the beginning all the way up to the present, the Soviet nuclear weapons project has served as foil for the American effort. It preoccupied Oppenheimer's minders in the Manhattan Project, fixated by the possibility of espionage; it drove his postwar attempts to achieve international control of atomic weapons just as surely as it lit up his opponents' mistrust of his stance on the hydrogen bomb. Throughout the arms race, Soviet progress vis-à-vis American weapons production helped drive the dynamic of the Cold War. More recently, Soviet and post-Soviet sources have inspired careful reanalysis of the dual courses of the two nations' nuclear projects. That effort has shed new light on the conditions under which Oppenheimer's wartime labors could find the conclusion they did.

The essays in this section come from two experts on the Soviet project who are also knowledgeable about the U.S. story. Focusing on the individuals at the center, David Holloway explores parallels between Oppenheimer and his Soviet counterpart, Iulii Khariton. Similarities in background and differences in role make Holloway's contrast instructive. At the scale of the national venture, Alexei Kojevnikov puts newly available material in the service of revisiting key aspects of the Soviet atomic project.

His account proposes some unexpected ways in which the American and Soviet efforts may been seen as converging— not just in their material basis, but their organizational structure and their consequences for the character of postwar science.

CHAPTER SIX

Parallel Lives?

Oppenheimer and Khariton

David Holloway

Introduction

Iulii Borisovich Khariton is sometimes called the "Soviet Oppenheimer" because he led the development of the first Soviet nuclear weapons. In 1995 the J. Robert Oppenheimer Memorial Lecture Committee invited Khariton to give the Oppenheimer Memorial Lecture in New Mexico. He very much wished to accept the invitation but ill health prevented him from doing so. He sent a letter to the Committee giving an account of some of his work, and in that letter he wrote: [1]

> Unfortunately I do not know very much about Robert Oppenheimer's personality, but what I do know compels me to regard him with great respect. Reading about his life, I have noticed several amusing coincidences in our biographies. Julius Robert Oppenheimer (his first name is the same as mine) was born in the same year as I was, 1904. His mother, like mine, was connected with the arts and evidently inculcated in him an interest in music, painting and poetry. In 1926 Oppenheimer spent a short time in Cambridge in Rutherford's laboratory, where I worked from 1926 to 1928. Unfortunately I do not remember him.

The parallels or coincidences can be continued. The two men were of course the first directors of their countries' nuclear weapons laboratories, at Los Alamos and at Sarov. Both came from cultured, assimilated Jewish families. And—if one is to believe recent revelations about Oppenheimer—both were members of the Communist Party. Khariton joined the Party in 1956, when he had already been scientific director of the institute at Sarov for ten years; if Oppenheimer did indeed join the Party, he was a member in the years 1938 to about 1942.

I am grateful to Barton Bernstein and Norman Naimark for comments on an earlier draft of this paper, and to Aleksei Semenov for many conversations on this topic. I am planning to write a longer essay on the issues raised in this paper.

1. "Pis'mo Iu.B. Kharitona v memorial'nyi komitet R. Oppengeimera," V.I. Gol'danskii, ed., *Iulii Borisovich Khariton. Put' dlinoyu v vek* (Moscow, 1999), 211.

There is another curious coincidence. In the United States the second nuclear weapons laboratory bears the name of Ernest Lawrence (the Lawrence Livermore National Laboratory); in Russia the second nuclear weapons laboratory, at Snezhinsk, is named for E.I. Zababakhin, who was its scientific director from 1960 to 1984. In spite of two resolutions passed by the State Duma, the institute at Sarov has not been named for Khariton, for reasons that are not wholly clear; but one reason sometimes given is that the Los Alamos National Laboratory does not bear the name of Robert Oppenheimer.

There are of course crucial differences in the biographies of the two men, apart from the fact that one lived in the United States, the other in the Soviet Union. The atomic bombs developed at Los Alamos under Oppenheimer's direction were dropped on Hiroshima and Nagasaki, but none of the Soviet nuclear weapons were used in war. After Hiroshima Oppenheimer became a public figure, associated around the world with the atomic bomb and the scientist's responsibility for entry into the atomic age. Khariton, by contrast, remained a very secret figure until the 1980s. In this essay I sketch some of the parallels and differences in the lives and careers of two men.

Oppenheimer's life and work have been examined in many books and articles, and he is soon to be the subject of an opera by John Adams (*Doctor Atomic*). He remains nevertheless an enigma, as the title of a recent book puts it, because of the complexity of his character. Khariton's life and work are difficult to study, for different reasons. He lived most of his life in a secret world. Yet the parallels and the differences between the two men suggest a number of questions that merit further reflection.

Biographical Background

Khariton was born in St. Petersburg on February 27, 1904, less than two months before Oppenheimer, who was born in New York on April 22. Khariton's father, Boris Iosifovich Khariton, was a well-known journalist, and his mother, Mira Yakovlevna Birens, an actress at the Moscow Arts Theatre. His parents separated when he was quite young, and his mother went to live abroad in 1910. She remarried in 1912 and lived in Berlin until 1933, when she and her new husband, Max Eitingon, went to Tel Aviv, where he had a position in psychiatry. (Eitingon was a close collaborator of Freud and founded the first psychoanalytical clinic in Berlin after World War I.)[2] Khariton's mother died in Israel in the late 1940s. Even before his parents separated, Khariton appears not to have seen a great deal of them. His mother was in Moscow at the Moscow Arts Theatre; his father was busy as editor of the newspaper *Rech'*, an organ of the Kadet (Constitutional Democratic) Party. His Estonian governess, Rozalia Loor, occupied therefore an important position in his life. She taught him and his two sisters to speak German almost as freely and easily as he spoke Russian.[3]

2. Biography of Max Eitingon, Deutsche Psychoanalytische Gesellschaft, at http://www.dpg-psa.de/in_ge_gesch_eitingon.htm.

3. I rely here exclusively on Khariton's own reminiscences, "Iz biograficheskikh zapisei Iu.B.

Oppenheimer's father was born in Hanau, Germany in 1871. He came to the United States at the age of seventeen and worked in his uncle's textile firm in New York; later he set up his own company, which imported linings for suits. In 1903 he married a young painter, Ella Friedman. The marriage was a happy one, and the family was prosperous.[4] Conditions of life were easier in New York in the first two decades of the twentieth century than they were in St. Petersburg, especially during World War I and the years after the October Revolution. Oppenheimer had a more comfortable family life than Khariton, but both grew up in a cultivated, non-observant, Jewish milieu where they acquired, especially perhaps from their mothers, a knowledge and love of European culture.

Khariton was educated at home until he was eleven years old. Restrictions on Jews made it difficult for him to enroll in a public gymnasium, and in 1915 he entered a not-too-expensive commercial school, where he managed to skip two years, graduating in 1919 at the age of fifteen. Students were not permitted to begin higher education until the age of sixteen, and he spent a year working as a mechanic in a railway telegraph workshop before entering the Faculty of Physics and Mechanics of the Polytechnical Institute in Petrograd in 1920. During his first year there, he took a course of lectures from A.F. Ioffe, the director of the newly created Institute of Physics and Technology, and decided to become a physicist. At the end of his first year N.N. Semenov invited him to work in his laboratory at Ioffe's institute while continuing his studies at the Polytechnical Institute. Evidently Semenov regarded Khariton as a very promising student. Khariton graduated from the Polytechnical Institute in 1925, after completing his five-year course.

In 1911 Oppenheimer entered the second grade at the Ethical Culture School, which had been founded in 1895 by upper-class New York Jewish families who found that the better private schools would not admit their children.[5] Oppenheimer graduated from the high school division in 1921. He distinguished himself as a brilliant student. He intended to begin his studies at Harvard in the fall of the same year but fell seriously ill on a summer trip to Bohemia, where he visited the old mines at Joachimsthal—famous as the source of uranium—in order to prospect for minerals; this was a youthful interest of his. He spent most of 1921-1922 recuperating, and it was not until the fall of 1922 that he started at Harvard, where he majored in chemistry, graduating *summa cum laude* in 1925.

Khariton's father, Boris Khariton, was active in the literary life of Petrograd after the October Revolution. The Kadet newspaper was closed down as the Bolsheviks brought the press under control, but Boris Khariton became director of the House of Writers, which the Union of Journalists opened with the permission of the authorities.

Kharitona (1978-1979 gody)," Gol'danskii (ref. 1), 16-41.

4. On Oppenheimer's early life see Alice Kimball Smith and Charles Weiner, eds., *Robert Oppenheimer: Letters and recollections* (Stanford, 1995), 1-10.

5. Jeremy Bernstein, *Oppenheimer: Portrait of an enigma* (Chicago, 2004), 10.

The younger Khariton recalled that hearing Mayakovsky read his own poetry there in 1920 created a powerful intellectual impression on him. In August 1922 the GPU (the political police) surrounded the building of the House of Writers and arrested the directing staff, including Boris Khariton, who was sent into exile later in the year as one of a group of about two hundred "ideologically alien" intellectuals. It was Lenin who decided on the policy of exile. The expulsion of this group indicated that the Bolsheviks would not tolerate independent-minded critical intellectuals in Soviet Russia.[6]

Cambridge

Khariton and Oppenheimer studied in Cambridge in the mid-1920s, but they did not meet each other. Oppenheimer was there for the 1925-1926 academic year. Rutherford did not at first admit him to the Cavendish Laboratory. The letter of recommendation from Percy Bridgman, Oppenheimer's physics professor at Harvard, may explain Rutherford's hesitation. "It appears to me," Bridgman wrote,[7]

> that it is a bit of a gamble as to whether Oppenheimer will ever make any real contributions of an important character, but if he does make good at all, I believe that he will be a very unusual success, and if you are in a position to take a small gamble without too much trouble, I think you will seldom find a more interesting betting proposition.

In spite of the initial rejection, Oppenheimer went to Cambridge and managed to find a place at the Cavendish Laboratory.

Oppenheimer was unhappy in Cambridge, where he suffered a nervous breakdown, for intellectual as well as personal reasons. He was assigned to work with J.J. Thomson, who had received the Nobel Prize in 1906 but was far removed from the latest developments in physics. "In my years at the Cavendish Laboratory," Khariton wrote later in his autobiographical notes, "Thomson had a small room there in which two or three young people worked. But what he was doing was somehow apart from the main direction of the physics of the time."[8] Despite his frustration with experimental work, Oppenheimer "began to develop his own taste in physics" during his year at Cambridge.[9] He attended seminars and went to meetings of the Kapitsa Club, an informal discussion group founded by the Russian physicist Peter Kapitsa. He came to realize that it was theoretical physics that interested him, and he accepted an invitation from Max Born, who would receive the 1954 Nobel Prize, to continue his work in Göttingen. Oppenheimer left Cambridge in the late summer of 1926.

6. Stuart Finkel, "Purging the public intellectual: The 1922 expulsions from Soviet Russia," *The Russian review, 62* (2003), 589-613.
7. Smith and Weiner (ref. 4), 77.
8. Gol'danskii (ref. 1), 33.
9. Smith and Weiner (ref. 4), 96.

It was Kapitsa, a close friend of N.N. Semenov's, who helped Khariton go to Cambridge. Kapitsa had gone to Cambridge with Ioffe in 1921, and Ernest Rutherford had agreed to take him into the Cavendish Laboratory. Kapitsa quickly established himself as an outstanding physicist and a great favorite of Rutherford's. During a visit to Leningrad in early 1926, Kapitsa heard a report from Khariton about the work he was doing in Semenov's laboratory on the oxidation of phosphorous vapors at low pressures. He suggested that Khariton come to Cambridge to do his Ph.D., and in the fall of 1926 Khariton traveled to England to spend two years there. Oppenheimer had left Cambridge shortly before Khariton arrived early in November.

Rutherford accepted Khariton at the Cavendish Laboratory, thanks no doubt to Kapitsa's recommendation. While he was there Khariton came across an article by the German chemist Max Bodenstein criticizing the experiments he had done on phosphorous while working in Semenov's laboratory. Bodenstein described as an experimental error the finding that oxidation did not take place below a critical pressure of the oxygen. Khariton wrote to Semenov to ask him to take this up. Further experiments by Semenov confirmed Khariton's results and led to the work on chain reactions for which Semenov received the Nobel Prize for Chemistry in 1956. While he was in Cambridge, Khariton worked with James Chadwick on the sensitivity of the eye to weak impulses of light and on alpha-radiation; he was awarded his Ph.D. in 1928. He was impressed by the simplicity and ingenuity of experimental work at the Cavendish, and that may have reinforced his own careful approach to experimental physics. He became close to Kapitsa and maintained friendly relations with him to the end of Kapitsa's life.

What might have happened if Oppenheimer and Khariton had overlapped at Cambridge? Would the two twenty-two-year olds have become friends? The Soviet Union and the United States were of course very different places, but the two young men had backgrounds that were similar in important respects. They were young physicists at a time when the community of physicists was becoming a genuinely international network, with Cambridge as one of the crucial hubs. Their interests in physics were somewhat different, and their personalities appear to have been very different. Oppenheimer was a complex and tortured individual, especially during his time at Cambridge. Khariton was much more self-possessed and self-reliant. But they both had a gift for friendship. Oppenheimer was able to get on well with people very different from himself—Ernest Lawrence is one notable example. Khariton too was close to people who were very different from him in personality—for example, Igor Kurchatov. Rudolf Peierls, who like Khariton was a brilliant and unassuming man, worked very well with Oppenheimer during World War II and admired him. So it is not impossible that, if they had met, Khariton and Oppenheimer would have become friends.

Research

Khariton returned to the Soviet Union in 1928, stopping in Berlin to see his mother and stepfather, and he later recalled that the growing strength of fascism had alarmed him during his visit. Shortly after his return to Leningrad he set up a laboratory devoted to the study of explosives at Ioffe's institute. This laboratory became part of the Institute of Chemical Physics, which was created in 1931 when Ioffe's institute was split into three. Semenov was director of the new institute. Khariton later claimed that his choice of this direction of research was motivated in part by his fear that the rise of fascism would lead to war. The laboratory did research on the physical and chemical processes involved in detonation and combustion. This had practical applications in mining and the oil industry, as well as in military affairs.[10]

Oppenheimer's trajectory as a physicist was quite different. He received his Ph.D. from Göttingen in the spring of 1927 for work on the application of quantum theory to transitions in the continuous spectrum. Like Khariton, Oppenheimer was dismayed by the political mood he found in Germany in the late 1920s. He recalled later "a very miserable German mood...bitter, sullen, and I would say, discontent and angry and with all those ingredients which were later to produce a major disaster."[11] Oppenheimer traveled back to the United States in the summer of 1927, but returned to Europe in 1928 on a nine-month fellowship, spending time at Leiden, Utrecht, and Zurich. In 1929 he went back to the United States to take up a joint appointment at the California Institute of Technology in Pasadena and the University of California in Berkeley, where he played an important role in building up theoretical physics in the United States.

Khariton was primarily an experimentalist and Oppenheimer a theoretician. They pursued different lines of research in the 1930s, but their interests began to converge after the discovery of nuclear fission. Oppenheimer was immensely stimulated by the discovery. "The U business is unbelievable," he wrote to a colleague in January 1939. He did not publish any papers on fission, but "he was constantly engaged in theoretical speculation and discourse."[12] He was very close to Ernest Lawrence, his colleague at Berkeley, and worked with him on the electromagnetic process for separating uranium-235. But Oppenheimer did not become directly involved in the work on fast neutron reactions until October 1941. In January 1942 he was asked to take charge of the fast neutron research at Berkeley, and in May he was given the task of coordinating all the work being done on fast neutron reactions under the auspices of what would soon be called the Manhattan Project. Towards the end of the year General Leslie Groves, who was now in charge of the Manhattan Project, appointed him director of the new laboratory at Los Alamos, where the atomic bombs were to be designed and fabricated. Los Alamos started work in the spring of 1943.

10. Ia. B. Zel'dovich, "Iulii Borisovich Khariton i nauka o vrzryve," A.P. Aleksandrov, ed., *Voprosy sovremennoi eksperimental'noi i teoreticheskoi fiziki* (Leningrad, 1984), 32-37.
11. Smith and Weiner (ref. 4), 103.
12. Ibid., 207.

Khariton too was stimulated by the discovery of fission. He collaborated with Iakov Zeldovich, who was also working at the Institute of Chemical Physics, on the conditions under which a nuclear chain reaction would take place in uranium. This was an obvious problem for them to tackle, given Semenov's work on chain reactions and their own interest in explosives. They made considerable progress in setting out the conditions for a nuclear chain reaction in a series of papers, some of which were published before the German attack, and some of which saw the light only many years later. These papers gave perhaps the fullest analysis of nuclear chain reactions at the time. In March 1940 Otto Robert Frisch and Rudolf Peierls, two refugees from Nazism who were then working in Britain, made a breakthrough by showing that a small quantity of pure uranium-235 would be enough to create an explosive nuclear chain reaction. The British Government set up a committee (the Maud Committee) to investigate the analysis made by Frisch and Peierls. This committee reported in July 1941 that 10 kilograms of uranium-235 would be sufficient for a bomb.[13] Khariton and Zeldovich, working with I.I. Gurevich, had reached the same conclusion in the spring of that year, though their paper was not published until long after World War II. Khariton and his colleagues did not know of the Frisch-Peierls memorandum or of the Maud Committee, but they did make use of an important paper on critical size that Peierls had published in 1939.[14]

The German invasion of June 22, 1941, brought Khariton and Zeldovich's research on nuclear chain reactions to a halt. They both worked on the powder fuel for the "Katyusha" rocket artillery, and Khariton later helped to develop anti-tank grenades and cheap surrogate explosive substances. Khariton became involved in the Soviet nuclear project in 1943, on a part-time basis, as the person responsible for research on fast neutron chain reactions. After the war he was drawn more and more into the project, and in the summer of 1946 he was made scientific director of Laboratory No. 2's branch at Sarov, where work on the bomb was to be done. (Laboratory No. 2, on the outskirts of Moscow, was the main center for nuclear research, and its director was Igor Kurchatov.)

Directors

In the late 1930s Oppenheimer became involved in left-wing politics and supported such progressive causes as the Spanish Republicans and the California farm workers. He helped to found a branch of the American Federation of Teachers at Berkeley, and he took part in a group that met regularly to discuss issues of the day. Oppenheimer said later that this was a discussion group, but another member of the group described

13. Margaret Gowing, *Britain and atomic energy, 1939-1945* (London, 1964), 45-89.
14. I.I. Gurevich, Ia.B. Zel'dovich, and Iu.B. Khariton, "Kriticheskie razmery i massa, neobkhodimye dlia tsepnogo deleniia iader neitronami," Gol'danskii (ref. 1), 119-149.

it as "a 'closed unit' of the Communist Party."[15] Oppenheimer never denied that he had had left-wing sympathies, but he consistently denied that he had ever been a Party member. There is recent evidence to suggest that he was indeed a member of the Party, but the evidence is not conclusive. Oppenheimer's activities attracted the suspicion of Army security officials, but Groves overruled their objections when he appointed Oppenheimer as director of Los Alamos.

Khariton's political background was, from a Stalinist perspective, also deeply suspect. After his expulsion from Soviet Russia, Khariton's father had spent a short time in Berlin before moving to Riga, where he worked for an émigré Russian newspaper. He remained in Latvia until the Soviet Union annexed the country in 1940. The NKVD arrested him in October 1940, and two months later he was condemned to seven years in the camps and three years without rights. He died in 1941 either on the way to the camp or in the camp itself. Khariton had lost contact with his father in the late 1920s or early 1930s; he did not have the right to correspond with him once he became engaged in secret research. Boris Khariton's fate would surely have been a black mark in Khariton's *kharakteristika* (personal dossier). Besides, by this time Khariton's mother was living in Palestine. And by the late 1940s, when the anti-cosmopolitan campaign reached its peak, Khariton's Jewishness could also have been held against him. Khariton's years in Cambridge would only have added to the suspicion with which the authorities must have regarded him.

In both the United States and the Soviet Union, men who did not seem to be especially well fitted for the task—either by temperament or by political background—were put in charge of the crucial task of building the bomb. Neither Oppenheimer nor Khariton had become scientists in the expectation that they would be called upon one day to direct large scientific-technical undertakings. Scientists of their generation might expect to head a research institute or become president of a university, but they could hardly have imagined bearing the responsibilities they eventually assumed. The atomic bomb gave science and scientists a new significance for the state. Oppenheimer and Khariton were in some ways unlikely candidates for the roles they played. Yet in each case the choice turned out to be an inspired one, for Oppenheimer and Khariton were widely acknowledged to be extremely effective directors of their respective laboratories.

Oppenheimer had had no administrative experience before becoming director of Los Alamos; he had not even been chairman of a Physics Department. Yet the qualities that made him successful at Los Alamos were qualities he had displayed before the war: "the quick and omnicompetent intellect, the capacity for hard work, the personal charisma, the thoughtfulness for others."[16] He also showed immense self-discipline in applying these qualities to the task of building the bomb. He quickly became a natural

15. Gregg Herken, *Brotherhood of the bomb: The tangled lives and loyalties of Robert Oppenheimer, Ernest Lawrence, and Edward Teller* (New York, 2002), 31.
16. Smith and Weiner (ref. 4), 221.

leader. "It was clear to all of us, whenever he spoke," Hans Bethe said at Oppenheimer's memorial service,[17]

> that he knew everything that was important to know about the technical problems of the laboratory, and he somehow had it well organized in his head. But he was not domineering, he never dictated what should be done. He brought out the best in us all, like a good host with his guests. And because he clearly did his job very well, in a manner all could see, we all strove to do our job as best we could.

Oppenheimer had the respect of the formidable group of scientists that he brought together at Los Alamos. He directed the whole complex undertaking to a rapid and successful conclusion. It was President Truman, of course, not Oppenheimer, who took the decision to use the atomic bomb against Japan, but Oppenheimer agreed with the decision and did not press for, and indeed advised against, alternative courses of action—a demonstration shot, for example—when he had the opportunity to do so. Oppenheimer stepped down as director of Los Alamos in October 1945, three years after he had been appointed to the position.

The Soviet context was very different. In the first place, Khariton worked very closely with Igor Kurchatov, scientific director of the whole nuclear project, who had no direct counterpart in the Manhattan Project. There were several scientists—notably the chemist James Conant and the electrical engineer Vannevar Bush—in leading positions in the Manhattan Project, but they were not as directly involved in management as Kurchatov was. Second, at Khariton's request, L.P. Beria, who was in overall charge of the Soviet project, appointed a "director" for the weapons laboratory at Sarov, while Khariton became the "scientific director." This relieved Khariton of some administrative responsibilities and allowed him to focus on the scientific and engineering tasks. Third, and perhaps most important, the Soviet Union in the last years of Stalin's rule was very different from the United States: being accountable to Beria, the head of the secret police, was certainly very much more stressful than answering to General Groves.

It is all the more interesting, therefore, that the reminiscences of Khariton's colleagues at Sarov emphasize qualities of character similar to those that the memoirs about Oppenheimer point to.[18] They stress in the first place his determination to get the science right. Like Oppenheimer at Los Alamos, Khariton was deeply knowledgeable about the work being done at the institute. He asked careful and probing questions. He had a motto, which was known to everyone: "We have to know ten times more than we are doing." Building the bomb was not merely a matter of fabricating a device: the project had to be based on a deep scientific understanding of the processes involved. Khariton's colleagues point also to his attention to detail. His experience at the Cavendish Laboratory may have reinforced the care with which he approached the work of the institute. Memoirs about Khariton emphasize also his *intelligentnost'*,

17. Cited in ibid., 264.
18. See, for example, Igor Tamm's remarks in Gol'danskii, ed. (ref. 1), 228.

his personal charm and gentleness, and his high moral qualities. The quiet-spoken Khariton, with his polite and modest manner, was very far from the standard model of a hard-driving Soviet manager.

The parallels between Khariton's style of leadership and Oppenheimer's—the deep understanding of the scientific and technical aspects of the work, and leadership through questioning and discussion rather than by *diktat*—spring as much from the nature of the enterprise as from similarities in personality. The two men were well regarded by their scientific colleagues for their sharp intellects and their firm grasp of the science and engineering of nuclear weapons. These qualities were evidently requirements of success in undertakings of this kind. But Oppenheimer and Khariton were also tough and resilient men, with a belief in the rightness of what they were doing; they could not otherwise have borne the pressures they had to work under.

Yet to talk of parallels is in a sense misleading. Khariton was following Oppenheimer, not moving alongside him. In the middle of 1941 Khariton's understanding of the feasibility of the atomic bomb may have been greater than Oppenheimer's. But the Manhattan Project moved forward with considerable speed, and the bombing of Hiroshima and Nagasaki in August 1945 made it clear to the whole world that the atomic bomb could be built and that it was very powerful. The Soviet Union had received remarkable intelligence from Los Alamos about the design of the atomic bomb, thanks largely to Klaus Fuchs. Kurchatov and Khariton decided early on that the quickest way to build a Soviet bomb would be to copy the design of the American plutonium bomb, or at least of the explosive charge. This is not to underestimate the work that needed to be done by Soviet scientists and engineers to build the bomb, nor to deny that it was a considerable achievement to make the bomb so quickly. But it does mean that after 1941 Khariton was following in Oppenheimer's footsteps.

Soviet efforts to acquire information about the American atomic project provide another curious link between Oppenheimer and Khariton. Late in 1942 Pyotr Ivanov, third secretary at the Soviet consulate in San Francisco, told George Eltenton, an English engineer working at the Shell Development Corporation, that he knew there was research connected with atomic energy going on at the Radiation Laboratory at Berkeley. Eltenton had spent a year in Leningrad working at the Institute of Chemical Physics at Khariton's invitation, but had moved from the Soviet Union to the United States in 1938. Ivanov asked Eltenton how well he knew Lawrence and Oppenheimer. Eltenton, who was sympathetic to communism, suggested to Ivanov that Haakon Chevalier, a Berkeley professor who was a friend of Oppenheimer's, should make the approach to Oppenheimer. Early in 1943 Chevalier had a brief conversation with Oppenheimer—or perhaps with Oppenheimer's brother Frank, or perhaps with both of them—in which Chevalier mentioned that Eltenton had means of getting technical information to the Soviet Union. Robert Oppenheimer and Chevalier later gave different versions of this conversation, but all versions agree that Oppenheimer made it clear that he would have nothing to do with such transactions. Oppenheimer did not immediately report this conversation to the security officials, and, when he did, he gave differing accounts that later proved damaging to himself and to others.[19]

The Eltenton link between Khariton and Oppenheimer seems to be little more than a coincidence. Eltenton was sympathetic to the Soviet Union and willing to help the Soviet consulate in San Francisco make contacts that might have intelligence value. But there is no evidence to suggest that Khariton had anything to do with Eltenton's being in the San Francisco Bay Area, or with his approach to Oppenheimer. Nor is there good evidence that Oppenheimer passed on information to the Soviet Union about the American project.

Contrasting Roles

The greatest difference between the two men was that, from 1945 until his death in 1967, Oppenheimer was a public figure, associated in the public mind with the beginning of the atomic age, while Khariton, who died in 1996, remained hidden from public view until the 1980s. Oppenheimer's face became well known, and his views on the bomb were widely solicited and read. He never regretted working on the bomb, nor did he believe that it was wrong to use the bomb against Japan. What did cause him great concern, however, was the danger that the bomb now posed to the human race. On October 16, 1945, his last day as director of Los Alamos, he put the problem starkly: "If atomic bombs are to be added as new weapons to the arsenals of a warring world, or to the arsenals of nations preparing for war, then the time will come when mankind will curse the names of Los Alamos and Hiroshima. The peoples of the world must unite, or they will perish."[20] The atomic bomb was born into a world in which large-scale war was a regular and apparently natural phenomenon. How could a nuclear world war be prevented in the future?

Oppenheimer was the guiding spirit behind the Acheson-Lilienthal Report on the international control of atomic energy, which was published in Washington in March 1946. The United States and the Soviet Union had agreed to engage in negotiations on the international control of atomic energy, and Secretary of State James Byrnes had set up a committee under Dean Acheson to advise him on the U.S. position at those negotiations. Acheson in turn created a Board of Consultants under the chairmanship of David Lilienthal, with Oppenheimer as one of the members. The report set out an ambitious plan for putting all "dangerous" activities under the control of an international agency, while "safe" activities were to be left to individual states. "Dangerous" activities were those that offered a solution to one of three problems: the supply of raw materials; the production of plutonium and uranium-235; and the use of those materials to make atomic bombs. Oppenheimer hoped that this would

19. Herken (ref. 15), 91-92, 111, 114, 160-162.
20. Lilian Hoddeson et al., *Critical assembly: A technical history of Los Alamos during the Oppenheimer years, 1943-1945* (Cambridge, 1993), 401-402.

remove the bomb from international politics, while making it possible to use atomic energy for peaceful purposes. The Baruch Plan, which the United States presented at the negotiations in June 1946, was a modified version of the Acheson-Lilienthal Report. The Soviet Union, which was determined to have nuclear weapons of its own, rejected the Baruch Plan. The negotiations on international control of atomic energy ended in failure.

Oppenheimer continued to exercise influence over U.S. nuclear weapons policy as chairman of the General Advisory Committee of the Atomic Energy Commission. In the last months of 1949, after the first Soviet atomic bomb test, an intense debate erupted within the U.S. government as to whether the United States should make a determined effort to develop the hydrogen bomb. At a meeting at the end of October, the General Advisory Committee opposed the development of the hydrogen bomb. The majority of the committee, including Oppenheimer, argued that the United States should not develop the hydrogen bomb, because it would be "in a totally different category from an atomic bomb," and might become a "weapon of genocide." Truman nevertheless decided to continue, and in effect to accelerate, the development of the hydrogen bomb. He announced this decision on January 31, 1950.[21]

Oppenheimer's opposition to the hydrogen bomb was bitterly resented by the advocates of the bomb—including Edward Teller, its most persistent proponent—and was one of the reasons why the Atomic Energy Commission filed charges against him in December 1953 for breaches of security regulations. In the spring of 1954 a personnel review board held hearings at which witnesses testified in support of the charges against Oppenheimer, as well as in his defense. The hearings exposed Oppenheimer's evasions about the Chevalier incident. Oppenheimer was deprived of his security clearance, and he took this very hard. He remained in his post as director of the Institute for Advanced Study at Princeton, but his influence in Washington was destroyed.

Khariton did not play the same kind of public role. He served as scientific director of the weapons institute at Sarov until 1992—forty-six years as director instead of Oppenheimer's two and a half. There is no doubt that he played an extremely important part in Soviet nuclear policy during those years, but little information is available about the advice he gave on such issues as weapons design, the comprehensive test ban, or arms control. It is clear, however, that he did not emulate Oppenheimer in opposing the hydrogen bomb. Andrei Sakharov moved to Sarov in 1950 to work on the hydrogen bomb and remained there until 1968. "In the 1940s and 1950s my position was much closer to Teller's, practically a mirror image," Sakharov writes in his memoirs. "Unlike Teller, I did not have to go against the current in those years, nor was I threatened with ostracism by my colleagues."[22] In other words, the Soviet nuclear weapons scientists were firmly in favor of developing thermonuclear weapons.

21. The quotations are taken from the majority's recommendation that the United States renounce the hydrogen bomb. That recommendation is reprinted in Herbert York, *The advisors: Oppenheimer, Teller and the Superbomb* (San Francisco, 1976), 156-157.
22. A.D. Sakharov, *Vospominaniia* (New York, 1990), 137.

Khariton was more or less completely absorbed by his work as scientific director of Sarov. He did, however, intervene on a number of issues affecting science in the Soviet Union. Physicists had acquired considerable prestige through their success in developing nuclear weapons, and they tried to use that prestige to improve the condition of Soviet science. In 1952 Khariton was one of a group of physicists who signed a letter to Beria protesting the publication of an article attacking relativity theory and quantum mechanics. Three years later, along with other physicists, he wrote to Khrushchev to complain about the damage that Lysenko was doing to science in general in the Soviet Union. In 1966 he wrote a letter to Brezhnev with N.N. Semenov and A.P. Aleksandrov to warn against efforts to rehabilitate Stalin at the XXIII Party Congress.

There were, however, limits on what Khariton could do, given his position. Sakharov, who had become Khariton's deputy as scientific director at Sarov, was forced to leave Sarov and to give up secret work in 1968, after his essay "Reflections on progress, peaceful coexistence, and intellectual freedom," had been published abroad. In 1973 Khariton signed a letter condemning Sakharov for his activities. This was a difficult decision for Khariton: it caused a rift in his family and he came to regret it later. Apparently he feared that if he refused to sign he would lose his position as scientific director of the institute where there were still things that he wanted to achieve.

Conclusion

I have sketched some points of connection between Oppenheimer and Khariton. Even this brief sketch raises a number of questions that I have been able to do no more than touch on here: how important was their Jewish identity to the two men, and what was the significance of that identity in the two societies? What was the relationship of science to policy, and scientists to policy-makers, in the two political systems? What is the role of public debate in understanding the issues presented by the bomb? What was the role of secrecy and espionage in two societies? How important is the distinction between leader and follower in the development of new technologies?

The cultural backgrounds of the two men were similar, and they became physicists at a revolutionary period in physics. They were members of a physics community that was becoming truly international, though the Soviet Union was cut off from that community from the mid-1930s to the mid-1950s. They chose different lines of research, but they were both excited by the discovery of nuclear fission. They were selected for extremely responsible and demanding positions in their own states, even though their backgrounds might have disqualified them from such positions. They were central figures in the most fateful scientific-technical development in the twentieth century.

I have said less about the obvious differences between the two men. They were not alike in personality. Oppenheimer was the more complex of the two, given to putting on masks and acting out charades, even if there was a fundamental seriousness beneath the surface. Khariton seems to have been more self-assured and more balanced. Their private lives reflected the difference in personality. Oppenheimer's was tempestuous; Khariton's was more stable. Oppenheimer moved, after the war, to focus on the political implications of the bomb. He sought to shape American policy in order to achieve international control of atomic energy, and he advised the government on the development of weapons and of the atomic stockpile. Khariton did not play a public role on these issues. There was of course no opportunity in the Soviet Union for the kind of public debate that took place in the United States. Khariton nevertheless carried a heavy responsibility in his position as scientific director at Sarov.

Both men confronted the question of the scientists' responsibility for technologies that they had helped to create. Oppenheimer was in a position to explain to political leaders and to public opinion the nature of the challenge posed by nuclear weapons, as he understood it. He could speak with great eloquence, though he was often elliptical in his statements. Khariton's opportunity to play such a role was much more restricted, though it would be interesting to know much more about the advice he gave to the political leadership and the way in which he framed the issues. It was Khariton's close colleague, Andrei Sakharov, who assumed the role of an independent public intellectual, addressing the nuclear issue in his 1968 essay. It is curious that Khariton witnessed, through his father's exile, Lenin's decision to curtail the autonomy of the intelligentsia and, through his colleague Sakharov, a decisive reassertion of that autonomy.

Khariton of course knew about Oppenheimer after 1945 and took a keen interest in the history of the Manhattan Project. Did Oppenheimer know about Khariton and his role in the Soviet nuclear project? There is no evidence that he did. It is a curious fact that Oppenheimer appears to have taken very little interest in the Soviet Union, even when he was sympathetic to its policies in the late 1930s.

CHAPTER SEVEN

The Making of the Soviet Bomb and the Shaping of Cold War Science

Alexei Kojevnikov

Preface: Oppenheimer and the History of the Soviet Bomb

Until relatively recently, it has seemed a rather thankless job to study the history of the Soviet atomic project. Some basic pieces of information were solid; most others relied on ambiguous sources such as censored official accounts and propaganda, conspiracy theories of espionage "experts," circumspect memoirs, banquet speeches, unverifiable rumors, or recollections and interviews recorded long after the fact. A few outside scholars, most notably David Holloway, were capable of navigating through this swamp, separating realistic from rhetorical and deceptive descriptions.[1] Yet from the professional historian's standpoint, the situation appeared frustrating—not because historians are unable to deal with a deficient source base, but because it was all too clear that incomparably richer collections of authentic primary documents existed somewhere very near, behind concrete walls, awaiting their time to become accessible. And it was hard to predict how long the wait might last.

Then things suddenly began to change—thanks in part to J. Robert Oppenheimer's ironic role in the infamous "Sudoplatov affair" of 1994. That affair has a history of its own, which in this context merits a brief excursion. The retired KGB lieutenant-general Pavel Sudoplatov had once supervised special operations, including the assassination of Leon Trotsky in 1940 and guerilla actions against the Nazi occupiers during the Second World War. After the arrest of his boss, Lavrenty Beria, a few months following Stalin's death in 1953, Sudoplatov was also arrested on accusations of mismanaging espionage networks. He spent fifteen years behind bars, and more than twenty years thereafter trying to rehabilitate himself in the eyes of the KGB and Soviet authorities. He finally

The following abbreviations are used: *Atomnyi*, *Atomnyi proekt SSSR* (6 vols., Moscow, 1998-2003); *VIET*, *Voprosy istorii estestvoznaniia i tekhniki* (*Russian journal for history of science and technology*).

1. David Holloway, *Stalin and the bomb: The Soviet Union and atomic energy, 1939-1956* (New Haven, CT, 1994). For reliable accounts by "insiders," see I.F. Zhezherun, *Stroitel'stvo i pusk pervogo v Sovetskom Soiuze atomnogo reaktora* (Moscow, 1978) and Arkady Kruglov, *Kak sozdavalas' atomnaia promyshlennost' v SSSR* (Moscow, 1994).

succeeded in getting cleared of those earlier charges only in 1992, when the Soviet Union was no more. In the new atmosphere of the criminal market, the collapse of old values, and the search for new kinds of profit, Sudoplatov's son, Anatoly, approached the American journalist Jerrold L. Schecter with a book proposal aimed at a Western publisher. And when Schecter and his wife, Leona, a literary agent, initially showed little enthusiasm—in their opinion, the proposal lacked in names familiar to Western readers—Anatoly responded with a document from his father that briefly mentioned Klaus Fuchs, Enrico Fermi, and Robert Oppenheimer as "sources" for the Soviet atomic project. These names, in particular Oppenheimer's, helped clinch the deal. Knowing what was required to make the memoir publishable, Anatoly questioned his father for additional details and, in collaboration with the Schecters, wrote the book.[2]

The historian Robert Conquest, who proclaimed the memoir in his introduction to be "perhaps the most important single contribution to our knowledge since Khrushchev's Secret Speech," called its chapter on nuclear espionage "the most striking and informative." In this chapter, a brief and unclear reference in the earlier document developed into full-blown charges of espionage against Oppenheimer, Fermi, Leo Szilard, and Niels Bohr, who reportedly "agreed to share information on nuclear weapons with Soviet scientists."[3] Just as the Schecters had predicted, the chapter became a media sensation, driving up sales—but it was also this very chapter that almost immediately proved unreliable.[4] The attention the book initially aroused in the American media quickly subsided; but, however short, it sufficed to produce an important effect in post-communist Russia. There, uncritical attitudes towards revelations contradicting official portrayals of the Soviet era also created a media hype around Sudoplatov's book, which quite unexpectedly produced a major crack in the concrete dams of official secrecy. A presidential decree of February 1995 ordered the Atomic Energy Ministry to launch a major declassification project and publish an official volume of authentic archival documents on the history of nuclear weapons.[5]

2. Pavel Sudoplatov and Anatoly Sudoplatov, with Jerrold L. Schecter and Leona P. Schecter, *Special tasks: The memoirs of an unwanted witness—a Soviet spymaster* (Boston, 1994), esp. xvii-xviii, 481.

3. Sudoplatov et al. (ref. 2), vii, ix, 172.

4. As a result, the entire book lost credibility, although most of its problems, it appears, are confined to the chapter on atomic spies, specifically designed to meet the wishes of literary agents. Professional historians quickly pointed out many details that were supposed to corroborate the main charge of espionage, but either contradicted well-established facts or included second-hand borrowings from published descriptions of events that happened elsewhere and involved different people. See Federation of American Scientists, *F.A.S. public interest report*, May/June 1994. One of Sudoplatov's stories—that Soviet intelligence sent a physicist to Copenhagen to interview Bohr about the uranium problem—found partial confirmation; but according to independent documents, the timing, goals, and results of the mission were very different. Bohr, in particular, had been careful to not provide any specifically classified information. A.V. Andreev and A.B. Kozhevnikov, "Kopengagenskaia operatsiia sovetskoi razvedki," *VIET*, 1994, no. 2, 18-21.

5. *Atomnyi proekt SSSR*, 6 vols. (Moscow, 1998-2003), esp. *I:1*, 3.

To be sure, historians who are not part of the official team are still deprived of direct access to primary documents on the Soviet atomic project. The work has proceeded along bureaucratic lines, with a ministerial commission reviewing existing collections, deciding what can be declassified, and preparing commented publications. The scale of the entire undertaking, however, has greatly exceeded initial expectations: as of the moment of this writing, volume I (documents prior to 1945) has appeared in two books; volume II (documents on the atomic bomb, 1945-1954) has come out in four books, altogether over 4300 printed pages. Volume III (the hydrogen bomb, 1945-1956) is still in preparation. Although some understandable restrictions apply, the project—along with an accompanying wave of publications by institutions and individuals of their own stories, documents, and recollections[6]—has reversed the situation: we, the historians on the outside, are no longer in a position to complain about the lack of sources, but find ourselves behind the curve of declassification, trying to catch up, digest, and comprehend the sheer quantity of newly available documents.

The following short essay cannot pretend to offer a new synthetic analysis, which will certainly require from historians much more time, space, and effort.[7] Elsewhere I have offered a brief narrative of the Soviet atomic bomb project that includes at least some of the new evidence,[8] while here I will concentrate on a few important interpretative points. The atomic spies issue would not qualify among them: it has been discussed far too often in the literature, especially that aimed at a general public, in comparison to its actual importance for the case. Other, historically more significant aspects of the story, on the other hand, have been largely ignored and need closer attention. Three are discussed below: first, the main difficulty that could have derailed the Soviet bomb; then the most important advantage that helped the project succeed, as well as its chief strategic dilemmas; and, finally, the contribution of nuclear weapons to postwar conservatism in science—on both sides of the Iron Curtain. Each aspect gives insight into the tasks of nuclear weapons construction and the nature of U.S.-Soviet competition, as they also came to shape Oppenheimer's era.

6. See, for example, *Sozdanie pervoi sovetskoi yadernoi bomby* (Moscow, 1995); *Sovetskii atomnyi proekt* (Arzamas-16, 1995); B.L. Vannikov, *Memuary, vospominaniia, stat'i* (Moscow, 1997); V.I. Shevchenko, *Pervyi reaktornyi zavod* (Ozersk, 1998); A. Emel'ianenkov, *Arkhipelag Sredmash* (Moscow, 2000); A.P. Zaveniagin, *Stranitsy zhizni* (Moscow, 2002).

7. For some of the recent studies contributing towards a new synthesis see: V.P. Vizgin, "Atomnyi proekt SSSR: Predvaritel'nye itogi izucheniia i novye materialy," *VIET*, 1996, no. 2, 86-93; *Science and society: History of the Soviet atomic project (1940s-1950s), Proceedings of the international symposium* (2 vols., Moscow, 1997-1999); Gennady Gorelik, *Andrei Sakharov: Nauka i svoboda* (Moscow, Izhevsk, 2000); Christoph Mick, *Forschen für Stalin: Deutsche Fachleute in der sowjetischen Rüstungsindustrie, 1945-1958* (Munich, 2000); Vladislav Larin, *Kombinat "Maiak"—Problema na veka*, 2nd edn. (Moscow, 2001); V.P. Vizgin, ed., *Istoriia sovetskogo atomnogo proekta: Dokumenty, vospominaniia, issledovaniia*, Vol. 1 (Moscow, 1998), Vol. 2 (St. Petersburg, 2002), esp. G.A. Goncharov, "Termoiadernyi proekt SSSR: predystoriia i desiat' let puti k vodorodnoi bombe," *2*, 49-148.

8. Alexei Kojevnikov, *Stalin's great science: The times and adventures of Soviet physicists* (London, 2004), chapt. 6.

The Critical Shortage

With the exception of a few astute observers whose statements in this regard remained largely unnoticed in the public perception,[9] most commentators hardly mention the single most important obstacle that severely delayed the Soviet post-Hiroshima drive to acquire atomic weapons—indeed, could have undermined it entirely. The newly available documentary evidence makes this omission hard to sustain, and the obstacle itself difficult to overstate. The matter in question is the shortage of raw materials, especially uranium.

Unlike most journalists and politicians (and many later historians), General Leslie R. Groves evidently understood that control over fissionable materials was more important for maintaining an American nuclear monopoly than anything called "the atomic secret." This consideration, according to his biographer, the late Stanley Goldberg, formed the reasoning behind Groves's optimistic postwar estimate of the length of time the Soviet Union would need to produce the atomic bomb. Indeed, one of Groves's first decisions as head of the Manhattan Engineer District was to try to acquire all the uranium from the world's richest and highest-quality deposit in the Belgian Congo. He subsequently managed to secure American control or exclusive rights over most identified resources of uranium in the world. It remained unknown to him, however, how much uranium might exist and be mined across the vast territory of the Soviet Union; his estimates in this regard must have been very approximate.[10]

As of 1945, Soviet authorities and experts themselves did not know much about the availability of uranium either, and their existing stock was negligible to the point of hopelessness. Prior to the war, uranium as a chemical element had almost no industrial use, except as a minor addition in dyes and photographic processing. There was no particular need to search for it underground, either, and most of the explored deposits were known because of the demand for radium, an element then regarded as much more useful and valuable. The geochemist Vladimir Vernadsky had initiated in 1911 a modest exploration of Russia's territory in search of radioactive ores under the auspices of the Radium Commission at the St. Petersburg Imperial Academy of Sciences. The Commission found a minor deposit of radium (and uranium) in Central Asia, and in 1918, with a subsidy from the Bolshevik government, the chemist Vitaly Khlopin, Vernadsky's collaborator, established a small factory in the Ural Mountains that in 1921 would extract the first Soviet radium from uranium ores.[11]

Uranium itself was largely seen as a by-product or even waste product of the radium industry until the discovery of fission in Germany in late 1938. Soviet physicists in

9. Holloway (ref. 1); Kruglov (ref. 1).
10. Stanley Goldberg, presentation at the conference "Wissenschaft und Krieg," Greifswald, Germany, May 1995; Robert S. Norris, *Racing for the bomb: General Leslie R. Groves, the Manhattan Project's indispensable man* (South Royalton, VT, 2002), 178-179; Holloway (ref. 1), 174.
11. *Atomnyi I:2*, 450-462; *II:2*, 313-316; Holloway (ref. 1), 175-177; B.V. Komlev, G.S. Sinitsina, and M.P. Koval'skaia, "V. G. Khlopin i uranovaia problema," *VIET*, 1982, no. 1, 63-75.

Leningrad learned the news about fission in a matter of weeks, but Vernadsky at his Radium Institute was alerted more than a year later, in June 1940, by a serendipitous *New York Times* clipping mailed by his son, George Vernadsky, professor of Russian history at Yale. Vernadsky and his collaborators contacted the Soviet government with a memorandum about the possible technical application of uranium in a chain reaction with slow neutrons. They emphasized geological exploration above all else and proposed to establish a State Uranium Reserve. Upon their urging, the Academy of Sciences set up a special Uranium Commission under Khlopin to coordinate research at various institutes. Disciplinary rivalries undermined the Commission's efforts, however, with physicists prioritizing research on fission proper rather than the search for raw materials. The Commission set as its goal an increase in uranium extraction to ten tons annually, but less than one ton was actually mined in 1940.[12] The entire activity stopped in June 1941, when German armies invaded the Soviet Union.

When work resumed at the end of 1942, it was physicists rather than chemists who played leading roles. Igor Kurchatov, the director of the newly established Laboratory for research on the atomic bomb, must have regretted his pre-war disagreements with Khlopin in the Uranium Commission. In early 1943 Kurchatov had less than 100 kg of uranium salts, which was not sufficient even for laboratory measurements. Necessity being the mother of invention, the coefficients of neutron diffusion were measured in so-called exponential experiments, in which scarce uranium was stretched out thinly in one dimension and the results of scattering and absorption measurements recalculated for the case of three dimensions.[13]

The virtual absence of fissionable material set the wartime Soviet project apart not only from the Manhattan Project, but even from the much smaller project in Nazi Germany, which at least had enough uranium for building a working reactor (though it failed at that task). In the rest of the world, only a few deposits of uranium had been explored prior to the war, and even fewer mined. The oldest and best-developed mines in Czechoslovakia remained under German control for the duration of the war. Almost the entire accumulated stock from the Belgian Congo was appropriated by General Groves. The Soviet atomic bomb, in contrast, required not only researching and designing the bomb itself, but also the additional task—actually much larger and more expensive—of building from scratch an entire industry of mining (and enriching) uranium. The resumption of geological exploration was ordered in November 1942; by mid-1943 the production of uranium salts reached two hundred kilograms per month, with ten tons as the envisioned (optimistic) target estimate for the year of 1944. At such a pace—Kurchatov estimated—at least five to ten years would be necessary simply to accumulate material for the first working uranium-graphite reactor

12. Holloway (ref. 1), 59-60; *Atomnyi I:1*, 113-130, 200-202.
13. A.P. Aleksandrov, ed., *Vospominaniia ob Igore Vasil'eviche Kurchatove* (Moscow, 1988), 276-277; Zhezherun (ref. 1); *Igor Vasilievich Kurchatov v vospominaniiakh i dokumentakh* (Moscow, 2003), 548-549.

(or a pile, as it was then called).[14] Indeed, any possibility of building a reactor, to say nothing of the bomb, was completely ruled out for the duration of the war.

The fist sizable amount of uranium came from occupied Germany in 1945. After the end of fighting, Soviet ministries dispatched several groups of engineers and scientists in military uniforms to look for valuable industrial equipment and information. In the strategically important fields of atomic and missile technologies they did not find much: most of the hardware, documentation, and personnel had already been moved to the Western occupation zones and came under the control of analogous programs on the American and British sides. Some German experts, however, remained in the Soviet occupation zone, and several groups agreed to move to the U.S.S.R. to work on military programs.[15] The most important find for the Soviet atomic project was the remainder of the German stock of uranium oxide—about a hundred tons—serendipitously discovered at a factory near Berlin where it was disguised as an ordinary chemical.[16] After subsequent reduction and purification, this amount allowed the construction of an experimental nuclear pile capable of sustaining a chain reaction. The first Soviet "boiler" F-1 used almost 35 tons of pure metallic uranium, 13 tons of uranium oxide, and 420 tons of pure graphite. It went operational in Kurchatov's Moscow laboratory on December 25, 1946.[17]

The actual bomb project, including a larger industrial-scale reactor for producing plutonium, required much more. The lack of fissionable material would remain the main bottleneck in the entire Soviet drive for the atomic bomb until 1949. During the initial postwar years, most of the uranium was coming from Eastern Europe, from the already severely depleted mines in Czechoslovakia and some deposits in East Germany and Bulgaria. Altogether, they provided almost eighty percent of the entire stock of uranium acquired by the Soviet project through 1949, when the first bomb was tested.[18] Meanwhile, in 1946 a geological party led by Antonina Likhtar' and Lidia Ivanova discovered a new uranium deposit in Ukraine. Additional sources of uranium were also found in Central Asia, and the overall balance started shifting gradually from imports toward domestic production.[19] As it turned out, uranium was not all that

14. *Atomnyi I:1*, 352, 364-365; 409; Kruglov (ref. 1), chapt. 10.

15. Mick (ref. 7); *Atomnyi I:2*, 249-250; *II:2*, 60-61, 319-321. By the end of 1946, 257 German specialists worked in the Soviet atomic project (64 scientists, 48 engineers, and 145 technicians and qualified workers). 122 of them were hired from occupied Germany; the rest were recruited from among prisoners of war. *Atomnyi II:3*, 594. The group included several prominent physicists and chemists: Manfred von Ardenne, Gustav Hertz, Nikolaus Riehl, Max Steenbeck, Peter Adolf Thiessen, Max Volmer, and a prewar émigré from Germany, Fritz Lange.

16. I.S. Drovenikov and S.V. Romanov, "Trofeinyi uran, ili istoriia odnoi komandirovki," Vizgin, *Istoriia* (ref. 7), *1*, 215-227. Other, higher estimates of the German bounty range up to 300 tons of uranium compounds, containing 150 to 200 tons of uranium; see Zaveniagin (ref. 6), 41-43; *Atomnyi, I:2*, 282-283, 287, 323-324, 423; *II:2*, 309-310.

17. Zhezherun (ref. 1); *Atomnyi II:1*, 631-632.

18. Kruglov (ref. 1), 263.

19. *Atomnyi II:3*, 264, 276-277, 523, 540-546, 686-688.

rare, and many more deposits would be eventually discovered. Soviet ores, however, were of a considerably poorer quality and required the invention of new methods of enrichment and extraction. Economically, developing the mining industry and the supply of uranium was the most challenging part of the entire Soviet atomic project, consuming the lion's share of total expenditures and manpower and determining the ultimate timetable for acquiring a bomb.

Most construction and mining work in the Soviet atomic industry until Stalin's death was done by convicts from labor camps, who worked in dismal conditions and suffered high mortality.[20] The atomic project was a terrible burden on the nation's economy, devastated as it was by four years of war, with half the country's cities lying in ruins and peasants starving in the villages, stripped once again of all their resources so the needs of speedy postwar industrial reconstruction could be met. The total human toll of producing the Soviet atomic bomb is hard to estimate, especially considering indirect casualties caused by the diversion of resources from human consumption into the defense projects of the escalating Cold War. Yet these sacrifices seemed justified to contemporaries, who had just barely survived the most destructive war in modern Russian history and did not need further proof of the consequences of a lack of military and technological preparedness in the face of a stronger adversary.

Socialist Big Science

Standard accounts of the Manhattan Project focus on elite nuclear physicists and their formulae. Thanks in large part to Thomas Hughes, a new view has started shifting the emphasis to the management of a gigantic industrial conglomeration with advanced research as its constitutive part.[21] A separate, long-existing historiographic tradition has described the Manhattan Project (sometimes in conjunction with other American undertakings during the war or slightly before) as initiating the transition to big science, a large-scale merger of advanced science, engineering, and the development of new, primarily military, technologies. A comparison with the Soviet experience can show the narrowness and short-sightedness of this traditional narrative as well. For a structurally very similar pattern of research and innovation had developed quite broadly in the Soviet Union beginning in the early 1920s. It was known there under a different name, the Soviet or socialist model of science—or perhaps not entirely different, as "socialist science" can also be translated into American political lingo as "big government science." This earlier Soviet history, unrecognized in the old historiography, made a difference for the atomic bomb. Prior experience with a similarly managed industrial-cum-R&D infrastructure made the task of imitating and replicating the Manhattan Project decisively easier for the Soviets.

20. Zh.A. Medvedev, "Atomnyi GULAG," *Voprosy istorii*, 2001, no. 1, 44-59.
21. Thomas P. Hughes, *American genesis: A century of invention and technological enthusiasm, 1870-1970* (New York, 1989), chapt. 8.

The idea of what later became known as the Soviet model of science goes back to the crisis of World War I, when Russian academics realized how isolated their research had been from the nation's economic and industrial needs. Abandoning the cult of pure science, they started looking for practical and military applications of their expertise. During the course of the war itself, relatively little could be accomplished to compensate for the previous almost total absence of links with, or often the very lack of, relevant industry. This inadequacy, exposed at a time of a major national crisis, stimulated plans and proposals by Russian academics for thoroughgoing changes in the goals and infrastructure of the country's scientific effort. Their drafts envisioned the recognition of science as a profession separate from teaching, with the creation of corresponding positions for researchers proper, the establishment of a network of research institutes for them, and an overall turn towards practical, applied investigations.[22]

In fact, these Russian proposals were not especially unique. Similar demands for "centrally planned and government-funded" science, public science policy, the opening of special research institutions, linking science to industrial and military needs, and an increased political role for scientists were also discussed during the war by academics in other belligerent nations in Europe and even in the United States.[23] Some of those initiatives failed, while others succeeded in part or temporarily, but most were abandoned after the war. In Russia, on the contrary, they were put into practice in a much more comprehensive and abrupt form because of the revolution and the ensuing Civil War of 1918-1920. Among a number of other modernist proposals for social change, the envisioned reform of science won support from the new Bolshevik regime, which helped give a material basis to the wartime plans, with some significant modifications, and assisted their further development during the subsequent period of peace.

The Bolsheviks were seeking cooperation with "bourgeois specialists"—their term for engineers, scientists, doctors, and other holders of special (including military) expertise—without whom, they understood, they were incapable of running the national economy. The resulting compromise between Bolshevik commissars and non-Communist experts ensured for scientists and engineers a historically unprecedented degree of access to state power, as the latter en masse took on responsible positions in government offices and played important, often decisive, roles in formulating and executing many state policies during the 1920s. "Specialists" also managed to secure

22. Alexei Kojevnikov, "The Great War, the Russian Civil War, and the invention of big science," *Science in context, 15* (2002), 239-275.
23. Frank M. Turner, "Public science in Britain, 1880-1919," *Isis, 71* (1980), 589-608; Andrew Hull, "War of words: The public science of the British scientific community and the origins of the Department of Scientific and Industrial Research, 1914-16," *British journal for the history of science, 32* (1999), 461-481; Zuoyue Wang, "The First World War, academic science, and the 'two cultures': Educational reforms at the University of Cambridge," *Minerva, 33* (1995), 107-127; Soraya Boudia and Xavier Roqué, eds., *Science, medicine, and industry: The Curie and Joliot-Curie laboratories*, Special issue of *History and technology, 13* (1997), 241-343; Ronald C. Tobey, *The American ideology of national science, 1919-1930* (Pittsburgh, 1971).

a correspondingly unprecedented level of state support, material and ideological, for the development of science and technology.[24]

As part of their mutual pact, scientists and Soviet officials quickly agreed on an organizational form: the research institute was the best and the most progressive way of organizing science. To Nikolai Gorbunov—a chemical engineer by training and the most influential early patron of science in Bolshevik circles—the concept looked so genuinely novel and revolutionary that he managed to insert a phrase about linking research to industrial production and establishing the "entire network of new scientific applied institutes, laboratories, experimental stations, and testing facilities" into the new Communist Party Program adopted at the 1919 Party Congress.[25] Sergei Ol'denburg, the permanent secretary of the Russian Academy of Sciences, summarized the mutual consensus a few years later: "while the eighteenth century was, for science, the century of academies, the nineteenth century became the century of universities, and the twentieth century is starting to become the century of research institutes."[26]

Most of the newly established institutes grew out of World War I proposals and research initiatives modified by revolutionary visions. By various estimates, some forty to seventy research institutes were founded in the country by the end of the Civil War. Some remained affiliated with universities and colleges, but more and more were institutionally separated from higher education. The favored form was the state-sponsored research institution directed simultaneously towards advanced research and utilitarian service, and usually multidisciplinary in character. The dominant organizational principles emphasized centralism, planning, and collectivism. The magnificent practical goal, even if it was not always realistically achievable, reflected the typical revolutionary combination of utopianism and utilitarianism. Hope was usually founded upon a radically novel theory or technology with revolutionary symbolism, such as radio, aviation, automobile, genetics, X-rays, or radioactivity.

One early and in many ways prototypical institute, the State Optical Institute or GOI, was founded in 1918 by physicist Dmitry Rozhdestvensky, chemist Ilia Grebenshchikov, and engineer Nikolai Kachalov on the basis of war-time collaboration between Petrograd University and the Imperial Porcelain Works for the production of optical glass and military optics. In the 1920s, GOI led the creation of the country's optical industry as the Soviet Union developed into an independent producer and later exporter of optical technology. The work done at the institute ranged from Vladimir Fock's fundamental theories of atomic spectra (the Fock-Hartree method)

24. Kojevnikov (ref. 8), chapt. 11.

25. M.S. Bastrakova, K.V. Ostrovitianov, et al., eds. *Organizatsiia nauki v pervye gody sovetskoi vlasti (1917-1925): Sbornik dokumentov* (Leningrad, 1968), 91; N.P. Gorbunov, *Vospominaniia. Stat'i. Dokumenty* (Moscow,1986), 14-15. Gorbunov's influence depended not so much on his relatively low formal rank in the party hierarchy as on his post as the secretary of the Soviet government, which provided him direct access to all important decisionmakers.

26. S.F. Ol'denburg, "Vpechatleniia o nauchnoi zhizni v Germanii, Frantsii i Anglii," *Nauchnyi rabotnik* 1927, no. 2, 88-101, on 89.

and quantum electrodynamics to design of the entire spectrum of optical devices—lamps, cameras, microscopes, telescopes, sights, etc.—for both civilian and military use. The balance between different aspects of research—fundamental and applied, military and civilian, classified and open—changed with time and often created tensions, but the basic principle of combining science with technology, and physics with chemistry and engineering, remained valid for the entire Soviet period. GOI's staff grew in the process from about 30 in 1919 to 240 in 1931 and 600 in 1936, and during World War II—now under the scientific leadership of Sergei Vavilov, the state plenipotentiary for the optical industry—the institute was responsible for supplying the Red Army with optical devices and technology.[27]

The association of special research institutes with experimental factories or workshops established the general Soviet model for industrial modernization and development of new technology, which was driven primarily by science rather than by market mechanisms. As a rule, the new institutes' tasks included production—manufacturing X-ray tubes, making optical glass, designing aircraft, or improving the sorting of grain—and many had experimental factories or production units closely associated with or administratively subordinated to research, rather than the reverse. The resulting Soviet system of research and development included most of the typical features traditionally associated with big science. Although not yet as gigantic as became common after the Second World War, it still constituted a quantum leap towards bigness in all quantitative measures—personnel, instruments, networks—and gave rise to a correspondingly more complex social structure and hierarchy. The Soviet R&D system essentially relied on and required increased governmental funding for science, which often came through military channels. It institutionally linked advanced research in fundamental science with production of new technology, rather than with teaching of undergraduate students, and established a new profession of "scientist-engineer."

In the U.S., the transition to the big-science "symbiosis [,]...a fusion of 'pure' science, technology, and engineering"[28] on the basis of state funding, came about mostly as a result of World War II. At first glance, the two developments, Soviet and American, appear independent, caused by similar challenges of war and military needs; yet a missing link between them may be found in the case of Great Britain. In the Europe of the 1930s, especially in France and Britain, influential leftist intellectuals saw certain advantages in the Soviet model of science, in particular a proportionally greater share of the state budget designated for science, the recognition of research

27. V.L. Chenakal, "Optika v dorevoliutsionnoi Rossii," *Trudy Instituta Istorii Estestvoznaniia i Tekhniki, 1* (1947), 145-167; D.D. Gulo and A.N. Osinovsky, *Dmitry Sergeevich Rozhdestvensky* (Moscow, 1980); Yu. N. Gorokhovsky, ed., *50 let Gosudarstvennogo Opticheskogo Instituta im. S. I. Vavilova (1918-1968)* (Leningrad, 1968); S.E. Frish, *Skvoz' prizmu vremeni: Vospominaniia* (Moscow, 1992).

28. Dominique Pestre and John Krige, "Some thoughts on the early history of CERN," Peter Galison and Bruce Hevly, eds., *Big science: The growth of large-scale research* (Stanford, 1992), 78-99, on 93.

as a separate profession, and the attention to placing science in service of the needs of society. Their demands for reforms contributed in France to the establishment of CNRS and a ministerial position for science in the Popular Front government of 1936. In Britain, starting in 1932, J.D. Bernal and other pro-Communist scientists working through the Association of Scientific Workers popularized Marxist views among British academics and the general public and demanded changes in the social role and standing of science. In 1937, the British government still turned down Bernal's memorandum written on behalf of the Parliamentary Science Committee proposing reform in the funding of research. As the war drew closer, however, even the Royal Society and the liberal opponents of Marxism agreed with the demand for government coordination of science, while Bernal and his supporters dropped their previous opposition to military research.[29]

With the start of the war, the Communist Bernal entered government service and played a key role in numerous military planning and research committees, seeing in the war-time mobilization of science a realization of his Marxist proposals of the 1930s. When British scientists working on radar decided to meet regularly for informal discussions with the users of their devices in different branches of the military, they called their gathering the "Sunday Soviet."[30] In this light, the wartime collaboration between Oppenheimer, a (formerly) left-wing scientist, with Groves, a (normally) anti-Communist military officer, does not look like an anomaly at all. On the contrary, it looks fairly characteristic of the wartime antifascist alliance. We know that Oppenheimer was involved in scientific and academic union drives in Berkeley in the 1930s. What still needs to be studied is how *au courant* he was with Marxist proposals for scientific reform advanced by the British and the American Associations of Scientific Workers and other socialist scientists, and what of this left-wing baggage and views he brought into the organization of the Manhattan Project.

In view of the desperate military situation, the wartime mobilization of science in the Soviet Union diverted scientists from uranium research and other uncertain long-term projects into work deemed relevant for immediate survival. In summer 1941 Yakov Zeldovich and Yuli Khariton abandoned nuclear fission for research on conventional explosives—despite having just discovered one of the two realistic scenarios for building the atomic bomb: in collaboration with Isai Gurevich, they estimated that a critical mass of some ten kilos of pure uranium-235 would suffice to produce a nuclear explosion with fast neutrons.[31] Even when nuclear research resumed in early 1943, the project remained a pilot undertaking and for the duration of the war did not

29. William McGucken, *Scientists, society, and the state: The social relation of science movement in Great Britain, 1931-1947* (Columbus, OH, 1984); Spencer Weart, *Scientists in power* (Cambridge, 1979); Maurice Goldsmith, *Sage: A life of J.D. Bernal* (London, 1980).

30. Guy Hartcup, *The challenge of war: Britain's scientific and engineering contributions to World War II* (New York, 1970), 24-25.

31. *Atomnyi I:2*, 469-495; Ya.B. Zeldovich and Yu.B. Khariton, "The mechanism of nuclear fission. II" [1941], *Soviet physics uspekhi, 26* (1983), 279-290.

achieve the scale of big science even by relatively modest Soviet standards. Kurchatov's wartime laboratory counted approximately a hundred workers, about twenty-five of them scientists. Concentration of resources and manpower, and their prodigal use on a scale anywhere close to the hundred-thousand-strong Manhattan Project, were simply ruled out in war-devastated and bombed-out Europe. The Soviet uranium project remained considerably smaller than even those in Germany and Britain proper, which all differed from the Manhattan Project in their limited size, the mostly academic style of their management, and the absence of large-scale industrial development and of a military general at the top.

This all changed for the Soviet atomic project two weeks after Hiroshima. On August 20, 1945, the government created a Special Committee for all work related to nuclear energy, with Lavrenty Beria at its head. As a candidate member of the Communist Party Politburo, Beria participated in making the most important political decisions; as a member of the State Committee of Defense and subsequently first deputy prime minister, he commanded the necessary branches of industry; and as former chief of state security, he continued to coordinate intelligence efforts and could mobilize the forced labor of thousands of prisoners. The productive alignment of all these resources in a gigantic military-industrial project was crucial for its eventual success, and Marshal Beria would prove himself a ruthless, cynical, and very efficient administrator. His task was quite certainly made much easier by the tradition of big science undertakings, long a familiar and recognized feature of the Soviet scientific and industrial landscape.

While the dimensions of the enterprise had to be rescaled, the basic organization remained typically Soviet—and also perfectly suited for replicating the Manhattan Project. Both projects shared the basic organizational principles of military-style management, hierarchical order of command, centralization of resources, and involvement of top academic scientists in the process of industrial development. The example of the Manhattan Project determined most of strategic directions of work in the Soviet case, but it also created dilemmas. The physicist Piotr Kapitza, a member of the Special Committee, argued that instead of squandering limited resources and imitating the American model in all its major aspects (working along a number of parallel routes, including plutonium and uranium-235, reactors with graphite and with heavy water, a variety of methods of isotope separation, etc.), a more narrowly defined, cheaper, and possibly more original solution should be explored.[32] The first option was more secure, but guaranteed massive overspending, while the second promised to be more economical but entailed a larger risk of mistakes. For Stalin and Beria at that particular juncture, as for General Groves earlier, the chief priority must have been time and security, rather than saving resources or satisfying scientists' egos. They chose the first option, also represented by the physicist Kurchatov, against

32. J.W. Boag, P.E. Rubinin, and D. Shoenberg, eds., *Kapitza in Cambridge and Moscow: Life and letters of a Russian physicist* (Amsterdam, 1990), 368-378; *Atomnyi II:1*, 613-623.

Kapitza's advice. In December 1945 Stalin met Kapitza's "request" and released him from membership in the Special Committee. Stalin received Kurchatov for the first time in February 1946 and urged him not to stint on resources, but to pursue the work on a "broad Russian scale." In a public speech a few weeks later, Stalin promised scientists that many new institutes would be constructed and offered them a slogan, "To catch up and to surpass!" The motto indirectly referred to the atomic bomb and reflected the same strategic choice of following the American path.[33]

Cold War Conservatism

Subsequent stages of work, in the Soviet Union as in the U.S., involved industrial development more than scientific research. Like General Groves before him, Marshal Beria ordered the work to be conducted in parallel, often even redundant fashion, and launched the construction of industrial facilities well before completion of the necessary research. The result, in both cases, was gross overcommitment of resources, but also an increased probability of overall success and some reduction in total time, which was the main priority. Beria distrusted Kurchatov and his team of scientists, but he also distrusted intelligence reports and documents smuggled from overseas, suspecting not merely mistakes, but deliberate disinformation. Typically, the security apparatus used espionage data to place alongside the Soviet scientists' research as a system of checks and counterchecks. As an additional precaution, many if not most concrete research and design tasks in the Soviet project were duplicated, with independent teams of scientists and engineers working in an indirect race. Often they became aware of each other's existence only when, towards the end of investigation, they had to meet face to face and argue whose product or solution was to be preferred and applied. Scientists in the Soviet atomic project learned the habits of working in remote secret locations, bound by the highest degree of compartmentalization and secrecy, living under surveillance and abiding military-style discipline. As a result of the atomic bomb, radar, missiles, and other "catch-up" military projects after the end of the war, big science in the U.S.S.R. became militarized to a much higher degree than it had been during the 1920 and 1930s, or even during the war itself.

In the U.S. and Great Britain, the transition to the Cold War destroyed the temporary alliance between socialists and liberals, and between left-wing scientists and the military.[34] Some of the consequences of their collaboration—those intended, such as the increased government support for science, and those unintended or expected to be temporary, such as its high militarization—became permanent fixtures in post-bomb science. On both sides of the Iron Curtain, militarization proceeded

33. *Atomnyi II:2*, 428-436; Stalin, "New five-year plan for Russia: Election address," *Vital speeches of the day, 12:10* (1 Mar 1946), 300-304.
34. Jessica Wang, *American science in an age of anxiety: Scientists, anticommunism, and the Cold War* (Chapel Hill, NC, 1999).

along the two main paths: the movement of scientists into secret, military-oriented research centers, and the increased penetration of massive military funding and classified research into ostensibly civilian institutions. Probably a large majority of academic physicists in both the U.S.S.R. and the U.S. during the postwar decade did at least some military research. Many moved between classified and academic settings and ran parallel projects in both, often within the same laboratory buildings and with the same or similar set of instruments (purchased with military funds for the needs of the classified part of research).

Such a hybrid lifestyle brought changes in the predominant mentality of physicists and the types of projects they considered most worth pursuing even in their non-secret, academic research. Paul Forman has described one of these changes as "gadgeteering," or the tendency to define and value scientific projects in terms of the special gadgets they produce rather than knowledge per se.[35] Another, somewhat related, consequence was a generally more conservative intellectual atmosphere in science—conservative not in the sense of denying progress, but in defining it in terms of extensive development (the famous "endless frontier") while preserving the basic conceptual foundations of the existing knowledge. In post-bomb physics, in vogue were technical virtuosity and the prediction and calculation of complex effects—rather than radical conceptual revolutions, the subversion and overthrow of science's most basic laws, the activities valued most highly by physicists in the first half of the century.

One of those who felt and disapproved of the change in dominant mood was David Bohm, a former student of Oppenheimer who received his Ph.D. in Berkeley during the war and worked there on a project marginally related to the question of isotope separation. In 1947 Bohm moved to Princeton as an assistant professor, and like many American physicists returning to academic positions from wartime projects, he faced a decision on what topics to choose for further research. Oppenheimer, among others, advised him to work on renormalization and nuclear physics, then the hottest and most career-friendly trends, simultaneously fundamental and appreciated by military patrons. Bohm ignored the well-meaning advice of senior colleagues: he was already developing an aversion to nuclear physics as a fundamentally boring and intellectually uninspiring field.[36]

In Bohm's view, the postwar physics community was becoming too conformist, mindful of hierarchy but driven by mindless fashion. Putting aside their ability to think

35. Paul Forman, "Into quantum electronics: The maser as 'gadget' of Cold-War America," Paul Forman and J.M. Sanchez-Ron, eds., *National military establishments and the advancement of science and technology* (Dordrecht, 1996), 261-326.

36. "Every time I hear the word 'nuclear physics,' it calls up to my mind an image of the most boring possible subject in the world. The surest way to discourage me from working in the quantum theory would be to continually remind me that it might be useful in nuclear physics." Bohm to Melba Phillips, late 1951 or early 1952, David Bohm Papers, Birkbeck College, London, C 46. See also Bohm, oral history conversations with Maurice Wilkins, Niels Bohr Library, American Institute of Physics, College Park, MD, 304-305, 320.

independently, physicists were behaving like sheep, jumping on the "bandwagon" of what was at the moment regarded as prestigious, and judging each other accordingly. Even Oppenheimer, the once revered teacher, has changed in that direction. Bohm commented: "I felt that was really very dull…I would have felt at the time…that if I had the qualities which enabled me, like Oppenheimer, [to] join the bandwagon and do what the majority were doing, then probably I would never have become a physicist. I would have become a furniture man. I would have become a businessman and perhaps one of the leading furniture dealers in Wilkes-Barre."[37]

Bohm preferred searching for unusual ideas, asking deep questions and thinking about foundations—a choice that a few years later led him to develop a causal interpretation of quantum mechanics. Bohm's theory explicitly contradicted the dominant Copenhagen interpretation, itself a revolutionary novelty a generation earlier, but a dogma during the early Cold War period. Running against the prevailing mood in the discipline cost Bohm dearly—his ideas were almost completely ignored for more than ten years—not so much because of the active opposition from the Copenhagen camp, but due to indifference on the part of his colleagues, who looked down upon such pursuits as unworthy of real professionals, preferring instead concrete results and calculations. Just as Bohm's Marxist beliefs and former Communist connections made him a political outcast in McCarthyite America, leading to the loss of his job and exile, his critical attitude towards the existing foundations of science made him marginal in the profession, his findings rejected by Cold War physics.[38]

As Bohm discovered with astonishment a few years later, a similar intellectual conservatism developed among the Soviet physics community, also in part because of the bomb. Initially, many Soviet physicists who went to work on the atomic project felt that they were sacrificing opportunities to accomplish more original research in fundamental science. Lev Landau was among those few who agreed to do the bomb work out of fear—in a precarious position after spending a year in prison during the Stalinist purges of the late 1930s, he felt that classified research provided additional protection. He still kept his involvement rather limited and withdrew at the first available opportunity after Stalin's death. Others, including Khariton, Kurchatov, Andrei Sakharov, Igor Tamm, and Zeldovich, participated out of a sense of moral and patriotic duty. Amounting to several years of complete engagement for some,

37. Bohm, oral history (ref. 36), 304-308.
38. On Bohm see David F. Peat, *Infinite potential: The life and times of David Bohm* (Reading, MA, 1997); Russell Olwell, "Physical isolation and marginalization in physics: David Bohm's Cold War exile," *Isis, 90* (1999), 738-756; Alexei Kojevnikov, "David Bohm and collective movement," *Historical studies in the physical and biological sciences, 33:1* (2002), 161-192. On Bohm's interpretation of quantum mechanics, see Olival Freire Jr., "Science and exile: David Bohm, the hot times of Cold War, and his lasting struggle for a new interpretation of quantum mechanics," paper presented at the workshop on "Migrant scientists in the twentieth century," Milan, June 2003; Olival Freire Jr., "Quantum controversy and Marxism," *Historia scientiarum, 7* (1997), 137-152.

for others becoming a lifelong commitment, no matter how strongly they felt about the necessity of the bomb work, they could not help regretting their diversion from fundamental physics. Not only did they view the atomic project as more technological than scientific, but also they found that the ultimate goal of the work consisted in large part of repeating and reproducing what had already been accomplished overseas.[39]

Kapitza's advice in 1945 to have it both ways—follow an original path and build the bomb—was rejected in favor of the strategy "to catch up and to surpass!" that minimized time and risks. Implementing this slogan in practice meant more often than not following the American lead, adopting many of the same strategic choices and goals, expending huge efforts and funds on catching up, usually trailing, while occasionally using chances to actually surpass, as happened with the hydrogen bomb of Sakharov's design. Some occasionally very heated discussions among Soviet physicists outside the atomic bomb project involved the same dilemma: whether to compete with American physics on its own terms or to seek more original directions in research. The latter choice entailed serious risks, which were particularly high in the Stalinist Soviet Union and were demonstrated there so disastrously by the 1948 Lysenko example in biology. Lysenko succeeded for a time being in persuading a large following and the political authorities that his theories were actually superior to foreign ones, but the majority of Soviet physicists saw him even during the period of his official dominance as a charlatan and imposter. For them then as for us now, Lysenko's rejection of international genetics meant great damage inflicted upon the whole of Soviet science.[40]

The combined positive example of the highly successful and appreciated atomic bomb project and the negative example of politically supported but ultimately disastrous Lysenkoist biology made Soviet scientists particularly wary of arguments in favor of seeking original socialist paths in science. Even after the completion of the main goals of the atomic project by mid-1950s, as military research was losing prestige and Soviet physicists were returning to civilian institutions and fundamental science, they brought back many of the new habits and attitudes acquired in the course of the bomb work.[41] Nuclear physicists became used to operating with industrial-scale resources and institutions and demanded a similar level of support for their civilian projects. They also extended the same "catch up and surpass" strategy as the right way to compete in Cold War science in general. The ubiquitous slogan worked not only for physics, but also in other branches of postwar science, military and civilian alike. Interiorized and uncritically accepted by the wider academic community, it not only survived Stalin's death in 1953, but determined the mentality of Soviet science for decades thereafter.

39. See, for example, Andrei Sakharov, *Memoirs* (New York, 1990).

40. See the most recent account of the Lysenko affair and citations to the earlier literature in Nils Roll-Hansen, *The Lysenko effect: The politics of science* (Amherst, NY, 2004).

41. Konstanin Ivanov, "Science after Stalin: Forging a new image of Soviet science," *Science in context, 15* (2002), 317-338.

In other European countries after the end of World War II, scientists faced the same dilemma as they contemplated strategies for coming to terms with the predominance of American science, methods, and money—and usually made the same choice of following overseas trends. The alternative—competing along different paths, as had usually been the case in scientific rivalries between leading European nations during the nineteenth and early twentieth centuries—was occasionally discussed, but more often than not deemed less promising of success or too risky.[42] For international science, the overall effects of the bomb and the Cold War resulted in a significant loss in diversity.

The bomb has long been seen as a milestone separating the two parts of the twentieth century with regard to governmental support for science, its militarization, and the overall dimensions of the scientific enterprise. It can also be seen as a milestone with regard to the trends discussed above: gadgeteering, homogeneity, and conservatism in attitudes towards the most basic conceptual foundations. These trends did not go unchallenged, in particular during the 1960s, but for the most part survived and continued to operate with fluctuating strength for the remainder of the century.

As does anyone who teaches the history of modern science, for my students I face the task of explaining why the first half of the twentieth century produced revolutions of basic scientific concepts, whereas the second, by comparison, was rich in devices, expenditures, and published papers. Helge Kragh, in his monumental history of twentieth-century physics, brings this curious phenomenon full circle.[43] His book opens with the often quoted view of the 1890s that the most basic laws of physics had already been established and were not expected to undergo major changes. Kragh's conclusion—five hundred pages, a great conceptual revolution, and over a hundred years later—evokes a similar feeling shared by the author and the majority of physicists at the end of the twentieth century. Can one not sense an irony here?

42. Pestre and Krige (ref. 28).

43. Helge Kragh, *Quantum generations: A history of physics in the twentieth century* (Princeton, 1999).

PART FOUR

Postwar Politics
The Oppenheimer Affair in Context

Without the infamous Atomic Energy Commission "inquiry" into Oppenheimer's loyalty and security status, our sense of the man would have been conspicuously different. Instead of a political martyr, he would have remained a scientific statesman and an elite Cold War liberal with powerful political connections. The Oppenheimer Affair was indubitably tied to his person and his moment. To understand it, however, we have to do more than assess the charges raised against him, or even trace his and other actors' moves in the framework of the hearings. We need to look at the role that Oppenheimer had assumed, the choices that put him under the microscope, and the poisoned political atmosphere that made the inquiry possible at all.

Much good scholarship now deals with the postwar political context. This body of work gives a careful account of scientific advising and consulting, as well as a fine-grained picture of loyalty-security measures. The program is carried forward by the essays in this section. First, James G. Hershberg looks at Oppenheimer's shifting position on the international control of atomic weapons in the later 1940s. After the Baruch Plan's failure, Oppenheimer remained an influential member of the American advisory elite, even as he cryptically signaled dissent. David Kaiser and Stephanie Young then turn our attention to the loyalty-security system. Kaiser examines rampant Red Scare stereotypes of theoretical physicists' propensity to pass atomic secrets, while Young delves into some of the earliest cases of AEC personnel security hearings, documented in the files of the security board's chairman in Berkeley. W. Patrick McCray's essay takes up one of Oppenheimer's run-ins with powerful military enemies in an ill-fated Caltech study assessing strategies for atomic warfare. Finally, Richard Polenberg reflects on the consequences of the 1954 security hearing and revisits a matter at its center, the political cast of Oppenheimer's mind.

CHAPTER EIGHT

"The Jig Was Up"
J. Robert Oppenheimer and the International Control of Atomic Energy, 1947-49

James G. Hershberg

In the crowded narrative of J. Robert Oppenheimer's intersection with early nuclear history, the international control of atomic energy and weapons generally makes a brief but dependable appearance. Like most Manhattan Project scientists, after Hiroshima Oppenheimer ardently believed that international control over all nuclear technologies and materials constituted the only hope of forestalling an arms race leading inexorably to catastrophe. Quoting Harry S. Truman's description of atomic energy as "a new force too revolutionary to consider in the framework of old ideas," Oppenheimer hailed the "injection of the spirit of the scientist into this problem of atomic weapons"—the "recognition that in these problems a common approach, in which national interests can play only a limitedly constructive part, will be necessary if a solution is to be found

I thank Cathy Carson for a thorough and splendid editing job to trim this essay down to publishable proportions. For access to their copy of Oppenheimer's declassified FBI file, I am grateful to Martin J. Sherwin and Kai Bird.

The following abbreviations are used: AEC, Atomic Energy Commission; *BAS*, *Bulletin of the atomic scientists*; DOE, Department of Energy Archives, Germantown, MD; FBI, Federal Bureau of Investigation; FHOP, Frederick H. Osborn papers, Harry S Truman Library, Independence, Missouri; *FRUS* 1947, *1*, U.S. Department of State, *Foreign relations of the United States, 1947,* Vol. 1, *General: The United Nations* (Washington, D.C.: Government Printing Office, 1973); GAC, General Advisory Committee; HFAP, Hamilton Fish Armstrong Papers, Seeley G. Mudd Manuscript Library, Princeton University, Princeton, NJ; JRO, J. Robert Oppenheimer; *JRO Hearing*, United States Atomic Energy Commission, *In the matter of J. Robert Oppenheimer: Transcript of hearing before Personnel Security Board and texts of principal documents and letters* (Cambridge, 1971); JROP, J. Robert Oppenheimer Papers, LC; LC, Manuscript Division, Library of Congress, Washington, D.C.; NARA, National Archives and Records Administration, College Park, MD; NDUA, National Defense University Archives and website, Washington, D.C.; RG, Record Group; *NYT, New York Times*; S/AE, Records of the Special Assistant to the Secretary of State for Atomic Energy Matters, 1944-1952, Series 1—General Records Relating to Atomic Energy Matters, 1944-1952, Lot 57 D 688, RG 59, NARA; UN, United Nations; UNAEC, UN Atomic Energy Commission.

at all."[1] As a consultant to the State Department group charged with drawing up the official U.S. position, he played a central role in developing the Acheson-Lilienthal Plan in early 1946. That plan envisioned a global Atomic Development Authority to promote civil uses of nuclear power—an alternative to rival secretive national programs, and equally to a system predicated on a "dreary, sterile, and policeman-like" program of international or multilateral inspections.[2]

However, over the course of 1946 Oppenheimer saw his dream dissipate. Acheson-Lilienthal's bright promise was, in his view, diluted and distorted by Truman's appointment of Bernard M. Baruch to present the U.S. program to the United Nations Atomic Energy Commission at Lake Success, and by the negotiator's stiffening of key provisions in line with hardening political currents in the Truman Administration and domestic politics as whole. Baruch seemed far more absorbed by the need for inspection, "condign punishment," removing the (Soviet) right to veto Commission actions, and protecting the prerogatives of private industry than he was impressed by the positive force a worldwide scientific enterprise could create, and the favorable impact that might have on international affairs.[3] When Soviet representative Andrei Gromyko countered Baruch's June 1946 presentation with a call for the immediate destruction of all (American) nuclear stockpiles but without any real acceptance of controls or inspections—in other words, a contemptuous rejection of the offer, however flawed and hedged, for the gradual relinquishing of the American atomic monopoly as an international agency went into operation—a deeply discouraged Oppenheimer tried to convince his colleagues (so he later recalled) that "the jig was up, because that was relevant to getting back to work." Even as he stood on trial for his security clearance and reputation in 1954, he remembered of the days in 1946 when hopes for international control ebbed, "I have seldom been as gloomy in my life; that even includes today."[4]

For those charting the trajectory of Oppenheimer's atomic career, the Baruch Plan's failure to gain Soviet support in the UN Commission's First Report of late December 1946 coincides neatly with Oppenheimer's assumption of the chairmanship of the General Advisory Committee of the new U.S. Atomic Energy Commission—and the GAC's rapidly reaching the reluctant but firm and unanimous conclusion ("I suppose with some melancholy," Oppenheimer told the AEC inquest[5]) that the Commission's

1. See JRO, "The new weapon: The turn of the screw," Dexter Masters and Katharine Way, eds., *One world or none: A report to the public on the full meaning of the atomic bomb* (New York, 1946), 22-25, on 22 (citing Truman's 3 Oct 1945 message to Congress), 24.

2. See JRO, "Atomic explosives," address of 16 May 1946, *The open mind* (New York, 1955), 2-17, on 10.

3. Donald A. Strickland, *Scientists in politics: The atomic scientists movement, 1945-1946* (Lafayette, IN, 1968), 121; McGeorge Bundy, *Danger and survival: Choices about the bomb in the first fifty years* (New York, 1988), 162; see also David E. Lilienthal, *The journals of David E. Lilienthal* (New York, 1964), *2*, 42-43, 54, 60, 69-71.

4. *JRO Hearing*, 45, 40.

5. Ibid., 69.

first priority must be building bombs, not developing power or cooperating with an international atomic development agency. With Cold War tensions deepening and the feared secret U.S.-Soviet nuclear arms race underway, international control largely vanishes from the Oppenheimer story, his subsequent actions rating at most a few sentences or pages, but usually nothing at all. "This ended Oppenheimer's formal attempts to influence government policy on the international control of nuclear weapons," Jeremy Bernstein writes typically, if inaccurately.[6]

However, Oppenheimer's post-Baruch Plan involvement with international control merits closer inspection. If Oppenheimer made few public comments drawing attention to his views, that was because of the conditions under which he and other members of the atomic elite operated. "I could not come out and say, 'This is a hopeless thing,'" he offered in 1954, "because I had some official connection with the Government until the Government had said so."[7] His evolving opinions—including his disagreements with one of his closest collaborators in atomic policy, James B. Conant, president of Harvard University and his "mentor in national policy matters"—illuminate important divergences in that atomic elite's thinking as hopes for international control faded.[8] This essay thus focuses on three aspects of Oppenheimer's shifting views in 1947-49: his strong secret opposition, beginning in early 1947, to further elaboration of American plans, in view of the Kremlin's refusal to play ball or open up Soviet society; his resistance, even while conceding the over-optimism of earlier predictions, to abandoning Acheson-Lilienthal's concept of the development of civil nuclear power; and his hints at his own quite acute private discontent, both with the direction of history and the conduct of U.S. policy, in the light of international control's failure. In view of the fact that he remained at this point a nuclear insider *par excellence*—chairing a supremely influential advisory panel and sitting on numerous secret government boards—and that, after his security hearing, he gained an only partially deserved reputation as an atomic dissenter, it is interesting to note the wary, cautious, sometimes cryptic, yet unmistakable manner in which he signaled his disagreements—and how those in power reacted.

6. Jeremy Bernstein, *Oppenheimer: Portrait of an enigma* (Chicago, 2004), 102; compare David C. Cassidy, *J. Robert Oppenheimer and the American century* (New York, 2005), 275-279. To my knowledge, the only Oppenheimer biography to consider his post-1946 international control activities is James W. Kunetka, *Oppenheimer: The years of risk* (Englewood Cliffs, NJ, 1982), 126-129. Kai Bird and Martin J. Sherwin note JRO's opposition in early 1947 to further pursuit of international control in their magisterial *American Prometheus: The triumph and tragedy of J. Robert Oppenheimer* (New York, 2005), 352-354.
7. *JRO Hearing*, 45.
8. Richard G. Hewlett and Francis Duncan, *A history of the United States Atomic Energy Commission*, Vol. 2, *Atomic shield, 1947-1952*, (University Park, PA, 1969), 378; James G. Hershberg, *James B. Conant: Harvard to Hiroshima and the making of the nuclear age* (New York, 1993). For this essay I have adapted some passages from the book.

Oppenheimer Turns Against International Control

The story opens in January 1947. His plan nominally approved by the UN Atomic Energy Commission, though in fact consigned to irrelevance by the Soviet abstention, Baruch resigned with a flourish, his post temporarily assumed by the U.S. representative to the world body, Warren R. Austin, a former Republican senator from Vermont. For the moment, U.S.-Soviet relations were relatively quiet, with no immediate crisis at hand—but the calm was deceptive, with looming disputes over Germany, Europe, and the Near East on the horizon. Despite the atomic impasse, it was presumed that negotiations at the United Nations would continue, in some form, and Washington remained committed, at least on paper, to its international control offer.[9]

For Oppenheimer, however, Moscow's conduct had already disqualified it as a partner. Even before the final vote at the UNAEC, Oppenheimer had made plain his private displeasure with the Russians' behavior. "I don't like the way they're acting," he told his friend the Caltech physicist Richard C. Tolman (who was advising Baruch) in late December 1946, in a telephone conversation intercepted and transcribed by the FBI. "I think we'll have to expect it, but I don't like it the least bit."[10] When friend and fellow theoretical physicist Hans Bethe visited Berkeley in January 1947, he was surprised by the vehemence with which Oppenheimer argued against further pursuit of an international control pact. Recounting some "quite long conversations" to the 1954 AEC security board, Bethe recalled Oppenheimer saying that "he had given up all hope that the Russians would agree to a plan which would give us security." Oppenheimer had little use for Moscow's insistence on first destroying all atomic stockpiles and then figuring out a system of controls, and gave Bethe a rationale that differed little from Winston Churchill's hard-line (and then still controversial) advocacy of firm measures to contain Soviet expansionism. "Particularly he [Oppenheimer] pointed out how much the Russian plan was designed to serve the Russian interests and no other interests, namely, to deprive us immediately of the one weapon which would stop the Russians from going into Western Europe; if they so chose, and not give us any guarantee on the other hand that there would really be a control of atomic energy, not give us any guarantee that we would be safe from Russian atomic attack at some later time."[11]

9. On the situation in the UN in Jan-Mar 1947, see Hewlett and Duncan (ref. 8), 264-267.
10. Consonant with Baruch's position, Oppenheimer now felt it "absolutely essential" to reject "anything in it [the international agency] corresponding to the veto" and warned against concessions in the interest of obtaining wider agreement that would amount to "buying fairly much of an illusion at a fairly high cost." Telephone conversation between JRO and Tolman, 26 Dec 1946, FBI file of JRO, #100-17828-120, Federal Bureau of Investigation, records at FBI Headquarters, Washington, D.C., courtesy of Martin J. Sherwin and Kai Bird.
11. Bethe testimony, *JRO Hearing*, 327. Bethe is confirmed to have been in Berkeley beginning on Jan 6, including a lunch with JRO on Jan 9 after the latter had returned from the East Coast the previous day. See materials in FBI file of JRO (ref. 10), #100-17828-127, esp. wiretap summaries of 5, 6, 8, 14, and 22 Jan 1947.

Though at that point neither Bethe nor many other atomic scientists had completely given up, Oppenheimer's gloomy forecast found echo in the classified confines of the GAC, which reunited Oppenheimer with wartime scientific titans like Fermi, Rabi, Conant, DuBridge, and Seaborg. "By 1947 it took more optimism than most anyone had to expect international control as proposed earlier," recalled Robert F. Bacher, a close friend of Oppenheimer who was the only physicist among the new U.S. Atomic Energy Commission's five commissioners (with Lilienthal as their chair).[12] After Bacher verbally briefed the GAC in early February on the size of the nation's nuclear stockpile and rate of production—numbers which all present found disconcertingly, even shockingly low—discussion ensued on "the relative emphasis on weapons and reactors," the minutes record, "with rather general final agreement that weapons were of first priority." After listening to his fellow luminaries, Oppenheimer, as was his wont, summarized the group's sentiments: "It seems to me that the heart of the problem has been reached with surprising speed: The making of atomic weapons is something to which we are now committed." He added plaintively: "It must be recognized that within our hearts we have been hoping that the world will be the world it was ten years ago. This is no longer possible and we must try in ourselves and in our colleagues to find some way of obtaining public service."[13]

Oppenheimer, Osborn, and the UNAEC

Though he kept his loss of faith in international control hidden from public view, Oppenheimer in March 1947 further displayed the intensity of his belief that continued negotiations in the UN were not only futile but potentially dangerous. The occasion was a crucial meeting with the newly-appointed U.S. delegate to the UNAEC (and deputy to U.S. UN ambassador Austin), Gen. Frederick H. Osborn, who had run the army's wartime information and education branch. The run-up to the Oppenheimer-Osborn meeting has previously been understood only cloudily, but now the course of events seems clear.[14] Acceding to a request from Osborn transmitted by

12. Robert F. Bacher, personal communication, 14 Jan 1982.

13. Minutes, GAC meeting, 2 Feb 1947, DOE; Hershberg (ref. 8), 305-306, 312-313.

14. The episode has been recorded in garbled form thanks to the vivid testimony Osborn gave in 1954 before the AEC security board (*JRO Hearing*, 342-346) suggesting that JRO contacted Osborn on his own initiative and came East on this self-chosen mission on Friday, March 7. The testimony has misled historians who have relied on it. See, e.g., Hewlett and Duncan (ref. 8), 268; Philip M. Stern with the collaboration of Harold P. Green, *The Oppenheimer case: Security on trial* (New York, 1969), 100-101; Joseph I. Lieberman, *The scorpion and the tarantula: The struggle to control atomic weapons, 1945-1949* (Boston, 1970), 390; Kunetka (ref. 6), 126; and Gregg Herken, *Brotherhood of the bomb: The tangled lives and loyalties of Robert Oppenheimer, Ernest Lawrence and Edward Teller* (New York, 2002), 180.

What in fact happened can be determined from contemporary documentation, supplemented by Oppenheimer's own 1954 testimony: On Monday, March 3, 1947, a wiretap

State Department official Joseph E. Johnson, Oppenheimer urgently made his way to Washington on an red-eye flight arriving Wednesday, March 5. "On the way" to see Osborn, Oppenheimer later testified at his security hearing, he "stopped at the State Department and [Under Secretary of State] Acheson showed me the President's [draft] speech on the Truman Doctrine. He wanted me to be quite clear that we were entering an adversary relationship with the Soviet, and whatever we did in the atomic talk [at the UN] we should bear that in mind."[15]

The day Oppenheimer reached the nation's capital was a grim one indeed, as far as the already moribund prospects for international control were concerned. In a speech to the UN Security Council that, in the words of one historian, "put an end to all optimism," Gromyko, in addition to maintaining Moscow's insistence on a total ban on atomic weapons prior to the institution of controls, brusquely rejected the Baruch Plan's principle of international atomic development as based on "fundamentally vicious premises" that were "incompatible with state sovereignty" and completely unreal.[16] In the background, too, as Oppenheimer noted, were the events leading to the pronouncement of the Truman Doctrine, with the emergent threat of a power vacuum in the strategically vital eastern Mediterranean, once the British had to abandon their efforts to prop up pro-Western governments in Greece and Turkey. All that week, Acheson had been deeply engaged in drafting Truman's momentous foreign policy speech, to be delivered to a joint session of Congress on Wednesday, March 12, 1947, on the need for the United States to aid countries threatened by communism. Acheson would later write, "This was my crisis. For a week I had nurtured it....[I]t was my task to bring it home."[17]

---

on Oppenheimer's home in Berkeley picked up an incoming telephone call originating at the Pentagon from "an individual identified only as JOE." According to the FBI summary, "JOE told OPPENHEIMER that OSBORN was anxious to get in touch with him and would like to know when OPPENHEIMER would come East." Oppenheimer responded that he expected to go east on March 22 or 23. But "JOE"—most likely Joseph E. Johnson, a State Department official dealing with atomic matters who accompanied Osborn to lunch that day with "the Interdepartmental Committee on Atomic Energy"—replied that "he did not think that would be soon enough and that OPPENHEIMER should be there in a very short time." On Tuesday, March 4, the wiretap disclosed, Kitty Oppenheimer told a friend—Barbara Chevalier—that Robert had already "left for Washington that evening" because Osborn "wanted to talk to OPPENHEIMER before he accepted the appointment." FBI wiretap summaries, 3 and 4 Mar 1947, FBI file of JRO (ref. 10), #100-17828. For Johnson's identity, see Osborn, UNAEC diary, Mon, 3 Mar 1947, FHOP, box 2, folder UNAEC diary. Johnson was then chief of the Department's Division of International Security Affairs.

15. *JRO Hearing*, 41.

16. Lieberman (ref. 14), 388-389; Hewlett and Duncan (ref. 8), 267; Thomas J. Hamilton, "Gromyko is bitter," *NYT*, 6 Mar 1947, accompanied by excerpts from Gromyko's speech.

17. Joseph Marion Jones, *The fifteen weeks* (New York, 1955), 148-155; Dean Acheson, *Present at the creation* (New York, 1969), 219.

On Thursday afternoon, March 6, after digesting Gromyko's speech, Acheson considered options for how Ambassador Austin should respond. For Austin simply to make a "'patience and good will'" speech would be too soft, especially in contrast to the "pretty strong and realistic statement" Truman would be delivering in a few days; but for Austin to tell the Council that now "the jig is up and there no longer is any hope of international control" would be too hard, implying, in conjunction with Truman's address, "a far-reaching and fundamental" stiffening of U.S. policy. But between these two alternatives, it would be "desirable" to keep debating the matter and thereby "expose fully the Soviet position" and alert the American public of the "great gulf" between the Kremlin and U.S. thinking on the issue. Austin might also note that Gromyko's statement had, alas, dispelled any lingering hope (fanned by a few ambiguous Soviet hints) that "the Soviet Government would after all sooner or later go along with the major conclusions of the Report."[18]

UN Ambassador Austin's speech needed to walk this fine line when he appeared before the Security Council on Monday. Accordingly, Acheson called Austin later on Thursday to outline the drift of his thinking. Austin consented to send Osborn, his new deputy representative, down from New York City to Washington for an all-day session Friday in the State Department "with Mr. Oppenheimer and others concerned."[19] Taking charge of the process, on Friday Oppenheimer himself "dictated" to Osborn and Johnson a draft statement for Austin to deliver as a *de facto* response to Gromyko. For authority, Osborn noted, Oppenheimer cited "the instructions of Dean Acheson with whom he had breakfast that morning" and told Osborn "to deliver it to Senator Austin as his [Acheson's] instructions" for the forthcoming Security Council speech.[20]

The draft speech bore many of the hallmarks of the arguments Oppenheimer would make in coming months. It opened by drawing attention to fact that Gromyko's speech had made clear "basic and vital" differences between the Soviet and American approaches. Moscow's latest proposals, Austin was to say, would both gut any effective international control plan and retard atomic energy development. In sum, they regrettably had "proved groundless" any hopes that the Soviet abstention on the Baruch Plan merely reflected disagreements on "relatively minor matters which could

18. Deputy Director of the Office of Special Political Affairs (John C. Ross) to Chief of the Division of International Security Affairs (Johnson), memorandum with notes on telephone conversation between Acheson and Ross, 6 Mar 1947, *FRUS* 1947, *1*, 426-427.

19. Acheson-Austin telephone conversation, 6 Mar 1947, USUN Files, *FRUS* 1947, *1*, 427 n. 5. Osborn's "UNAEC diary," apparently compiled shortly after the events, corroborates both Oppenheimer's presence in Washington on March 7 and his collaboration with Acheson to influence the American position. Osborn, UNAEC diary (ref. 14), 7 Mar 1947.

20. Osborn, note attached to "Draft of statement by Senator Austin," FHOP, box 2, folder UNAEC Diary: "This letter was dictated by Robert Oppenheimer in Joe Johnson's office in the presence of Johnson and myself." This and subsequent paragraphs are based on the draft statement.

be negotiated." Before moving to the "grave" issue of the "reappraisal...of the task before us," Austin would also strongly rebut Gromyko's contention that Washington sought an atomic monopoly. He would explain that the U.S. nevertheless could not accept the Soviet proposals because they lacked the "minimum and essential" features necessary to provide security, and would instead lead to "a system which at best would give no security, and at worst would be a constant incentive to distrust and war." The Soviet objections to the Baruch Plan, Austin was to declare, "go to the very heart of the proposals in that they deny the necessity, if security is achieved, for an international cooperative development in the field of atomic energy."

Exactly a year after Oppenheimer had collaborated with Acheson on an optimistic, idealistic plan for international control, the two men were now proposing that the U.S. representative to the United Nations formally shelve the idea. Despite reaching a mutual accord on some "trivial" aspects, Austin's draft speech declared, the "basic issues dividing" the Soviet Union and the U.S.-led nations "preclude agreement on this matter so vitally affecting the peace of the world." Thus while expressing readiness for further exploration of such fundamental disputes, "we...should admit that unless these can be resolved, progress in detail will be neither likely nor significant." Consequently, the U.S. would withdraw its resolution calling on the Commission to elaborate further its blueprint for international control.

This was Osborn's introduction to Oppenheimer's thinking. After the drafting session in Washington, the two men spent an intense two days exchanging ideas. They traveled to New York for the weekend, meeting in Manhattan on Saturday and driving to Osborn's country house near West Point, where Oppenheimer remained until the two returned to New York City on Sunday afternoon. Recounting his international control experiences before Osborn came on the scene, Oppenheimer described the sad fate of the Acheson-Lilienthal Report, including the disastrously poor relations between that group and Baruch's team. But his main point was that continuing to seek international control at this juncture no longer made sense—partly because the Acheson-Lilienthal provision for an international atomic development authority "to conduct all intrinsically dangerous operation in the field" had been so watered down by both American and Soviet statements as to become "meaningless," but mostly because of a justified distrust of the Russians.[21] That much Osborn distinctly recalled before the AEC Personnel Security Board in 1954: [22]

> Oppenheimer came on to tell me that there were two very serious dangers in continuing negotiations. The general background was that he was now certain... that the Russians had no intention of accepting any plan for the control of atomic energy—international control of atomic energy—which would mean lifting the Iron Curtain....[T]o do so would mean the end of the regime....Yet he felt certain that, if the Iron Curtain was not lifted, any plan of international control would

21. Osborn, notes on conversation with JRO, dated "March 9, 1947," S/AE, box 64, folder 15D.
22. *JRO Hearing*, 344.

> be exceedingly dangerous....[W]e could accept compromises which would put the United States in a very dangerous position of not really knowing what was going on in Russia, whereas the Russians would know all about what was going on here.... He also felt that this continuation of negotiations was something that the Russians would be glad to use the United Nations as a medium for propaganda...So he was for discontinuing the negotiations.

As Osborn noted in his secret "UNAEC diary," Ambassador Austin's delivery of the draft speech Oppenheimer had dictated in Johnson's office on March 7 would have "prevented progress in drafting specific proposals until agreement was reached with the Russians." However, he acknowledged, "I did not understand the proposition clearly at the time."[23] After driving back from Garrison with Oppenheimer to New York City and seeing off the physicist, Osborn met Austin at the ambassador's apartment late on Sunday afternoon, March 9, and endorsed Oppenheimer's suggested speech breaking off U.S. participation in the international control talks. But after a "considerable discussion," Osborn reported, Austin "very specifically decided not to use the State Department draft which I gave him and urged on him."[24]

On Monday at the Security Council, to Oppenheimer's disappointment, Austin presented a revised (and, Osborn felt, "rather weak") new version of the earlier U.S. resolution on international control—rather than withdrawing it entirely, as Oppenheimer's draft envisioned.[25] In strongly rejecting the "defamation and innuendo" in Gromyko's statement, as well as its specific ideas, the U.S. representative did use passages from Oppenheimer's draft rebuffing the charge that Washington plotted to retain an atomic monopoly and characterizing the flaws of the Soviet proposals.[26] But Austin then inserted an entirely different conclusion. Already viewed by some in the administration, including Acheson, as an inveterate optimist, he rejected the idea of walking away from the talks altogether. Instead he put forward a resolution that merely called on the UNAEC "to continue its inquiry into all phases of the problem," to "develop as promptly as possible...specific proposals," and "in due course to prepare and submit to the Security Council" draft agreements and to hand in a Second Report by the time the General Assembly convened on September 16, 1947. And so it would happen. Unanimously, the Council approved tossing the matter back to the Commission.[27]

23. Osborn, note attached to "Draft of statement by Senator Austin" (ref. 20).
24. Osborn, UNAEC diary (ref. 14), 9 Mar 1947. Baruch and another of Austin's deputies, Herschel V. Johnson, were also on hand.
25. Osborn, record of telephone conversation with Joe Johnson, 1 Apr 1947, FHOP, box 2, folder UNAEC Diary.
26. Compare JRO draft (ref. 20) with Austin, advance text of 10 Mar 1947 speech to the Security Council, U.S. UN delegation press release, Warren Austin Papers, Bailey-Howe Library, University of Vermont, Burlington, VT, carton BV, folder 358A; see also Thomas J. Hamilton, "Atom plan is sent to U.N. Commission," *NYT*, 11 Mar 1947.
27. Austin, advance text of 10 Mar 1947 speech to the Security Council (ref. 26).

Oppenheimer had clearly established a degree of rapport with Osborn so obviously absent from his relationship with his crusty predecessor Baruch. Charmed by Oppenheimer's "fine" visit, Osborn thanked him for providing "an initial point of view which has enabled me to envisage immediate problems with far greater clarity."[28] Nonetheless, through the spring of 1947, the UN negotiator showed that he did not feel bound to accept the physicist's advice. In a late March meeting including Bacher and Conant, he told Oppenheimer that cutting off talks would be regarded as an "act of bad faith" by friendly nations, who had promised to take the lead in promoting proposals acceptable to Washington. Eager to stay on good terms with Osborn and perhaps reassured by his skeptical assessment of the likelihood of agreement with Russia—but not really altering his views—Oppenheimer told him that "this changed the situation in his mind and we should go ahead along these lines."[29] In any case, Osborn's inclination to press ahead for these and similar reasons soon hardened into policy. At a high-level meeting in early April, senior administration officials, including Acheson, War Secretary Forrestal, and Lilienthal, favored continuing the talks, if mostly to lay the blame squarely on Moscow when they failed.[30] And despite more indications of a deepening U.S.-Soviet rift, friendly countries on the Commission still vehemently urged Osborn to reject Oppenheimer's plea to terminate the talks. Doing so before finishing the proposals, the Canadian, British, French, and Belgian representatives told him, would "[c]ast grave doubts on [American] good faith," leave the West unprepared should a Kremlin policy shift suddenly make international control attainable, and hand the Soviets a propaganda victory.[31] In early May, Osborn publicly signaled his determination to press forward by formally adding prominent consultants to his delegation, including Conant and Groves, to a roster that already included Tolman.[32]

28. And upon hearing that Osborn planned to see Bacher at the end of the month in order to put together a small core group to formulate the American position, Oppenheimer responded with his second transcontinental trip in three weeks. Osborn to JRO, 13 (quote) and 21 Mar 1947, S/AE, box 77, folder 20.103.

29. Osborn, UNAEC diary (ref. 14), 29 Mar 1947; see also entry of 1 Apr 1947, in which Osborn alludes to JRO's views in a conversation with Austin.

30. Lilienthal (ref. 3), 166-167, describing meeting of 3 Apr 1947; Hewlett and Duncan (ref. 8), 268-269; Osborn, UNAEC diary (ref. 14), 3 Apr 1947. Austin even urged those present to eschew "defeatism" regarding the possibility of eventual agreement. Osborn went along with Austin, Forrestal recorded, "except that he did not breathe the same degree of optimism." Walter Millis, ed., *The Forrestal diaries*, 3 Apr 1947 (New York, 1951), 259.

31. Osborn to Acheson, unsent draft, shown to Joseph Johnson, 2 May 1947, S/AE, box 23, folder 15F. Osborn later recalled that "if we embarked on a serious attempt to write specific proposals…we would be doing something constructive. The Russians might say that we were blocking progress, but…it is hard to bark against facts all the time." Osborn, "United Nations Atomic Energy Commission," transcript of remarks to National War College, Washington, D.C., 14 Sep 1947, secret, 9-10, S/AE, box 9, folder 8.

32. Nancy MacLennan, "U.S. to push moves for U.N. atom plan," *NYT*, 7 May 1947.

Oppenheimer worked closely and cooperatively with Osborn's staff to help fashion proposals for the UNAEC's Second Report, and on April 29 he briefed the Commission's Control Committee ("Committee 2") at Lake Success as an "expert witness" on technical aspects of the plans for an international control agency and the prospects for nuclear power.[33] While sidestepping extended references to the disputes ripping apart the wartime Grand Alliance, he reiterated the rationale of the international atomic authority he had proposed a year earlier, to promote both world security and—the real focus of his presentation—the "fundamental principle of international cooperative development of atomic energy" for peaceful purposes.[34] He had little success, however, in overcoming the skepticism of the Soviet delegate, the physicist Dmitrii Skobeltzyn, over how the proposed agency would distinguish between dangerous activities (over which it would maintain a monopoly) and other activities related to nuclear research and development (which it would license or monitor and which would be non-secret).[35]

Though Oppenheimer publicly concealed his belief in the futility of further negotiations,[36] Osborn fully understood that the two viewed the matter differently. The day after his UNAEC appearance, the physicist spent several hours with the U.S. delegation reviewing its positions. "Went over all our work with Dr. Oppenheimer," Osborn recorded. "Got enormously valuable advice on next matters in subject list and on research and mining. Dr. Oppenheimer thinks we are making a mistake to go on with our work in Committee No. 2, but recognizes that [that] is the decision and has been enormously cooperative and helpful in assisting us to prepare it."[37] To senior State Department officials, Osborn profusely praised Oppenheimer's contributions even while alluding to their ongoing divergence: "I may say that while Dr. Oppenheimer disagrees with one part of what we are doing, he has given us his unstinted help in seeing that we do that part well, and I am deeply grateful to him. He is a remarkable man."[38]

33. A transcript of JRO's closed-door appearance, including exchanges with delegates, can be found in "Verbatim record of the seventeenth meeting, Committee 2 of the Working Committee—Atomic Energy Commission" 29 Apr 1947, restricted, S/AE, box 77, folder 20.103.

34. A.M. Rosenthal, "Atom work urged by Oppenheimer," *NYT*, 30 Apr 1947.

35. "[W]e are lost in words," JRO said in frustration after several attempts to convince Skobeltzyn that little more than explosions, uranium mining, and a very few potentially dangerous developmental activities would be under the agency's exclusive purview. "I understand better, but I am not satisfied," the Russian finally concluded, warily (and wisely) stressing that he was reserving any firm views on JRO's presentation until he received instructions from Moscow. "Verbatim record" (ref. 33), 51-63, 73-74.

36. For one pessimistic private assessment, see JRO to Lincoln Gordon, 21 Apr 1947, JROP, box 194, folder UN/AEC, April-May 1947.

37. Osborn, UNAEC diary (ref. 14), 30 Apr 1947.

38. Osborn to Acheson, unsent draft, 2 May 1947 (ref. 31).

Oppenheimer, International Control, and Civil Nuclear Energy

When a sanitized version of what the *New York Times* dubbed Oppenheimer's "specific blueprint" for an international atomic agency was released in early June, what garnered attention was his relatively optimistic forecast for civil nuclear power. "Predicting an atom-powered civilization," the *Times* reported, Oppenheimer had envisioned a demonstration within five years (and "possibly considerably less than that") of the feasibility of nuclear reactors for generating electrical energy, and in one, or certainly "less than two decades," the application of nuclear power to "certain specific highly critical problems of energy generation," such as ships or rockets. It would take, he estimated, thirty to fifty years for atomic energy to "in any substantial way supplement the general power resources of the world." His estimate, the United Press pointed out, was "the most optimistic that has yet been made by physicists who must do the actual work of solving atomic energy problems." Like his depiction of a thriving international agency's research and development laboratories, his words and arguments still reflected the spirit that had animated Acheson-Lilienthal.[39]

But as he clung to his vision of an international scheme predicated on atomic energy research and development—even while convinced none could be adopted or implemented for the foreseeable future—Oppenheimer could not help but note an increasing number of alternative proposals. Citing both the desperate need to entice the Soviets into an agreement and far more pessimistic estimates for civil nuclear power, these other proposals envisaged forgoing the original concept of an international development agency intensively engaged in mining uranium and building and operating power reactors. Instead, to simplify the complications involved in such enterprises, they contemplated largely or completely forswearing the peaceful, civil uses of atomic energy indefinitely—if that was what it took to safeguard against atomic war.[40]

In the spring of 1947, Oppenheimer encountered two such proposals. One came from a Harvard Law School professor, David F. Cavers, who privately circulated a memorandum recommending a moratorium on developing nuclear power. Cavers's plan would have permitted use of isotopes for scientific and medical research, but natural uranium would remain unmined and reactors unbuilt. Such a change, he argued, would greatly ease the task of inspection. It would also avoid the problem of handing over U.S. nuclear factories to international control, since they would be

39. U.S. Representative to the UNAEC, press release, containing JRO, statement to 17th meeting of Committee 2 on "Functions of the international agency in relation to research and developmental activities" (Item A.2 (a), Document AEC/C.2/16), 29 Apr 1947, JROP, box 285, folder United Nations AEC—Committee 2—4/29/47. A "slightly condensed version" was printed as JRO, "Functions of the international agency in research and development," *BAS, 3:7* (Jul 1947), 173-176, 197; see also George Barrett, "Oppenheimer plan in U.N. atom study," *NYT*, 8 Jun 1947, and UP report, "Atom-driven generators seen in 5 years," *Washington post*, 8 Jun 1947.
40. See Hershberg (ref. 8), chapt. 18.

destroyed, or supporting the building of reactors in Russia, since they would not be needed. Nor would the world have to worry about their seizure in times of crisis or war. Mankind would sacrifice the benefits of atomic power, but Cavers did not see this as a great loss.[41] Such arguments, brought to Oppenheimer's attention, failed to impress.[42] Though fully sharing Cavers' "grim, gloomy" prognosis, he found himself "in rather total disagreement with the remedy." His main criticism of such an "enormous retrenchment" was that it would relegate the international control organization to a mission dominated by "the grim machinery of purely negative policing and inspection." In other words, an agreement along these lines would require an "improbable and, in my opinion, inachievable inspection apparatus if it were to give any satisfaction at all." He further lamented the plan's lack of any positive element of "cooperative enterprise on any significant scale for which we might look for an eventual betterment of the relations between the powers."[43]

Oppenheimer had considerably more respect for a scheme advanced by Cuthbert Daniel and Arthur M. Squires in the *Bulletin of the atomic scientists* to prohibit large-scale power production for a generation. The two physicists contacted him after they were disturbed to see him quoted in the newspapers as calling their plan "a sour note" and adding, "I can't possibly see how reliable controls can be established, unless there is something for an agency to do besides snoop." Their detailed, technically sophisticated letter stressed agreement with Oppenheimer's contention that an international agency must have attractive developmental duties, explaining how these were built into their plan, "save only the prohibited activities of constructing and operating large-scale nuclear plants." But the persistent danger of political leaders who might succumb to the "irresistible temptation" to use atomic energy for conquest, still far outweighed whatever benefits civil nuclear power might offer in the next twenty or so years.[44] In a brief, friendly response, Oppenheimer did not disavow the quotations attributed to him, but also added that he would "not, in fact, raise strong objections

41. See David F. Cavers, "Draft of article on the international control of atomic energy," JROP, box 194, folder UNAEC April-May 1947.
42. JRO's contact was Harvard Business School professor Lincoln Gordon, a former Baruch adviser and now a part-time U.S. UNAEC consultant with whom he was on friendly terms. Gordon to JRO, 15 Apr 1947, enc. Cavers (ref. 41), JROP, box 194, folder UNAEC April-May 1947.
43. JRO to Gordon, 21 Apr 1947, JROP, box 194, folder UNAEC April-May 1947. While *Fortune* magazine rejected Cavers's manuscript, he eventually published a version later in 1947. David F. Cavers, "Atomic power versus world security," *BAS, 3:10* (Oct 1947), 283-288, 302; documents in Cavers Papers, Langdell Library, Harvard Law School, Cambridge, MA, Atomic Energy Control box 1, folder Bulletin of Atomic Scientists; and Hershberg (ref. 8), 327, 346-347, 841-842 n. 25-26, 847 n. 108.
44. Cuthbert Daniel and Arthur M. Squires, "The international control of safe atomic energy," *BAS, 3:4-5* (Apr-May 1947), 111-116, 135; Daniel and Squires to JRO, 9 May 1947, JROP, box 64, folder Su-Sq, including quotations from Peter Kihss article, *New York herald tribune*, 30 Apr 1947.

to modifications along these lines were that necessary to the willing adoption of an effective international cooperation" (even while finding, he remarked, their division of activities related to development and production technically "unsatisfactory"). But the crux was that he saw no evidence that the Soviets would find the Daniel-Squires notion any more acceptable than the Baruch Plan, given their own evident nuclear ambitions, nor (in the unlikely event they did agree) did he "believe that any measure of genuine international cooperation can be undertaken without recognizing and facing up to the risks which a collapse of such cooperation would bring to the peace of the world."[45]

Oppenheimer's ambivalent, even warily positive response to Daniel and Squires, as well as his calculation that large-scale power production was probably at least three decades off, hinted at a greater willingness to consider scaling back the Acheson-Lilienthal emphasis on peaceful civil uses. By early June, he had apparently even teamed up with Conant to promote a "Conant-Oppenheimer Plan" (as one of Osborn's advisors dubbed it in an internal memo) that "would in effect restrict the use of nuclear fuel to research and limited development." In general the American negotiators now began to move toward a less expansive vision of the power-related activities of a proposed international atomic energy.[46]

But if Oppenheimer had started to inch away from his earlier enthusiasm for power development as a centerpiece for international control, in late June Conant began to move well beyond the U.S. delegation—and Oppenheimer—in his willingness to discard atomic energy altogether. A confidential meeting on June 18 with the physicist Sir George P. Thomson, a British delegate to the UN talks, may have accelerated this process.[47] Thomson informally conveyed to Conant, Tolman, and Osborn (and Oppenheimer surely learned soon thereafter) what he described as his own, personal idea to bring the atomic energy enterprise to a screeching halt: a plan under which all nuclear power development would be barred for, say, twenty years, or until small-scale experimentation proved atomic energy's economic value. At least, Thomson reasoned, in this scenario many contentious issues could be substantially or indefinitely deferred.[48] Encouraged by such views and discouraged by the latest,

45. JRO to Daniel and Squires, 20 May 1947, JROP, box 64, folder Su-Sq.
46. Paul Fine to Osborn, 10 Jun 1947, S/AE, box 22, folder 15F. At least one fellow U.S. UNAEC consultant seemed to approve of "the Conant-Oppenheimer scheme": Bell Telephone head Chester I. Barnard. Barnard to Osborn, 19 Jun 1947, S/AE, box 12, folder 15A.3.
47. Thomson helped trigger a secret diplomatic crisis by relaying word that the British government, in order to secure Moscow's acquiescence, was considering putting forward its own international control alternative based primarily or solely on inspection. Shocked and alarmed—"if the British ball this up it will have ramifications which would go round the world," Acheson tartly told Osborn—the State Department launched an urgent mission to dissuade London from going forward. Osborn, UNAEC diary (ref. 14), 11 Jun 1947. On the U.S.-U.K. contretemps, see *FRUS* 1947, *1*, 489ff.
48. The Americans gathered that Thomson had little faith that his plan would actually work,

clearly unacceptable Soviet proposals—apparently feeling "that since it was obvious that Russia wouldn't agree to our proposal we must find something else they might agree to, and this might be it"—Conant adopted the moratorium plan as his own. On June 28, his ideas popped up during a conversation between Acheson and Lilienthal. Lilienthal had learned of Conant's plan from Bacher, "who didn't care for it," but had heard that Oppenheimer "rather" favored it.[49]

Such high-level secret doubts about the "sunny side of the atom" (as an hour-long CBS radio broadcast hosted by Edward R. Murrow that summer was called) ran sharply counter to public pronouncements being regularly made by journalists, social commentators, and some scientists. And altogether shunning the development of nuclear power, as Conant advocated, constituted a virtual heresy, especially given the philosophy of Lilienthal and his fellow AEC commissioners of capturing public imagination and congressional support by promoting atomic energy's purportedly bright future—in power, medicine, transportation, agriculture, and other areas.[50] Though not going so far in his willingness to turn away from nuclear power in the international control proposals, Oppenheimer shared Conant's view that the public needed to be disabused of the wildly overoptimistic notions then swirling around. The rest of the GAC agreed. When the advisors met in Washington on July 28-29, they shocked Lilienthal and his fellow commissioners by drafting a statement for the Commission to issue. It would state bluntly that the current status of nuclear power efforts was "not fully understood," that it "does not appear hopeful" to use uranium as a power source, and that the costs of nuclear reactors rendered them "economically prohibitive." Even in the most favorable projections, the group warned in their draft statement, "decades will elapse" before atomic power could significantly contribute to the industrialized world's needs.[51] "Had quite a blow today," a shaken Lilienthal recorded in his diary. "...This pessimism didn't come from nobodies, but from a top group—Oppenheimer, Conant, Rabi, Seaborg, etc."[52]

The scene was set for the first full-dress meeting of the U.S. UNAEC consultants at Osborn's offices on the 23rd floor of the Empire State Building on Thursday morning, July 31. As the group gathered, the world situation, reflected at the United Nations, gave scant cause for optimism. The Soviet Union a few weeks earlier had brusquely

but felt it offered "the only hope" for inducing Moscow to sign a treaty. Johnson to Secretary of State [Marshall], 20 Jun 1947, *FRUS* 1947, *1*, 526-527; Osborn, UNAEC diary (ref. 14), 18 Jun 1947.

49. Lilienthal (ref. 3), 216.

50. See, e.g., Paul Boyer, *By the bomb's early light: American thought and culture at the dawn of the atomic age* (New York, 1985), 96-97, 294-298; Brian Balogh, *Chain reaction: Expert debate and public participation in American commercial nuclear power, 1945-1975* (Cambridge, 1991), 78-79.

51. Draft statement on atomic power, appended to minutes, GAC meeting, 28-29 Jul 1947, DOE. Further draft GAC statements on atomic power from Sep and Nov 1947 may be found in JROP, box 292, folder 1947.

52. Lilienthal (ref. 3), 229.

rejected participation in the Marshall Plan offer of economic aid to Europe and forced its "allies" to walk out on the East-West talks in Paris called to consider the program. Moreover, that month's issue of *Foreign affairs* featured an article on "The sources of Soviet conduct" by a government official identified only as "X" (probably known to Oppenheimer to be his State Department friend George F. Kennan) outlining the need for a long-term policy of containment. Nor was the situation at the UNAEC any brighter. On June 11, Gromyko had presented a new set of Soviet proposals that, despite some more forthcoming language, struck the Americans as "wholly inadequate" because of its lack of effective safeguards against cheating. Helping to guarantee its immediate dismissal by Washington, Gromyko continued to insist on the destruction of all (i.e., all American) atomic stockpiles as the first step in any plan and the retention of Soviet rights to veto inspections by an international agency.[53]

Yet, despite the emerging clarity on the international front, the U.S. atomic negotiator found his advisors sharply divided on how the delegation should proceed in the talks—or if it should proceed at all. Attending the secret meeting, in addition to Osborn and his top staff, were some of the leading lights of the atomic establishment: Oppenheimer, Conant, Groves, Tolman, Bell Telephone head and Acheson-Lilienthal consultant Chester I. Barnard, and former Groves aides Maj.-Gens. Thomas Farrell and Kenneth D. Nichols. At the outset, Osborn told the influential group that if it could settle on a policy line, it would undoubtedly be officially adopted. But it could not: Osborn afterwards described the long day of discussions to Bacher (invited but unable to attend) as "full of ideas and fireworks," adding, "You really missed something."[54]

At the meeting's outset, Osborn handed copies of the latest American draft proposals, prepared in collaboration with friendly delegations, to Oppenheimer and the other consultants. The six documents, all highly classified, contained nothing startling, largely devoted as they were to setting the agenda for the upcoming fall UN debate and to delineating U.S. positions on the proposed atomic agency's structure, financing, staffing, hierarchy, functions, inspection rights, and sanctions. Their significance lay not so much in their content as in their signal that the United States was continuing negotiations. They occasioned few strong objections, and no less a perfectionist than Oppenheimer told Bacher that despite "some bad defects, especially in the raw materials document, of which Osborn and his staff are fully aware," he was on the whole "pleasantly astonished that such sensible things could have been written under the circumstances."[55]

Real arguments flared, however, over what future direction U.S. international control policy should take. As Osborn recounted afterward, the discussion pitted a

53. Osborn to Austin, 1 Aug 1947, S/AE, box 38, folder 20.5; Even the ardently pro-control *Bulletin of the atomic scientists* sternly responded that the "Gromyko Covenant" could not be taken seriously. "If the UN Atomic Energy Commission fails," *BAS*, *3:7* (Jul 1947), 1; see also Lieberman (ref. 14), 391-392.

54. Osborn to Bacher, 5 Aug 1947, S/AE, box 22, folder 15F.

55. JRO to Bacher, 6 Aug 1947, JROP, box 18, folder Bacher.

makeshift alliance of Oppenheimer and Groves against Tolman and Barnard, "with Conant taking an independent line."[56] Conant's impassioned support for not only pressing ahead with detailing an international control plan but sharply cutting back the level of Atomic Development Authority activity and specifically committing the U.S. to the destruction of nuclear fuels predictably horrified Groves—Russia might cheat while Washington dutifully got rid of its hard-earned materials—but the idea's sponsor viewed leaving uranium underground as a "great contribution to humanity" and charged that Groves was "merely arguing against any treaty."[57]

Conant probably anticipated resistance from Groves, who was famously covetous of nuclear ores, but then ran into opposition from an unexpected quarter—Oppenheimer. With no real prospect of agreement, Oppenheimer believed, it was time to terminate the charade. To proceed with U.S. plans for collaborating with the U.S.S.R. on atomic controls, he argued, would be dishonest—and worse. The "real danger," he feared, was the unlikely possibility that the Russians might actually *accept* the American proposal at a time when neither the U.S. government nor its people favored international control. Conant, too, saw the danger in disingenuously advancing an unserious proposal—but his solution was to advance a *more* serious one.[58]

According to the fragmentary notes of what was clearly a contentious day-long session, the conversation gradually followed Osborn's lead in the direction of approving the draft reports and going ahead to finish the international control plan. Even Oppenheimer grudgingly joined in voting to move forward with completing the proposed treaty, stopping short only of spelling out precisely the stages that would be involved as the plan went into operation. Going ahead, he rationalized, might "advance hopes of agreement" and impress world opinion.[59]

But "much thought" over the weekend changed his mind, and late on Monday afternoon he sought out Osborn in his office to emphasize that he was "very disturbed" about the idea of continued participation in the talks.[60] Osborn listened with con-

56. Osborn to Bacher, 5 Aug 1947 (ref. 54).
57. Minutes of meeting of advisers and consultants, 31 Jul 1947, S/AE, box 64, folder 15D; see also JRO to Bacher, 6 Aug 1947 (ref. 55), and Osborn, UNAEC diary (ref. 14), 31 Jul 1947. The next American proposal, Conant insisted, should include what Acheson-Lilienthal and Baruch lacked—explicit provisions for stages and strategic distribution (and destruction) of nuclear assets. Conant to Osborn, 3 May 1947, S/AE, box 39, folder 20.29.
58. Minutes of meeting, 31 Jul 1947 (ref. 57). According to the minutes, JRO also poured cold water on the idea, floated by some, that merely exploring the feasibility of nuclear power might be a way to avoid large-scale production of militarily useful isotopes. Citing Fermi, and echoing his objections to Cavers's plan, he pointed out that a "real pilot plant to learn about denatured fuel power production would require large dangerous amounts." On balance, the minutes show, JRO preferred the destruction of plutonium rather than maintaining exclusive U.S. possession or worldwide redistribution.
59. Minutes of meeting, 31 Jul 1947 (ref. 57).
60. Osborn, UNAEC diary (ref. 14), 4 Aug 1947; Osborn to Matthew B. Ridgway, 6 Aug 1947, S/AE, box 22, folder 15F.

siderable chagrin as Oppenheimer warned that it would be "unwise, not altogether honest, and exceedingly dangerous to the American position" to keep negotiating.[61] As he recounted afterward to Bacher, Oppenheimer told Osborn:[62]

> In the present situation, with the rapid past and probable future deterioration of relations, the deep and fully justified suspicion with which the people in this country regard Russia, the sharp even if temporary alignment, and the always present threat of war in the not remote future, it is not possible to undertake a cooperative enterprise of the kind that we originally had in mind. No other alternative, certainly not that of Conant's [sic], appears to me either hopeful or salable either to us or to the Russians. If we believe, as I still do, that there is a long term hope in ambitious international cooperative enterprise like the ADA [Atomic Development Authority], we can only prejudice the chances of their ultimate realization by continuing these negotiations as though they had reality. Almost everyone is aware of the extraordinary situation which we should find ourselves in should the Russians wholeheartedly agree themselves, say, to the committee reports. Almost everyone is aware how difficult it then would be to spell out a reasonable system of floating this ship. It seems to me that to continue negotiations under these conditions is an act of bad faith, and that in time this bad faith will catch up with us.

Oppenheimer recommended that Osborn explain that the world situation that compelled the United States to discontinue negotiations and "record our willingness to resume, bi[-] or multi-laterally within or without the United Nations, in the Security Council or the Commission discussions of the possibility of removing the threat of atomic war; but that in the absence of agreement on basic issues, and in the present state of hostility between major powers, the future detailed elaboration of proposals seemed to us wrong in principle." While Oppenheimer acknowledged that his suggestion entailed an extremely difficult assignment for Osborn, he fretted that the negotiator's plan to continue elaborating the proposals "can only be prejudicial to any long term hopes and that it is likely to make us out both fools and swindlers if the Russians play their cards somewhat more skillfully than they have in the past."[63]

Oppenheimer, according to Osborn, claimed that Bacher "would have agreed with him" had been at the meeting, and that Conant supported his position.[64] But Osborn doubted that Conant really went along with Oppenheimer's negative approach, and his skepticism was vindicated the next morning when he received a letter from the Harvard president.[65] Precisely contradicting Oppenheimer, and to Osborn's relief, Conant expressed his "hope that people in Washington will take an optimistic view even in this grim world and *complete* the American proposal. I think you understand,"

61. Quoted in Osborn to Tolman, 5 Aug 1947, S/AE, box 42, folder 20.132.
62. JRO to Bacher, 6 Aug 1947 (ref. 55).
63. Ibid.
64. Osborn to Austin, 5 Aug 1947, *FRUS* 1947, *1*, 589-590.
65. Osborn to Ridgway, 6 Aug 1947 (ref. 60).

he added, "that my specific position is that the United States made a proposal, but an incomplete one."[66]

While no more optimistic than Oppenheimer about the chances of an agreement, Osborn still rejected his arguments because of his keen awareness that an American walkout would be a diplomatic blunder. "Strong propaganda by Russia that we had never intended the offer seriously" would be the result, he told Austin, suggesting that if Washington desired a clean break with Moscow, it should take that step somewhere other than the UN Commission. As for Oppenheimer's fear that the United States, by staying in the talks, might end up signing a disadvantageous accord, Osborn assured Austin that it should be easy to stall consideration of a detailed staging process for destroying or handing over the U.S. stockpile.[67]

When Osborn spoke to the U.S. UN ambassador, he was pleased to find that Austin "agrees strongly with me and not with Dr. Oppenheimer."[68] Rather than continue the argument, the atomic negotiator merely sent Oppenheimer a brief note enclosing Conant's letter, sympathizing with him for having to endure "Princeton's tropical August," and promising to give a call after a vacation and a visit to Washington, where he hoped to get a better handle on official policy toward Russia.[69] And work at the U.S. UNAEC delegation continued on the Second Report.

Oppenheimer, Osborn, and *Foreign Affairs*

Such sharp private arguments over whether the struggle to control the bomb was really dead or merely comatose, critical, and on death's door spurred Oppenheimer's thinking. Pondering international control's sad fate for more than just a long weekend that summer, he penned a lengthy essay on the subject for *Foreign affairs* which would appear in the journal's January 1948 issue.[70] Essentially an obituary for international control—"We are beyond advocacy, and yet not far enough for history"—Oppenheimer's piece drew particular notice ("a brilliant piece of analysis," commented the *Washington post*) for its critical appraisal of American handling of the issue and its imputation that Washington, and not Moscow alone, bore some responsibility for international control's ultimate demise.[71] "To answer simply that we have failed because of non-coöperation on the part of the Soviet Government is certainly to give a most essential part of a true answer. Yet we must ask ourselves why in a matter so

66. Conant to Osborn, 4 Aug 1947, copy enclosed with Osborn to JRO, 7 Aug 1947, JROP, box 54, folder Osborn.

67. Osborn to Austin, 5 Aug 1947 (ref. 64); see also Osborn to Tolman, 5 Aug 1947 (ref. 61).

68. Osborn to Ridgway, 6 Aug 1947 (ref. 60).

69. Osborn to JRO, 7 Aug 1947 (ref. 66).

70. JRO, "International control of atomic energy," *Foreign affairs, 26* (1948), 239-252.

71. "Atomic future," *Washington post,* 8 Jan 1948.

overwhelmingly important to our interests we have not been successful; and we must be prepared to try to understand what lessons this has for our future conduct."[72]

Oppenheimer's *Foreign affairs* article is a remarkable document, bringing into the open some of this highly placed advisor's pointed criticisms. Read in this light, it delicately strikes a balance between criticizing so harshly as to alienate Truman Administration officials and imperil his own insider standing, and failing to reflect adequately the agonizing doubt and second-guessing ("hand-wringing," Truman and Baruch would probably snort) Oppenheimer shared with so many atomic scientists. The essay identified alleged deficiencies in the U.S. approach to international control even as it qualified their significance. Echoing the complaints of other administration critics (and foreshadowing those of critical historians), its author faulted Washington for pursuing international control through the United Nations, on a commission with many states, large and small, instead of in intense bilateral negotiation with the great power (and its dictator) that really mattered if there were to be any real chance of agreement. Oppenheimer frankly and fully admitted the grave and justified doubts on the part of American leaders that the Kremlin would countenance a rupture from past practices of police state secrecy internally and extreme suspicion internationally—yet he believed, and deeply regretted, that the effort, in his view, had never seriously been made to find out. "The later relegation of problems of atomic energy to discussions within the United Nations, where matters of the highest policy could only be touched upon with difficulty and clumsily, would appear to have prejudiced the chances of any genuine meeting of minds."[73]

Thus while praising the U.S. government for going as far as it did to endorse "a quite new approach to international problems" in its UN atomic energy proposals, Oppenheimer's article expressed regret that it did not make comparably forthcoming and revolutionary efforts to establish cooperative relations with Moscow. Moreover, the essay implied that opportunities to test Soviet readiness for cooperation had been frittered away due to inattention, incompetence, or a lack of comprehension. "[I]n our preoccupation with determining and clarifying our own policy," it suggested quite pointedly,[74]

> We have given far too little thought to attempting to influence that of the Soviet Union on the only plane where such an influence could be effective....Our internal effort has cost delay, it has cost confusion, it has cost the injection of some irrelevant and some inconsistent elements in our policy with regard to atomic energy. Above all, it has cost a sort of schizophrenic separation of our dealings in this field from our dealings in all others.

72. JRO, "International control" (ref. 70), 239-240. Oppenheimer declined to use the words "*the* most essential" surely not by accident.
73. Ibid., 246.
74. Ibid., 247-248.

As mitigation, these criticisms were coupled with the bleak recognition that the Kremlin's ingrained tendencies and actions gave "little support to the view that by prompter, clearer and more magnanimous action we might have achieved our purposes."[75]

Oppenheimer's January 1948 *Foreign affairs* piece is well known, of course, to his analysts and biographers, for whom it serves as a paradigmatic public expression of his discontent. But placing it in the context of his secret work with Osborn in the summer of 1947—when he drafted the essay—offers opportunities to derive new insights. In addition to reading the article alongside Oppenheimer's classified arguments with Osborn and his consultants as well his private objections to alternatives proffered by Cavers, Daniel-Squires, and Conant, we can exploit the little-known fact that more than four months before its publication, in late August 1947—just a few weeks after his secret dispute with Osborn over the desirability of further talks—Oppenheimer arranged to have *Foreign affairs* send Osborn a copy of the galleys for the article.[76]

Viewed against this backdrop, the essay proves even more revealing of the balancing act that Oppenheimer sought to carry off. "My concern," the physicist wrote Osborn, "is not that the paper misrepresents my own views, nor even perhaps that yours will be so very different. It is only that the publication of such views at this time by someone with a quasi-official connection with our delegation could conceivably prove of embarrassment. That is the last thing I want to do."[77] Oppenheimer even offered Osborn the option of vetoing the article's appearance and solicited his "desires" in "complete frankness." Osborn's scribbled marginal notations (and those of an unidentified aide), as well as discrepancies between the galleys, Oppenheimer's own various drafts, and the published version, prompt some observations on official reaction to the article as well as Oppenheimer's own evolving views.

To go back, then, to the article's origins: these dated to late May 1947, when *Foreign affairs* editor Hamilton Fish Armstrong solicited from Oppenheimer a "summing up" on atomic energy, not only its scientific but particularly its social and political ramifications. It was intended for the journal's twenty-fifth anniversary issue that October. "I had a long talk on this matter yesterday with Dean Acheson," the editor explained, "and he agrees both as to the value the right sort of article would have and as to your being the right person to write it."[78] Accepting the assignment, Oppenheimer worked on the piece over the summer and sent Armstrong the completed manuscript on August

75. Ibid., 248.

76. JRO, galleys, 21 Aug 1947, of "International control" (ref. 70), S/AE, box 77, folder 20.103.

77. See JRO to Osborn, 23 Aug 1947, JROP, box 257, folder Foreign Affairs Article—Jan. 1948—Correspond.

78. Armstrong to JRO, 29 May 1947, JROP, box 257, folder Foreign Affairs Article—Jan. 1948—Correspond. Armstrong had first asked Acheson to write the piece upon his retirement as Under Secretary that summer, but Acheson declined and suggested Oppenheimer instead. See also Armstrong to Acheson, 20 May 1947, HFAP, box 48, folder 5.

15, on the heels of his secret disagreement with the U.S. UN negotiator and some of his fellow consultants. If *Foreign affairs* wished to publish the article, Oppenheimer told Armstrong, "we should give Mr. Osborn an opportunity to pass on whether its publication would in any way embarrass him or prejudice his position."[79] The editor, delighted to have this "ornament" to his anniversary issue, promptly agreed to send the negotiator a copy of the galleys. After speaking with Osborn by telephone, Armstrong forwarded Oppenheimer the proofs and in a hand-written P.S. tried to stiffen his spine: "I hope you will not agree to make any changes which you do not yourself consider sound or necessary."[80]

Reading over the galleys, Osborn was taken aback by some of Oppenheimer's implicit and explicit criticisms—probably a mild indication of the irritation that more hard-line members of the Truman Administration, less personally fond of the physicist, must have felt when the article eventually appeared. Osborn did not appreciate, for example, a section in which Oppenheimer lamented what he considered the "prevalent view" at the top of the U.S. government at the end of World War II that international control could be used as a tool of what would come to be called atomic diplomacy: "The prevalent view, that is," the draft suggested, "saw in the problems of atomic energy not an opportunity to allow the leaders of the Soviet state to carry out a policy of international cooperation, of openness, candor and renunciation of violence to which they were already committed; rather, it saw an opportunity to cause a decisive change in the whole trend of Soviet policy." Next to this description Osborn wrote: "* all seems unfortunate." And next to the final words, about the desire to coerce a "decisive change" in Soviet conduct, he wrote: "Why say this? Is it really so?" At another point, Oppenheimer acerbically juxtaposed the U.S. government's rhetorical advocacy of proposals for international control and its simultaneous coveting of the "military advantages which we only reluctantly would lay down"—a contradiction which, Oppenheimer judged, would have produced American resistance "were there more reality to the [international control] plans" and which had contributed to the "lack of reality of all discussions to date." An unhappy Osborn wrote "Confusing" and "Out" beside this imputation of inconsistency, if not hypocrisy.

Osborn also found other remarks in the galleys objectionable: a description of U.S. proposals as "sketchy" (next to which he wrote, "Out"); a lament that secrecy prevented the revealing of much relevant information (prompting an "Out"); and an observation that the history of international atomic energy control efforts was "one of high and provably unreasonable hopes, and of failure" (Osborn wished the words "and of failure" cut, or at least supplemented by the phrase "to date"). Osborn also scribbled a marginal quibble besides Oppenheimer's most trenchant and specific criticism of Truman's policy, that he had dropped international control in the lap of the United

79. JRO to Armstrong, 15 Aug 1947, "JROP, box 257, folder Foreign Affairs Article—Jan. 1948—Correspond. This correspondence is duplicated in HFAP, box 48, folder 5.
80. Armstrong to JRO, 20 and 25 Aug 1947, JROP, box 257, folder Foreign Affairs Article—Jan. 1948—Correspond.

Nations: "Byrnes & Attlee & MacKenzie King tried to in Moscow"—a reference to the fact that the Soviet leadership, despite agreeing to the notion of creating a UN commission, had declined to engage in substantive conversations on international control during the December 1945 Council on Foreign Ministers meeting.[81]

Beyond all these specific qualms, Osborn evidently did not relish the prospect of his most famous advisor publishing such a gloomy and critical piece in so prominent a journal just as the United Nations prepared to consider the revised proposals on which he had worked for so many months. As world diplomats prepared to converge on New York for the annual session of the General Assembly, the UNAEC approached a final debate and vote on its Second Report to the Security Council against a backdrop of mounting tensions between Washington and Moscow. An article by Oppenheimer in *Foreign affairs* on international control was not only liable but likely to attract considerable press attention, risking a public relations brouhaha Osborn obviously desired to avoid—especially one involving criticism, however gentle or implicit, of the U.S. position by one of its most prominent advisors. Months of secret disputes with American allies, especially the British, over Washington's sincerity could only have exacerbated such trepidation.

Accordingly, on August 26, the day after receiving the galleys, Osborn telephoned the editor to request that he drop Oppenheimer's piece at the last minute from the anniversary issue.[82] Armstrong evidently did not give up hope until Oppenheimer himself confirmed his acquiescence to Osborn's wishes. "You can imagine what a disappointment it was to me to learn that you felt you shouldn't go ahead with the plan to publish your article in our autumn issue," the editor wrote Oppenheimer on September 9, after the issue had gone to press. "I talked at length about it with Fred Osborn and had to agree that the considerations which he felt so important would at any rate be properly conclusive with you."[83]

However, he continued, Osborn had assured him that he enthusiastically backed the idea of publishing Oppenheimer's piece in the next issue, too late to interfere with the Second Report's safe disposition. Not only would it be a shame for Oppenheimer to "waste the work" he had put into the article, Osborn believed, but it

81. Conant had been added to the American delegation at the last minute precisely for such a contingency, and despite a few intriguing brief atomic exchanges with Stalin and Molotov, found them uninterested in any serious talks on the subject. Hershberg (ref. 8), 250-257. See also David Holloway, *Stalin and the bomb: The Soviet Union and atomic energy, 1939-1956* (New Haven, 1994), 158-159.

82. "8/26/47 Oppenheimer article dropped at request of Osborn," Armstrong wrote at the bottom of Armstrong to JRO, 25 Aug 1947, HFAP, box 48, folder 5. That Osborn raised the fate of the UNAEC documents then under discussion in his argument to postpone *Foreign affairs'* publication is confirmed by Osborn to Armstrong, 26 Aug 1947, HFAP, box 48, folder 12.

83. Armstrong to JRO, 9 Sep 1947, JROP, box 257, folder Foreign Affairs Article—Jan. 1948—Correspond.

would be worthwhile to reap the "advantages" of such an "authoritative piece" on the controversy with the Russians over international control. Returning to the chase, Armstrong warmly invited Oppenheimer to okay the article's appearance, with any necessary revisions, in the next issue.[84]

Intervening events apparently served to remove Osborn's—and therefore Oppenheimer's—hesitations. On September 11, with only the Soviets voting *nyet* (Poland abstained again, as it had on the Baruch Plan), the UNAEC approved the American-backed Second Report on the structure and program of the proposed international atomic energy agency, and eight days later the Commission's Canadian delegate formally presented it to the Security Council (chaired that month, ironically enough, by Gromyko).[85] By early October, reassured that Osborn no longer objected, Oppenheimer telephoned Armstrong to confirm his willingness to have the article appear in *Foreign affairs'* winter issue.[86] Late that month, he sent in a slightly modified manuscript with an insistence on no further editorial revisions[87] and a characteristically apologetic conclusion: "Let me again tell you how sorry I am for the trouble that this poor paper has been to you and to your journal. That seems to be a characteristic of everything that has to do with the atom."[88]

Oppenheimer explained that he had made "only minor changes of substance" in the text, attempting to "clarify some sentences" and adding a few paragraphs "in order that the balance should not be distorted." A close comparison suggests that at least some of the modifications stemmed from his efforts to mollify Osborn.[89] For instance, to his assertion that Washington at the close of the war had perceived an opportunity to exploit its atomic advantage to press for a wholesale transformation of Soviet policy,

84. Ibid., Armstrong to JRO, 25 Sep 1947, JROP, box 257, folder Foreign Affairs Article—Jan. 1948—Correspond.
85. "10 nations approve atom control plan," *NYT*, 12 Sep 1947; "Second atom report: U.N. Commission submits its findings to Council," *NYT*, 20 Sep 1947.
86. Armstrong to JRO, 6 Oct 1947, JROP, box 257, folder Foreign Affairs Article—Jan. 1948—Correspond.
87. "I believe," JRO explained in typically elliptical fashion, "that only in this way can we be sure that the intent of the author will really be purveyed and such blame as attaches to the obscurity of style will rightly be ascribed to him and no one else." JRO to Armstrong, 27 Oct 1947, JROP, box 257, folder Foreign Affairs Article—Jan. 1948—Correspond. Annoyed at such authorial impertinence, a *Foreign affairs* aide insisted on making minor alterations "just because nobody should have the right to tell us that we have to print anything without change of a word. (We publish magazine articles, not documents.)" B.D. to Armstrong, memorandum, 29 Oct 1947, HFAP, box 48, folder 5.
88. JRO to Armstrong, 27 Oct 1947 (ref. 87).
89. It remains unclear whether Osborn explicitly asked Oppenheimer to change language to which he objected, or whether Oppenheimer, sensing Osborn's discomfort, made the revisions completely on his own. In addition to their talks in New York City in late September, the negotiator visited the physicist twice in Princeton in the two weeks before Oppenheimer sent Armstrong the revised draft. See Osborn, UNAEC diary (ref. 14), 25 Sep, 11, 22-23 Oct 1947.

Oppenheimer added a softening phrase that implied understanding, perhaps even a degree of approval (added words in italics):[90]

> The prevalent view, that is, saw in the problems of atomic energy not an opportunity to allow the leaders of the Soviet state to carry out a policy of international cooperation, of openness, candor and renunciation of violence to which they were already committed; rather, it saw an opportunity to cause a decisive change in the whole trend of Soviet policy, *without which the prospects of an assured policy were indeed rather gloomy, and which might well be, if accomplished, the turning point in the pattern of international relations.*

He also added a new paragraph to ease the impact of another section which had disturbed Osborn. In it, the physicist had noted the "lack of reality" surrounding "all discussions to date" of international control, given the incongruity of U.S. rhetorical promotion of the idea while avidly developing atomic energy for military purposes. In the final article there would appear an observation absent from the August galleys: "Yet despite these limitations the work of the United Nations Atomic Energy Commission has established one point: through many months of discussion, under circumstances of often dispiriting frustration, and by delegates not initially committed to it, the basic idea of security through international cooperative development has proven its extraordinary and profound vitality."[91] On the other hand, Oppenheimer left unaltered many of the statements that had irked Osborn, as well as all of the essential criticisms of the Truman Administration's handling of the atomic energy issue.

A final go-round between Armstrong and Oppenheimer in November yielded one more noteworthy modification.[92] In the published version, one can discern dimly Oppenheimer's secret debates over whether continuing U.S. participation in the UN talks risked a Soviet maneuver to embrace international control while secretly sabotaging it to the detriment of American security. With "the world as it now is," the article stated, it was unreasonable to modify the U.S. stand in order to entice the

90. Compare JRO, galleys (ref. 76), with JRO, "International control" (ref. 70), 245.
91. Ibid., 249-250. Citing their eventual acceptance by most other UNAEC members, JRO also commented approvingly on the "vitality of the fundamental ideas of the United States position" during a closed-door appearance at the National War College. JRO, "Atomic energy as a contemporary problem," 17 Sep 1947, NDUA; an essentially identical version of JRO's remarks, absent the Q & A session, appeared in JRO, *The open mind* (ref. 2), 21-41.
92. This last revision was occasioned by Armstrong's query about Molotov's comment at the U.S.S.R.'s thirtieth anniversary celebration that the secret of the atomic bomb "has long ceased to exist." The *Foreign affairs* editor invited JRO to add a comment to the article at the final proof stage. JRO declined but made another change instead. Armstrong to JRO, 7 Nov 1947, and JRO to Armstrong, 8 Nov 1947, JROP, box 257, folder Foreign Affairs Article—Jan. 1948—Correspond. For Molotov's remarks and reactions to it, see the *NYT*, 7 Nov 1947, and Edward A. Gullion, Special Assistant to the Under Secretary of State (Lovett), memorandum "Molotov's remarks of November 6, 1947 on atomic bomb 'secret,'" *FRUS* 1947, *1*, 861.

Soviets into "intermediate solutions involving, perhaps, a formal concession with regard to access to atomic facilities on the part of international inspectors," or to seek "palliative solutions" that would foster "intensified suspicions and intensified rivalries" while sacrificing substantial American advantages. The initial version had been more transparent. However, when Oppenheimer sent Armstrong his final revised proofs on November 8, he deleted a significant passage (here in bold face) in which his concrete disagreement with alternatives promoted by Conant, most prominently, as well as by Cavers and others, emerged far more clearly:[93]

> It is my own view that none of the other proposals that have been made have any elements of hopefulness in the short term. It does not seem to me that by offering to widen the areas of cooperation we would at this stage increase the likelihood of acceptance on the part of the Soviet Government. **It does not seem to me that arrangements from which the Soviet Government is excluded, even if only by its own decision, bear directly on the maintenance of security. It does not seem to me that the essential difficulties of agreement can be resolved by reducing the scale of atomic activities contemplated for the future, or by any other device to make the abrogation of sovereignty appear less extreme than it must in fact be if cooperation is to be achieved in this field.** In fact, it appears most doubtful if there are now any courses open to the United States which can give to our people the sort of security they have known in the past. The argument that such a course *must* exist seems to me specious; and in the last analysis most current proposals rest on this argument as their best advocate.

Although in a covering note Oppenheimer indicated that he had removed twelve lines of text because they were "not fundamentally clear or explicit enough to be useful,"[94] he may have concluded that the deleted comments, originally drafted prior to mid-August, could even be read as a veiled slap at the current official U.S. position.

Indeed, a close scrutiny of the Second Report's lengthy contents revealed the success of Osborn's advisors in changing U.S. policy on expectations for nuclear power development. In contrast to its statement nine months earlier, which had spoken optimistically of the "straightforward development" of peaceful atomic energy "competitive" with other sources of power such as coal, the new report echoed the GAC's private assessments (and Oppenheimer's April 29 testimony) in conceding that nuclear power "probably cannot become an important factor in the world's existing power resources for some decades."[95] And, in a provision that attracted little public notice, it

93. JRO made the deletion in the proofs he enclosed with JRO to Armstrong, 8 Nov 1947 (ref. 92). Compare JRO, galleys (ref. 76), with JRO, "International control" (ref. 70), and see draft manuscripts in JROP, box 257, folder Foreign Affairs Article—Jan. 1948—Correspond.
94. JRO to Armstrong, 8 Nov 1947 (ref. 92).
95. UNAEC, *Official records, special supplement, Report to the Security Council* (Lake Success, NY, 1946), 31 Dec 1946, 27-28; UNAEC, *Official records, special supplement, Second report to the Security Council*, 11 Sep 1947 (Lake Success, NY, 1947), 26.

actually mandated that the international agency limit the production of nuclear fuel to "the minimum required for efficient operating procedures necessitated by actual beneficial uses, including research and development."[96] The new U.S. position, adopted at Osborn's urging, actually *did* seek to enhance the prospects for eventual agreement (as Oppenheimer's original draft had stated) "by reducing the scale of atomic activities contemplated for the future." Though this "entirely new provision" seemed completely to escape public notice, as Osborn pointed out in secret remarks to the National War College, for his most prominent scientific advisor publicly to disdain its import, even indirectly, might have appeared unseemly.[97]

Armstrong, for his part, probably ruing that he had asked Oppenheimer to trim a few lines to save space, regretted the mysterious deletion but had by now grown accustomed to his author's delicate position. "I must say I am rather sorry to lose the explicit passage on the last galley," he wrote, "but I presume you left it out for calculated reasons, and so won't urge you to reconsider."[98]

Oppenheimer vs. Conant

In the autumn of 1947, the East-West impasse grew even more obvious, with the Soviets following up their rejection of the Marshall Plan with the creation of a "Cominform" to coordinate policies of European communist parties, as the world divided into what Stalin's lieutenant Zhdanov termed the "two camps" of aggressive capitalist imperialism and "people's democracies." At Moscow's behest, European communists militantly sought to disrupt governments moving closer to Washington—which in turn was rapidly organizing for rearmament and Cold War competition through military integration and the creation of the CIA.

In such a context, the idea of attaining an agreement on international control appeared utterly outlandish to most U.S. atomic policymakers, including Oppenheimer, at least for the foreseeable future. Keeping quiet in public, for the moment, Oppenheimer related his views to an elite audience in a "restricted" speech to the National War College on September 17 (a few days after Osborn appeared). There he presented some of the same wistfully pessimistic views that characterized the *For-*

96. UNAEC, *Second report* (ref. 95), 16.
97. "The final provision is one which is entirely new and has had almost no publicity....The provision is this; that the production of nuclear fuel will be at the minimum necessary to supply power plants actually under construction. That means the production of nuclear fuel would stop short all those great plants in the United States, the plants in Canada and England. They would all shut down because power is forty years away, and there won't be any power plants under construction for a long while to come....That is the most revolutionary thing, but nobody had noticed it yet." Osborn, "United Nations Atomic Energy Commission," transcript, 14 Sep 1947 (ref. 31), 14.
98. Armstrong to JRO, 11 Nov 1947, JROP, box 257, folder Foreign Affairs Article—Jan. 1948—Correspond.

eign affairs piece then awaiting publication. Reaffirming his belief in international control's ultimate desirability, if not its foreseeable attainability, he told his audience that only through a genuinely cooperative global atomic development agency could America restore absolute security, comparable to that which it enjoyed before World War II—"the only way in which we will be able to live with bad governments, with new discoveries, with irresponsible governments such as are likely to arise in the next hundred years."[99] Yet, he admitted, in line with the views he had privately expressed all year to Osborn, any realistic prospects for an acceptable pact had been foreclosed by U.S.-Soviet conflicts and Moscow's persisting patterns of secrecy, dictatorship, and communist ideology. It might happen some day, he supposed, but not as a consequence of the debates at the UN.[100]

In view of this absence of mutual confidence, so necessary for an unprecedented control scheme to function, it struck Oppenheimer as "a kind of schizophrenia which can only lead to very bad political confusion" to believe seriously that any prospect existed to set up an atomic development authority at any point in the next year and a half. Nor—in contrast to what he had told Osborn earlier in 1947—did he any longer worry that Gromyko or the Soviets might suddenly reverse course and accept the American plan in order to sabotage it, since the U.S. proposals contained safeguards against "very bad faith" and "presuppose a very large measure of peaceful intention, of cooperation, of confidence and candor before they can get started."[101]

But Oppenheimer knew that Conant, his close associate and one of the nation's most prominent figures in atomic policymaking, still clung to the idea that the United States should consider a serious policy shift—a long-term, perhaps permanent moratorium on nuclear power development—in order to make a last-ditch attempt to draw Moscow into an international arrangement. And, aware that Conant would soon be addressing this same distinguished audience, Oppenheimer regretfully but firmly expressed his view that such "cheap solutions" to the atomic stalemate were illusory:[102]

> [A] number of people of thoughtful character and undoubted devotion to this problem are somewhat less certain that we might not settle for less, [and believe] that in these negotiations we have set our sights far too high. We have asked for things that cannot be realized; and in view of the privileged position of a police state in the possession of weapons of this kind, we might do well to make a cheaper bargain. I have had many discussions with Dr. Conant about this. I know he is going to talk to you in the next week; and I believe he will speak about this question. He has some suggestions, which he thinks have not been taken seriously enough, and which come down to a much more modest proposal for internationalization, based essentially on leaving nuclear fuel in the ground, and cutting back the scale of technical activity to

99. JRO, "Atomic energy as a contemporary problem" (ref. 91), 5-6.
100. Ibid., 6-7.
101. Ibid., 8.
102. Ibid., 9.

> a point where very much less interference with sovereignty, and thus very much less international cooperation, would be involved. I myself am confident—I am most reluctantly confident—that the arguments that were given in the Acheson-Lilienthal report are correct arguments, and that such cheap solutions do not exist. But this is a possibility which should not be casually dismissed.

Sure enough, when Conant appeared before the group on October 2, he ardently advocated a nuclear moratorium as a plausible way to escape what he called the "Age of the Superblitz," making clear that eschewing atomic energy for peaceful uses would be a bargain indeed if it safeguarded the species from oblivion in a nuclear war.[103] Conant devoted much of his remarks to the need for intensive military planning for the coming period once Moscow got the bomb, and assessed the pros and cons of a "nuclear alliance" or "limited international scheme." But while acknowledging that a satisfactory international control pact could be ruled out "in the immediate future," he refused to relinquish hopes for an accord before a Russian nuclear stockpile came into existence.[104] He postulated that a window of three to twelve years for diplomacy might still remain before the Soviets got the bomb, and, terming himself "a hardy optimist," declared that if U.S.-Soviet relations stabilized in the next five years or so, "we can develop a scheme which stands a very high chance of working even in an unfavorable international climate and as the scheme works it will change the climate." So desperately did Conant desire to avoid the grim "Superblitz" age that he made two assertions that startled many in his audience: first, that it was "in the interest of the military security of the United States to enter into an agreement even if the scheme breaks down during the operation"; and second, that "the United States would from a military point of view be very much better off if there were neither atomic bombs nor their equivalent [i.e., separated uranium or plutonium] in existence anywhere in the world." Considering their "rather explosive" nature, Conant promised not to press his "idealistic" and frankly "iffy" proposals in public—yet, believing (as he had told the National War College group a year earlier) that "even a rickety bridge out of the shadow of the Superblitz is better than none," he saw no harm in positing even radical alternatives as the world verged on the new age of global enemies armed with atomic weapons.[105]

103. See James B. Conant, "The atomic age: A preview 1947 edition," and "The atomic age: Discussion," 2 Oct 1947, NDUA. All quotations that follow are from these documents. For a summary, see also Hershberg (ref. 8), 339-343.

104. In public speeches that autumn, Conant had spoken out against the notion that war with Russia was inevitable, and even predicted that once Moscow came to its senses and understood the dangers involved in nuclear war, the international control of atomic weapons could become feasible "in a relatively short time." See Hershberg (ref. 8), 338-339.

105. Conant suggested advancing a three-stage international control proposal: first, Washington would "trade all its [nuclear] industrial know how and remaining secrets about the bomb" for an "unimpeded" two-year inspection of Russian territory; second, the U.S. would swap more data for a trial run at a "rascal proof" regime of international inspection and ownership and

Upon reading Conant's War College talk, Oppenheimer sent him a letter that was a model of diplomatically couched criticism. Terming the speech "wonderfully stimulating and rich," he could "well understand why it should have attracted attention in the Big City" and, predictably, endorsed its calls to "re-enlighten" secrecy policy and provoke fresh thinking.[106] More warmly than Vannevar Bush—who told Conant he was "decidedly pessimistic" but "unwilling to dismiss the possibility" of an atomic agreement should the Soviets turn more moderate[107]—Oppenheimer welcomed his "spirit of hopefulness and open-mindedness with regard to the ultimate prospects of international control." But he also registered several protests. To start, he highlighted their disagreement on nuclear power's potential, still believing as he did in the Acheson-Lilienthal theory that "constructive uses of atomic energy" could ultimately contribute "strength and vitality" to international control. He also chided Conant for blurring the distinction between a general international control authority as envisioned in Acheson-Lilienthal and a "limited nuclear alliance" in which the United States would share control over atomic weapons with its partners. On balance, still, Oppenheimer favored a limited nuclear alliance, with the rather wan hope that it would be a step toward full internationalization.

Oppenheimer dryly added one final, "very minor" objection, flowing from their months-old rift over whether a flawed control setup was better than none. "It has often been our concern," he wrote, "that if we embarked on international undertakings and they collapsed, that could only mean war. In my own fumbling way I am an advocate of postponing war, at least for a little while. I should be worried about embarking on a scheme whose collapse is probable and...would face us with the necessity for immediate military action."[108]

By the time Osborn's consultants assembled once more on December 12, 1947, most spoke of the UN talks as an exercise in futility. Only Conant pressed for a serious American effort to spell out further details of the international agency's proposed operations, including its phased implementation, although he opted not to resuscitate his nuclear power moratorium scheme. Once again, in opposition to Conant, Oppenheimer teamed up with Groves (along with Bacher and Gordon Arneson) to urge against any further development of the proposals. Oppenheimer also rejected Conant's willingness to "leave final decisions on stages to the Security Council by majority vote,"

---

management of all uranium mines; and, finally, U.S. nuclear assets would be "dumped into the sea or 'denatured' rather than handed over to an international agency as envisioned in Acheson-Lilienthal." For the "rickety bridge" quotation, see Conant, "The atomic age," speech to the National War College, 16 Sep 1946, cited in Hershberg (ref. 8), 275; "rather explosive" is from Conant to JRO, 22 Oct 1947, JROP, box 27, folder Conant.

106. JRO to Conant, 29 Oct 1947, JROP, box 27, folder Conant. Quotations that follow are from this letter.

107. Bush to Conant, 21 Oct 1947, Vannevar Bush Papers, LC, box 27, folder Conant.

108. Conant, in response, still clung to hope, and in the medium rather than the long term. Conant to JRO, 1 Nov 1947, JROP, box 27, folder Conant.

remarking that such concepts lacked "reality." Oppenheimer and Conant agreed, however, with Osborn's proposition that Washington could not relinquish the bomb so long as the Soviets remained poised to "over-run Europe." And in the end, again, all agreed that Osborn's delegation should leave the heavy lifting on a Third Report to friendly countries, let the staff flesh out existing proposals without venturing into sensitive areas, yet not break off the talks.[109]

As 1948 opened, and his *Foreign affairs* article finally saw print, Oppenheimer's private divergence with Conant on international control persisted. Though Conant thought no wide-ranging settlement with Moscow possible for at least several years, he could still, remarkably, speak of negotiating an international control agreement—even of accepting a *Soviet* plan (if sufficiently improved), an astonishing heresy. Oppenheimer, in contrast to his friend's willingness to walk even a "rickety bridge" with the Russians, if only they would accept strict inspection, insisted that pervasive U.S.-Soviet hostility precluded any full-fledged cooperative venture. Osborn wrote in his diary that Oppenheimer "thinks that the substitution of the Russian plan—however strengthened and improved—for the majority would be impossible, dangerous and unacceptable in the present state of the world. He feels that the weakness of the Acheson-Lilienthal Report is that it did not sufficient[ly] define the state of the world necessary for any effective plan of control. If an ideal state of the world were attained—unlikely in our generation—then he might go along with Conant. Otherwise, [Oppenheimer considers] Conant's position absolutely untenable."[110]

A "Closed Chapter"

In the ensuing weeks, international events rendered their argument moot—or rather, added further evidence to bolster Oppenheimer's contentions. In Central Europe, a hard-line communist coup toppled the Czechoslovak government, ending hopes it might remain an oasis of neutrality between East and West; in Berlin, increased tensions prompted fears of a Soviet clampdown, triggering a full-fledged war scare in Washington and presaging the blockade that the Red Army imposed a few months later. By the summer of 1948, against the backdrop of a U.S.-U.K. airlift to the besieged West Berliners, the Cold War was permeating American election-year politics, as the House Un-American Activities Committee held hearings on charges that former State Department official Alger Hiss had been a Soviet spy. Meanwhile, the Truman Administration speeded secret negotiations toward the establishment of a military treaty linking the United States and Western Europe and accelerated preparations for

109. "Meeting of staff consultants" and the more formal "Memo of meeting of Board of Consultants," memoranda dated 17 Dec 1947, S/AE, box 64, folder 15D; see also Osborn to Tolman, 19 Dec 1947, S/AE, box 42, folder 20.132.
110. Osborn, UNAEC diary (ref. 14), 18 Feb 1948.

the creation of a West Germany, the Federal Republic of Germany, rather than rely on attaining agreement with Moscow on that country's postwar unification.

By September 1948, when he made his third consecutive annual pilgrimage to the National War College for a secret speech, even Conant conceded that international control represented a "closed chapter" in history.[111] It seemed merely to be stating the obvious when Oppenheimer, upon returning from travels in Europe the following month, told reporters that international control could not happen "while the two great powers are fighting on almost every front....It cannot be done as long as one country makes a fetish of secrecy and doesn't even let her own people know what the situation is."[112] In matters atomic, Oppenheimer's priorities focused even more sharply on the GAC's secret efforts to promote the Atomic Energy Commission's development of a panoply of fission weapons and increase their production rate and military utility, especially as the danger of possibly imminent conflict with the Soviets hovered over Berlin. Still, Oppenheimer (like Conant) kept up his periodic consulting for Osborn, whose delegation plodded along in preparing more reports even as history passed their efforts by. "There was really very little left to do," one scholar has noted. "Meetings of the commission occurred most infrequently. The organs of the group atrophied, and a form of diplomatic senility oppressed debate whenever it occurred."[113] Oppenheimer tried to stay on cordial terms with the American negotiator, congratulating Osborn and his staff on a "very fine" report in May and interceding with the *Bulletin of the atomic scientists* when Osborn was disturbed by an editorial.[114] In September 1948, the UN General Assembly, meeting in Paris, received the UNAEC's Third Report, which outlined the proposed international agency and the impossibility of its creation due to Soviet opposition, and suggested that the group's work be suspended indefinitely. The world body approved the report by an overwhelming but meaningless majority (since the Soviet bloc cast its *nyet*), but insisted that the Commission continue meeting (albeit to no apparent end).[115]

With the UN Commission little more than a forlorn reminder of the bright hopes of three years earlier, Oppenheimer in early 1949 published an article in the *Atlantic monthly*, like his *Foreign affairs* piece full of forebodings about the future and not without misgivings about the past.[116] Once again, the physicist sent an advance text to Osborn. While effusively praising the "excellence of our proposals" for interna-

111. James B. Conant, "The impending atomic age: 1948 preview," 14 Sep 1948, NDUA.
112. "Oppenheimer bids U.S. to guard atom," *NYT*, 14 Oct 1948.
113. Lieberman (ref. 14), 392.
114. Osborn to JRO, 4 May 1948, and JRO to Osborn, 7 May 1948, S/AE, box 77, folder 20.103.
115. Lieberman (ref. 14), 393; JRO, *The open mind* (ref. 2), 48.
116. JRO, "The open mind," *Atlantic monthly, 183:2* (Feb 1949), 221ff.; *The open mind* (ref. 2), 45-57. The piece originated as a talk to a Rochester, New York, foreign policy audience on 11 Dec 1948. See "The open mind: Draft of talk given to The Rochester Institute of International Affairs," 11 Dec 1948, S/AE, box 77, folder 20.103.

tional control, Oppenheimer did little to conceal his distress that the effort had failed, and again dropped a number of cryptic hints that one could not entirely blame the Soviets. Osborn was now disturbed not so much by specific criticisms of U.S. policy or policymakers—for Oppenheimer made none—but by some of the phraseology and tone: by the physicist's laments about the United States racing along a "course of action in some important respects inconsistent with our original purposes," those being foreign policy aims which "cannot in any real or enduring way be achieved by coercion"; by his misgivings over the nuclear weapons being developed; and by his admission that he could not suppress an "uneasy" or "troubled" conscience over the path his country had taken, committed to "a future of secrecy, of coercion, and to an immanent threat of war." The negotiator urged Oppenheimer to delete "troubled conscience" and the repeated references to atomic "coercion." In a last-minute, half-hearted placatory gesture, the physicist wired the magazine's editor on the evening of December 24 to make a few such revisions, while leaving other comparable phrases unaltered. "Have managed to change one conscience and two coercions," he telegraphed Osborn. "Merry Christmas."[117]

The UNAEC kept meeting, of course, in line with the General Assembly's injunction—which might well appear, so Oppenheimer suggested, "no more than an indication of unwillingness on the part of the Assembly to accept as permanent the obvious past failures of the Commission to fulfill its mandate." But Osborn could not dismiss it, so instead of closing up shop, in March 1949 he recalled Oppenheimer and the rest of his consultants to review documents and strategy one last time. Now the group reached consensus that the only useful thing to do was to make the strongest case for the record that the effort had failed due to Soviet obstruction on the issues of access and openness, and seek to go out of business in order to "[g]et us completely out of an impossible situation."[118] Finally, in the summer of 1949, shortly before the first Soviet atom bomb officially opened the next phase of the nuclear arms race, the UNAEC majority approved an American resolution stating that "the differences are irreconcilable at the Commission level, and that further discussion...would tend to harden these differences and would serve no practical or useful or useful purpose." Over Soviet objections, the UN Security Council and—after word of Joe One in September—the UN General Assembly finally put the Commission out of its misery.[119]

117. Osborn, handwritten notations on first page of JRO, "The open mind: Draft" (ref. 116); JRO to Edward Weeks, 24 Dec 1948, telegram, JROP, box 280, folder Rochester Talk; Publcn. Atlantic Monthly; JRO to Osborn, 24 Dec 1948, telegram, JROP, box 280, folder Rochester Institute—Correspondence. After the piece appeared, the negotiator assured him that it "was perfectly bully and a very real contribution, with or without the coercions." Osborn to JRO, 7 Dec 1949, S/AE, box 77, folder 20.103.
118. Osborn, memorandum on 8 Mar 1949 meeting of consultants, 10 Mar 1949, S/AE, box 13, folder 15A.
119. Lieberman (ref. 14), 393.

Aftermath

Ironically, on the same weekend back in March 1947 that Oppenheimer was first warning Osborn against Kremlin perfidy, J. Edgar Hoover secretly forwarded to David Lilienthal a "rather voluminous file" of "derogatory information" on the physicist, meant to cast aspersions on Oppenheimer's fitness to hold a security clearance on the basis of suspicions regarding possible pro-Soviet sentiments. An appalled AEC chairman hurriedly organized a quiet campaign to preserve the physicist's standing. After a suspenseful delay, during which Oppenheimer's fate hung in the balance, Truman sent word to Lilienthal in early April 1947 that he had no desire to pursue the matter.[120] Philip M. Stern has noted the incongruity of the FBI's insinuations when contrasted with Oppenheimer's actual conduct that March weekend.[121]

When Oppenheimer was called before the AEC personnel security board in 1954, his lawyer, Lloyd K. Garrison, led the physicist through the story of the evolution of his views on international control in order to demonstrate his solidly anti-Russian views and conduct. Osborn reaffirmed that view during his own brief appearance.[122] Yet—although those determined to revoke his clearance appear to have been impervious to rational argument or a balanced consideration of his security foibles—the story presented here suggests that his attorneys might have made more of the fact that Oppenheimer, in his secret advice on international control policy in 1947-49, not only showed himself to be firmly anti-Soviet, but took a consistently harder line in his counsel than did most of the other prominent consultants—so much so, as has been shown, that he found himself allied with, of all people, Leslie Groves, and opposed to his close friend and collaborator, James Conant.[123] As an ardent supporter of international control in principle, he could quite easily and reasonably have, like Conant, pressed for substantial modifications or even concessions in order to make the U.S. control plan more negotiable, or "salable," to the Soviets. Instead, he clearly and

120. The secret crisis of Mar 1947 can most conveniently be traced in Lilienthal's 1954 testimony, with relevant documents, *JRO Hearing*, 374-381; see also, e.g., Hershberg (ref. 8), 316-319, and Herken (ref. 14), 177-180. For a sanitized copy of the FBI material forwarded to Lilienthal, see Hoover to Lilienthal, 8 Mar 1947, with a 12-page, single-spaced attachment, in FBI file of JRO (ref. 10), #100-17828-136.

121. Stern (ref. 14), 101. "Oppenheimer's pressing, self-appointed mission [to Osborn]," Stern writes, "might have surprised some of the FBI's 'confidential sources,' whose accusations of pro-Russian sympathies on Oppenheimer's part were contained in the investigative file." Though Oppenheimer's mission was not, in fact, "self-appointed," as Osborn's later testimony suggested, Stern makes a valid point.

122. See *JRO Hearing*, esp. 36-45, 342-346.

123. When he testified, Conant spoke in definite terms about JRO's opinions of Soviet conduct in his atomic energy consulting for both Baruch and Osborn. However, Conant never specifically mentioned, nor was he asked about, JRO's explicit opposition in 1947-48 to further U.S. participation in international control talks. Likewise, Groves neither discussed nor was asked about JRO's international control activities. *JRO Hearing*, 384-394; 163-180.

consistently warned of the dangers of falling victim to Kremlin plots, and endorsed flatly the contention that the United States needed to retain atomic weapons so long as the threat of a Soviet military assault on Western Europe remained. If anything, as Robert Gilpin once observed, while most atomic scientists favored to some extent the doctrine of containing the U.S.S.R., Oppenheimer (like Edward Teller) belonged to the harder-line, more anti-Soviet wing of that group.[124] Substantially supplementing already available testimony from the 1954 hearing, the declassified record of Oppenheimer's record of advising the U.S. government on the international control of atomic energy in 1947-49 further refutes William L. Borden's absurd contention, in his November 1953 letter to J. Edgar Hoover that led to the AEC security hearings, that "More probably than not, he has acted under a Soviet directive in influencing United States military, atomic energy, intelligence, and diplomatic policy."[125]

More importantly, the story of Oppenheimer's post-Baruch Plan involvement with U.S. international control helps restore balance and complexity to our understanding of a life that has often been mythologized and caricatured by both his admirers and detractors. Demonized on the right as a naïve communist sympathizer, if not an actual Soviet spy, hailed on the left as a nuclear dissenter, martyr, or even pacifist, Oppenheimer emerges from this narrative as a committed, only slightly left-of-center cold warrior—"hard headed, realistic, and thoroughly anti-Soviet," as Conant put it at the security inquest.[126] While agonized about the atomic arms race his efforts had failed to forestall, he remained firmly convinced that the United States had to reserve the potential to wield and, if necessary, use such weapons to safeguard its expanded national security interests in its competition with Soviet-led communism. While lamenting certain aspects of U.S. nuclear policy, he hardly felt impelled to endanger his status as a nuclear insider by openly trumpeting his disagreements, even if he was willing to drop hints of unease to select audiences. Viewed in this light, Oppenheimer's ultimate expulsion from the nuclear and cold war establishment as a potential security risk appears even more ironic.

124. Robert Gilpin, *American scientists and nuclear weapons policy* (Princeton, 1962), 71.
125. Borden to Hoover, 7 Nov 1953, *JRO Hearing*, 837-838.
126. Ibid., 385.

CHAPTER NINE

The Atomic Secret in Red Hands? *American Suspicions of Theoretical Physicists During the Early Cold War*

David Kaiser

Visions of Science and Scientists in Postwar America

The early years of the Cold War were not a pleasant time to be an intellectual in the United States, especially if he or she happened to have a past or present interest in the political left. Following pathbreaking work by historians such as Ellen Schrecker on academics' ill treatment, scholars have begun to examine the toll of McCarthyism—which began earlier and lasted longer than Joseph McCarthy's fevered tenure in the Senate—on American scientists. Through special scrutiny, unfair treatment, blacklists, and quiet removals from classrooms and laboratories, American scientists bore the brunt of "loyalty-security" investigations during the Cold War era. Thanks to the efforts of many historians, we now know a great deal about how scientists fared from the late 1940s through the mid-1950s.[1]

This essay originally appeared in *Representations, 90* (2005), 28-60. My thanks to Bart Bernstein, Luis Campos, Harry Collins, David Engerman, Michael Gordin, David Hollinger, Alexei Kojevnikov, John Krige, Donald MacKenzie, Shawn Mullet, Ellen Schrecker, Steven Shapin, Rebecca Slayton, and Jessica Wang for helpful discussions; to Kieran Downes and Shane Hamilton for research assistance; and to Rebecca Press Schwartz for sharing several of her unpublished dissertation chapters with me.

The following abbreviations are used: AEC, Atomic Energy Commission; *BAS*, *Bulletin of the atomic scientists*; FAS, Federation of American Scientists; HUAC, United States Congress, House Committee on Un-American Activities; JCAE, United States Congress, Joint Committee on Atomic Energy; *NYT*, *New York Times*; *SHPMP*, *Studies in history and philosophy of modern physics*; TBL, The Bancroft Library, University of California, Berkeley; *VSD*, *Vital speeches of the day*.

1. On McCarthyism and American higher education see esp. Ellen Schrecker, *No ivory tower: McCarthyism and the universities* (New York, 1986); Sigmund Diamond, *Compromised campus: The collaboration of universities with the intelligence community, 1945–1955* (New York, 1992); James Hershberg, *James B. Conant: Harvard to Hiroshima and the making of the nuclear age* (New York, 1993); and John McCumber, *Time in the ditch: American philosophy and the McCarthy Era* (Evanston, IL, 2001). On American scientists in particular, see esp. Jessica Wang, "Science, security, and the Cold War: The case of E.U. Condon," *Isis, 83* (1992), 238-269; Jessica Wang, *American science in an age of anxiety: Scientists, anticommunism, and the Cold War* (Chapel Hill,

Paradoxically, by talking generally about "scientists," most of these previous studies have been at once too general and not far-reaching enough. There is a finer-grained story to be told about *which types* of scientists received the most attention at the time, and about the shifting assumptions about science that undergirded this scrutiny. In short, theoretical physicists emerged as the most consistently named whipping-boys of McCarthyism: repeatedly subjected to illegal surveillance by the Federal Bureau of Investigation (FBI), paraded in front of the House Un-American Activities Committee (HUAC), charged time and again in the media as well as in federal courts with being the "weakest links" in national security, and widely considered to be more inherently susceptible to Communist propaganda than any other group of scientists or academics. Focusing more squarely on allegations against this subgroup of American scientists, and the representations repeatedly made of them in congressional committees and broad-market media outlets, I will attempt to illuminate broader changes in assumptions both about scientists and about the nature of science after World War II.

More than a dozen theoretical physicists suffered publicly—and often over long durations—in the vise-grip of postwar anti-Communism: Edward Condon, Joseph Weinberg, Giovanni Rossi Lomanitz, David Bohm, Max Friedman, I. David Fox, Byron Darling, Bruce Dayton, Philip Morrison, Bernard Peters, Wendell Furry, Albert Einstein, and J. Robert Oppenheimer. Their cases often remained in the news for months at a time, resurfacing as a result of surreptitious leaks by their political enemies.[2] Thousands of scientists and engineers were affected by McCarthyism; only

---

NC, 1999); Silvan S. Schweber, *In the shadow of the bomb: Oppenheimer, Bethe, and the moral responsibility of the scientist* (Princeton, 2000); Lawrence Badash, "Science and McCarthyism," *Minerva, 38* (2000), 53-80; Naomi Oreskes and Ronald Rainger, "Science and security before the atomic bomb: The loyalty case of Harald U. Sverdrup," *SHPMP, 31* (2000), 309-369; David Kaiser, "Nuclear democracy: Political engagement, pedagogical reform, and particle physics in postwar America," *Isis, 93* (2002), 229-268; Fred Jerome, *The Einstein file: J. Edgar Hoover's secret war against the world's most famous scientist* (New York, 2002); and Gregg Herken, *Brotherhood of the bomb: The tangled lives and loyalties of Robert Oppenheimer, Ernest Lawrence, and Edward Teller* (New York, 2002).

2. On Condon, see Schrecker (ref. 1), 274-275; Wang, "Science, security" (ref. 1); and Wang, *American Science* (ref. 1), chapt. 4. On Weinberg, Lomanitz, Bohm, Friedman, and Fox, see Schrecker (ref. 1), 130-148; Russell Olwell, "Physical isolation and marginalization in physics: David Bohm's Cold War exile," *Isis, 90* (1999), 738-756; and Shawn Mullet, "Intelligence files and the unlearning of atomic espionage: Joseph Weinberg and Giovanni Rossi Lomanitz" (unpub. msc., 2003). On Darling and Dayton, see Schrecker (ref. 1), 207-209, 277-279. On Philip Morrison and Bernard Peters, see Schrecker (ref. 1), 148-160; and Schweber (ref .1). On Furry, see Schrecker (ref. 1), 197-204; and Catharine M. Hornby, "Harvard astronomy in the age of McCarthyism" (A.B. thesis, Harvard University, 1997), chapt. 2. On Einstein, see Jerome (ref. 1). The literature on Oppenheimer's case is vast; see esp. Philip Stern, *The Oppenheimer case: Security on trial* (New York, 1969); Barton Bernstein, "'In the matter of J. Robert Oppenheimer,'" *Historical studies in the physical sciences, 12* (1982), 195-252; and Charles Thorpe, *Oppenheimer: The tragic intellect* (Chicago, forthcoming), chapt. 7. Many additional theorists were subject to

a few became headlines in the newspapers or suffered repeated hounding by Congressional committees. More theoretical physicists were called by HUAC, for example, as ("unfriendly") witnesses between 1948 and 1953 than members of any other academic specialty: more than twice the number of chemists or historians; almost three times the number of biologists; nearly four times the number of economists or philosophers, and so on. The committee devoted no fewer than twenty-seven hearings to investigations of theorists and their "Communist infiltration" of weapons projects and educational institutions during this period—an average of nearly five per year, twice as many as those involving any other academic field.[3] Nor was HUAC alone. When *Life* magazine ran a two-page spread in April 1949 on a "red rumpus" of "dupes and fellow travelers" who purportedly gave aid to Communist-front organizations—featuring photographs of fifty individuals—twice as many theoretical physicists as members of any other academic field were included.[4] The only other scientists to receive comparable, sustained scrutiny from the likes of HUAC and the national media included astronomer Harlow Shapley and physical chemist Linus Pauling—both politically outspoken senior scientists—and experimental physicist Frank Oppenheimer, Robert's younger brother. Chemists and engineers were certainly implicated at times during the early Cold War. Yet theorists became cultural stand-ins for postwar intellectuals, their portrayals fitting a common pattern with remarkable consistency.

These dozen or so theoretical physicists surely came under fire for many reasons; with hindsight their troubles appear almost overdetermined. Most were Jewish; several had been active in labor organizing before or during the war; a few had flirted with the Communist Party in their youth; many were active in other left-wing political organizations after the war. Perhaps most important, however, they had close and long-standing ties with Robert Oppenheimer, all but Darling and Einstein having been trained by him as graduate students or postdoctoral fellows. (When he became director of the Institute for Advanced Study in Princeton in 1947, Oppenheimer even became Einstein's "boss," although Einstein had been under intense FBI surveillance long before then.) As previously classified documents make clear, Oppenheimer had

---

behind-the-scenes harrassment, including FBI wiretaps and mail covers, although their cases did not erupt into the media spotlight: Barton Bernstein, "Interpreting the elusive Robert Serber: What Serber says and what Serber does not explicitly say," *SHPMP*, *32* (2001), 443-486.

3. In addition, eight mathematicians and five experimental physicists were called as witnesses before HUAC between 1948 and 1953, totaling less than three-quarters and one-half the number of theoretical physicists, respectively. Information on witnesses before HUAC hearings (in both public and executive sessions) is most conveniently available via the "Congressional universe" search engine: http://web.lexis-nexis.com/congcomp. Similarly, Schrecker's meticulous account in *No ivory tower* (ref. 1)—which ranges beyond HUAC to include several other congressional, state, and university committees—includes twelve physicists who were adversely affected (eleven of whom were theoretical physicists), more than any other academic specialty.

4. "Red rumpus: Dupes and fellow travelers dress up Communist fronts," *Life* (5 Apr 1949), 42-43.

enemies within the military and intelligence branches of the government from his earliest days in the wartime Manhattan Project; he was under surveillance even before most of his students and colleagues were. Debate continues to this day over the extent of Oppenheimer's sympathies—and possible affiliation—with the Communist Party.[5] It is certainly plausible that this circle of theorists garnered undue attention as "collateral damage," caught in the escalating cross fire between Oppenheimer's growing list of enemies and the central figure himself, whose personnel security hearing in spring 1954 marked the climax of McCarthyist attacks on scientists and public intellectuals. Yet even if connections to Oppenheimer explain why these particular people were singled out, the pedagogical links alone do not explain the types of representations formed about theory and theorists as a whole. Petty politics might have fixed the victims, but not the charge.

The portrait of theoretical physicists that eventually solidified into a recognizable (and oft-repeated) ideal type relied on three separate elements. None was beyond question; some were openly challenged in their day. Changing political exigencies, however, eventually snuffed out the dissenting voices, and by 1948–49 the formula had been set. The first piece of the Cold War equation was that theorists had single-handedly built the atomic bomb. The wartime Los Alamos laboratory had been headed by the theorist Oppenheimer, and soon after the war lack of information due to government secrecy led to a skewed picture of how the bombs had been built. Second, a widespread presumption emerged that "atomic secrets" existed whose possession would allow other countries immediately to build their own bombs. By the late 1940s, if not immediately in the wake of the bombings of Hiroshima and Nagasaki, many came to believe that these "secrets" consisted of purely theoretical information, perhaps even single equations that only theorists understood and that could be smuggled out of the country on tiny scraps of paper. As soon as journalists and politicians concurred that text-based atomic secrets existed and that theorists were their special keepers, it seemed to follow that theorists required special scrutiny. The anti-Communist backlash struck theorists particularly hard, coalescing in the third and final element of the Cold War formula: reporters, senators, and federal judges alike began to assert that theoretical physicists were inherently a breed apart, more susceptible to Communist influence than any other group of people. By the late 1940s, theorists routinely were cast as doubly dangerous proto-spies: of all scientists, theorists held the most important information, and they were most likely to give it away.

Lurking behind this formulaic depiction of the postwar theorist lay a specific idea of how science works, not just how various scientists behave. By tracing discussions of

5. On the early and sustained surveillance of Oppenheimer, see Bernstein (ref. 2); and Herken (ref. 1). Recent exchanges in the ongoing debate over Oppenheimer and Communism include Daniel J. Kevles, "The strange case of Robert Oppenheimer" (review of Herken, *Brotherhood of the bomb*), *New York review of books, 50* (4 Dec 2003), 37-40; and Gregg Herken, "The Oppenheimer case: An exchange," with a reply by Kevles, in *New York review of books, 51* (25 Mar 2004), 46-48.

purported "atomic secrets" and their "red" keepers—as these unfolded in such varied venues as the *New York Times*, *Newsweek*, *Time*, *Life*, *Fortune*, the *Saturday evening post*, *Reader's digest*, and the *Bulletin of the atomic scientists*, as well as in congressional hearings, government reports, and oratory reprinted in *Vital speeches of the day*—we learn about the shifting cultural and political valence of "tacit knowledge" within postwar America.[6] Historians and sociologists of science have long been interested in tacit knowledge: craftlike skill and artisanal "know-how" crucial to the practice of science, yet incapable of being transmitted by texts alone. Earlier figures such as Thomas Edison had been publicly valorized precisely on the strength of their storehouses of tacit knowledge rather than formal book-learning. Recent innovative studies by sociologists and anthropologists, meanwhile, have emphasized how critical tacit knowledge has been and remains for designing and building nuclear weapons.[7]

As I argue here, such evaluations of tacit knowledge have a history—a history closely intertwined with changing ideas about theory and theorists in Cold War America. Naturally certain kinds of textual information could aid other groups in their efforts to build atomic bombs, mainly by helping them avoid "blind alleys" and focus their replication efforts on techniques that had been shown to work. What I find fascinating is the way in which congressional, judicial, and journalistic discussions collapsed all these issues, beginning in the late 1940s, to focus narrowly around "secret formulas" that somehow held the key to the entire bomb. During the early Cold War years, the importance of tacit knowledge fell out of discussions, leaving only text-based information as the seeming "secret" of the bomb—with deleterious effects for many theoretical physicists.

6. Relevant articles were found by examining all entries for 1945–55 in the *Reader's guide to periodical literature* indexed under the following keywords: "atomic bomb," "atomic secret," "atomic espionage," and "physicists."

7. Donald MacKenzie and Graham Spinardi, "Tacit knowledge, weapons design, and the uninvention of nuclear weapons," *American journal of sociology*, *101* (Jul 1995), 44-99; Laura McNamara, "Ways of knowing about weapons: The Cold War's end at the Los Alamos National Laboratory" (Ph.D. dissertation, University of New Mexico, 2001), esp. chapt. 4; and Hugh Gusterson, "A pedagogy of diminishing returns: Scientific involution across three generations of nuclear weapons science," in David Kaiser, ed., *Pedagogy and the practice of science: Historical and contemporary perspectives* (Cambridge, 2005), 75-107. On "tacit knowledge" in science studies more broadly, see esp. Harry Collins, *Changing order: Replication and induction in scientific practice*, 2nd edn. (Chicago, [1985] 1992); Harry Collins, "Tacit knowledge, trust and the Q of sapphire," *Social studies of science*, *31* (2001), 71-85; Trevor Pinch, H.M. Collins, and Larry Carbone, "Inside knowledge: Second order measures of skill," *Sociological review*, *44* (1996), 163-186; H.M. Collins, G.H. de Vries, and W.E. Bijker, "Ways of going on: An analysis of skill applied to medical practice," *Science, technology, and human values*, *22* (1997), 267-285; Steven Shapin and Simon Schaffer, *Leviathan and the air-pump: Hobbes, Boyle, and the experimental life* (Princeton, 1985); Myles Jackson, *Spectrum of belief: Joseph von Fraunhofer and the craft of precision optics* (Cambridge, 2000); and David Kaiser, *Drawing theories apart: The dispersion of Feynman diagrams in postwar physics* (Chicago, 2005).

The Work of Many People

Nuclear weapons have always been "the work of many people." Edward Teller introduced the phrase in 1955 when recalling the range of efforts and contributions—from specialists in many different fields—to the American hydrogen-bomb effort.[8] (Many of Teller's critics found the maneuver a disingenuous effort to deflect criticism after his withering performance at Oppenheimer's personnel security hearing mere months before, and his less-than-generous sharing of credit throughout the early 1950s.[9]) Whether made in good faith or not, Teller's remark was accurate and applies equally to the wartime efforts to build atomic bombs. At its peak, the wartime Manhattan Project employed more than 125,000 people at more than thirty sites, ultimately spending more than $2 billion. Huge factory towns sprung up at Oak Ridge, Tennessee, and Hanford, Washington; university facilities were converted to top-secret laboratories at Berkeley, Columbia, the University of Chicago, and elsewhere; and a new laboratory was created at Los Alamos, New Mexico.

In all these facilities, scientists, engineers, and technicians of many stripes contributed their expertise to turn nuclear weapons into a reality. The overall scientific directors of the wartime project were Vannevar Bush and James B. Conant, an electrical engineer and a chemist, respectively. The theorist Oppenheimer served as scientific director of the Los Alamos laboratory, where he was joined by a nine-person governing board that included one fellow theorist alongside four experimental physicists, two chemists, and two experts in ordnance. Several divisions were created at Los Alamos, focusing on metallurgy, chemistry, ballistics, ordnance, electrical engineering, experimental physics, and theoretical physics. The laboratory's wartime organization chart shows the groups arranged in a circle and connected by spindly links; no group appeared on top directing the others.[10] A few months after the war, the War Department awarded eleven "Medals for Merit" to scientific staff of the Manhattan Project for their contri-

8. Edward Teller, "The work of many people," *Science, 121* (25 Feb 1955), 267-275.

9. See, e.g., Herbert York, *The advisors: Oppenheimer, Teller, and the superbomb*, 2nd edn. (Stanford, [1976] 1989), 133-134; and Hugh Gusterson, "The death of the authors of death: Prestige and creativity among nuclear weapons scientists," Mario Biagioli and Peter Galison, eds., *Scientific authorship: Credit and intellectual property in science* (New York, 2003), 281-307, on 283-284.

10. See esp. Richard Hewlett and Oscar Anderson, Jr., *The new world, 1939-46*, Vol. 1. *A history of the United States Atomic Energy Commission* (University Park, PA, 1962); David Hawkins with Edith Truslow and Ralph C. Smith, *Project Y: The Los Alamos story* (New York, [1947] 1983), esp. 30 on the Governing Board; Lillian Hoddeson et al., *Critical assembly: A technical history of Los Alamos during the Oppenheimer years, 1943-1945* (New York, 1993); Andrew Pickering, "Cyborg history and the World War II regime," *Perspectives on science, 3* (Spring 1995), 1-48; Peter Hales, *Atomic spaces: Living on the Manhattan Project* (Urbana, IL, 1997); Peter Galison, *Image and logic: A material culture of microphysics* (Chicago, 1997), esp. chapt. 4; and Rebecca Press Schwartz, "Manhattan beyond the mesa" (unpub. msc., 2004). The Los Alamos wartime organization chart appears on the cover of Galison, *Image and logic.*

butions to the atomic bomb project; a little more than one-quarter went to theoretical physicists, while half went to chemists and engineers.[11]

Of all the work done at wartime Los Alamos, the challenge of implosion—detonating a sphere of conventional explosives to compress fissionable material into a superdense critical mass—best exemplified the melding of many groups' efforts, and the central place of tacit knowledge in making bombs.[12] From the metallurgy of plutonium, to the design and fabrication of sufficiently homogenous shaped-charge lenses, to the chemistry of radioactive "initiators," to the physical measurements of spontaneous fission rates and reaction cross sections, to the challenges of wiring the detonation circuits to fire simultaneously, to the mathematical studies of in-going shock waves and the hydrodynamics of matter under extreme conditions, the implosion device was a product of varied—and harried—contributions. The Theoretical Physics or "T-division" remained the smallest of all divisions at Los Alamos, soon dwarfed by such groups as the Engineering Ordnance division.[13] The T-division played some important roles throughout the project, estimating critical masses, trying to calculate optimal shapes for the conventional-explosive lenses, and so on; in some instances, their contributions proved critical, while in others the T-division struggled to play catch-up to the other groups.[14] Throughout the war, theorists remained hamstrung, having no self-consistent theory of nuclear forces and no reliable way to calculate properties of nuclear matter in bulk, even under idealized conditions—problems that would haunt theoretical physicists for the next decade and a half.[15] Members of Los Alamos's T-division developed some "good enough" approximation methods, learning clever ways to "get the numbers out," in Sam Schweber's telling phrase. But even these approximations rarely drove design or manufacturing efforts; most calculations relied upon tedious hand-cranked or punched-card mechanical calculators, and the nonlinear nature of the equations made extrapolation from one curve to another treacherous. At no time could the T-division *calculate* the properties of the bomb from first principles;

11. See the following War Department press releases: "Medal for Merit presented to eight scientists for work on atomic bomb," 4 Mar 1946; "Atomic bomb scientists awarded Medal for Merit," 13 Mar 1946; and "Harvard scientist awarded Medal for Merit for atom bomb work," 9 Apr 1946, all available in Paul Kesaris, ed., *Manhattan Project: Official history and documents* (Washington, D.C., 1977), 12 microfilm reels, on reel 1, part 6.
12. The implosion device was first tested at the so-called "Trinity" test near Alamagordo, New Mexico, on July 16, 1945, and a comparable bomb destroyed Nagasaki on August 9, 1945; the bomb that destroyed Hiroshima on August 6, 1945, used a different technique to trigger the explosive chain reaction.
13. Hoddeson et al. (ref. 10), 93.
14. See esp. Hawkins, Truslow, and Smith (ref. 10); Hoddeson et al. (ref. 10); and Anne Fitzpatrick, "Igniting the light elements: The Los Alamos thermonuclear weapon project, 1942-1952" (Ph.D. dissertation, Virginia Institute of Technology, 1999), available as report LA-13577-T from Los Alamos National Laboratory.
15. Kaiser (ref. 7), chapts. 5-9.

atomic bombs have never resulted from the mere "application" of explicit formulas, let alone full-blown theories.[16]

News of the surprise bombings of Hiroshima and Nagasaki in August 1945 was the first most people in the United States and abroad had heard of atomic bombs. The only information released about technical aspects of the bombs came from a specially prepared government report, released to the press on the evening of August 11. The two-hundred-page document with the long-winded title—*A general account of the development of methods of using atomic energy for military purposes under the auspices of the United States Government, 1940-1945*—quickly became known as the "Smyth Report," named for its author, the Princeton physicist and Manhattan Project consultant Henry DeWolf Smyth. As Rebecca Press Schwartz comments in her stunning analysis of the Smyth Report's composition, "What it included, and what it omitted, had a great impact on what the Manhattan Project meant to the American people." Security concerns dominated what Smyth could and could not include within his report. Early on, Smyth, General Groves, and advisors agreed that only information that had already been published in the open literature, that was widely known to working scientists, or that had "no real bearing on the production of atomic bombs" was fit for release. Little of the messy combination of chemistry, metallurgy, engineering, and industrial-scale manufacturing met these criteria; these aspects of the sprawling project, crucial to the actual design and production of atomic bombs, remained closely guarded. Instead, Smyth focused narrowly on ideas from physics, pushing theoretical physics in particular to the forefront—only such material was deemed safe enough to publicize. Ironically, most people read in Smyth's Report the lesson that theorists had built the bomb (and by implication, had won the war).[17] A supplemental docu-

16. On theorists' wartime efforts to "get the numbers out," see Silvan S. Schweber, *QED and the men who made it: Dyson, Feynman, Schwinger, and Tomonaga* (Princeton, 1994), esp. chapt. 3. On the nontheoretical elements involved in nuclear weapons design and manufacture, see also Hawkins, Truslow, and Smith (ref. 10); Hoddeson et al. (ref. 10); Fitzpatrick (ref. 14), chapt. 2; MacKenzie and Spinardi (ref. 7); Hugh Gusterson, *Nuclear rites: A weapons laboratory at the end of the Cold War* (Berkeley, 1996); and McNamara (ref. 7).

17. Rebecca Press Schwartz, "The making of the history of the atomic bomb: The Smyth Report and the historiography of the Manhattan Project" (Ph.D. dissertation, Princeton University, in preparation), quotations from chapt. 2, pp. 3, 11. The security regulations governing what could and could not be included in the Smyth Report are also quoted in Arvin Quist, *Security classification of information*, Vol. 1, *Introduction, history, and adverse impacts*, 2nd edn. (Oak Ridge, TN, 2002 [1989]), chapt. 5, p. 2 (available at http://www.fas.org/sgp/library/quist). In 1946 Princeton University Press took over the printing of the Smyth Report (the Government Printing Office had become overwhelmed by all the requests for copies), releasing it with a slightly more manageable title: Henry DeWolf Smyth, *Atomic energy for military purposes* (Princeton, 1946). Not all of its readers agreed that the Smyth Report overplayed theoretical physics at the expense of engineering and industrial contributions. Early in 1947, one Manhattan Project veteran complained that the report gave too much attention to the industrial contractors, and that the report had been cavalier in revealing "technological aspects of the

ment released a few months later by the Senate's new Special Committee on Atomic Energy, entitled *Essential information on atomic energy*, reinforced the lesson, depicting atomic bombs as the natural outcome of continuous developments in atomic theory. A "chronological table" at the end of the report extended the narrative as far back as 400 BC to the ancient Greek atomists![18] Official reports on the Manhattan Project painted a consistent—if consistently inaccurate—picture: theoretical physicists had built the bomb. True or false, the first part of the Cold War formula was in place.

The Making of a "Secret"

The notion that there existed a single "secret" to the atomic bomb was not hatched all at once with the bombings in August 1945 or the release of the Smyth Report. It emerged over time, unfolding in lockstep with changing political tidings. In the weeks after news of the bombs broke, rumors of legislative efforts (in the form of the May-Johnson bill) that would have extended the wartime military regime into peacetime spurred groups of scientists and engineers at Los Alamos, the University of Chicago, Oak Ridge, and elsewhere to speak out about the nature of atomic weapons. The new bill would have kept all aspects of atomic energy under strict military control, perpetuating the wartime secrecy protocols. In response, the fledgling "atomic scientists' movement" launched a vigorous public-relations campaign, putting mimeograph machines into motion to counter the proposal.[19] Their pamphlets made it into the hands of editors at the *New republic*, who quoted from them approvingly in the October 8, 1945 issue, adding italics for their own emphasis: "'*There is no secret to be kept.*'" The "principles" required for the explosive release of atomic energy, the editors continued (still quoting from the scientists' memoranda), "'have been the common property of scientists throughout the world for the last five years.'" President Truman sounded a similar note that same week when he delivered his first speech to Congress on atomic policy matters. Like the atomic scientists, Truman insisted that "the essential theoretical knowledge upon which the discovery [of the atomic bomb] is based is already widely known. There is also substantial agreement that foreign research can come abreast of our present theoretical knowledge in time."[20] That autumn, scientists and journalists

---

production of fissionable materials and manufacture of bombs." E.R. [Eugene Rabinowitch], "Editorial: The atomic secrets," *BAS, 3* (Feb 1947), 33, 68, on 33; see also "Smyth disclosure amazed savants: Chicago atomic scientists say they were 'dumfounded' [*sic*] at report; writer defends it," *NYT*, 18 Feb 1947.

18. United States Senate, Special Committee on Atomic Energy, *Essential information on atomic energy* (Washington, D.C., 1946), esp. 21-31, 41-44.

19. See esp. Alice Kimball Smith, *A peril and a hope: The scientists' movement in America, 1945–47* (Chicago, 1965); and Wang, *American science* (ref. 1), chapts. 1-2.

20. "The bomb is no secret," *New republic* (8 Oct 1945), 451-453, on 451; Harry S. Truman, "Atomic energy: Domestic and international control," *VSD, 12* (15 Oct 1945), 8-10, on 9-10.

developed a second, related response: the only secret about the bomb was whether or not it would work—not how to design or produce one—and that "secret" had disappeared with the destruction of Hiroshima and Nagasaki.[21] No secrets, no need for military control.

Beyond these immediate denials that there were any secrets, speculation continued as to whether "atomic secrets" existed and what they might be. Scientists, journalists, and politicians proposed at least eight distinct candidates, each put forward as *the* "atomic secret" between 1945 and 1955. The candidates hardly emerged at random; two distinct phases marked the "secrets" discussions.[22] Between 1945 and 1948, most coverage focused on raw materials and industrial infrastructure as the keys to producing atomic weapons, emphasizing nontextual "know-how" rather than textual "knowledge" or "information." Beginning late in 1948 and accelerating through the mid-1950s, the weight of discussion among politicians and journalists shifted, focusing instead on textual and theoretical "information" as the essential "secret" of the atomic bomb, rather than experimental skill or industrial capacity. Many now claimed that specific, esoteric formulas—the *x*'s and *y*'s of theoretical physics—contained the true secrets of the atomic bomb.[23]

Three main contenders for the "secret" received the most attention between 1945 and 1948: (1) raw materials and their handling; (2) production plants and industrial methods; and (3) technical details of design and manufacture. Editors at *Newsweek* entered the fray first, proclaiming "A new era: The secrets of science" in their August 20, 1945, issue. No doubt drawing on the just-released Smyth Report, *Newsweek* heralded uranium isotope separation as "the No. 1 secret of the war, known to only a few men in the 'Manhattan Project.'" President Truman similarly called for the need "to establish control of the basic raw materials essential to this power" in his speech to Congress on October 3, 1945.[24] One month earlier, a vice president at the University

21. Reuben G. Gustavson, "Story behind the atomic bomb: Teamwork among scientists," *VSD, 11* (1 Oct 1945), 762-767, on 766; "The bomb is no secret" (ref. 20), 451; A.K. Solomon, "Physics of the bomb," *Fortune* (1 May 1946), 114-122, 173-176, on 174. Later commentators sometimes returned to this theme: E.R. [Eugene Rabinowitch] (ref. 17), 33; and "The Russians knew," *Time* (2 Jan 1950), 54. See also the Congresional testimony from Oct 1945 by Manhattan Project scientific veterans Leo Szilard and Harold Urey, as quoted in Quist (ref. 17), chapt. 5, p. 30.

22. Cf. William A. Reuben, *The atom spy hoax* (New York, 1955). Reuben did not focus on the implications of the "secrets" discussions for theory or theorists, and he periodized the debate slightly differently than I do here (Reuben saw the Soviets' detonation of Joe I as the main pivot between two phases of discussion), but his book nonetheless remains a prescient study worthy of renewed attention.

23. Looking back at the troubled decade in October 1954, for example, physicist Ralph Lapp lamented that in August 1945 "public attitudes had not congealed into their present state of rigidity. Then the notion of '*the* secret of the atomic bomb' was only a notion and not a fixation. And it had not become a plank in a political platform." Ralph E. Lapp, "Atomic candor," *BAS, 10* (Oct 1954), 312-314, 226, on 312.

24. "A new era: The secrets of science," *Newsweek* (20 Aug 1945), 34-42, on 37; Truman (ref.

of Chicago emphasized industrial contributions, "working on a gigantic scale," as the essential ingredients in the bomb's development. *Time* magazine reassured its readers in mid-September 1945 that there were "no immediate dangers" that other nations could threaten the United States with their own atomic bombs, "because at this stage of the bomb's development huge production plants (which exist in the U.S. alone) are necessary." Throughout the remainder of 1945 and much of 1946, conservative members of Congress—alarmed to learn the extent of foreign-born scientists' contributions to the Manhattan Project—repeatedly heralded "American industrial and technological superiority" as the real secret to the project's success. One representative went so far as to declare, "I am not giving the principal credit for the production of the bomb to the scientists," who (he continued) would have been unable to do anything useful had it not been for "the trained fingers of the technician...—in other words, American know-how produced the atomic bomb."[25] As late as October 1948, a reporter for the *New York Times* emphasized that "engineering and technical know-how"—embodied in teams working in "gigantic facilities"—was "of equal or greater importance" than any "theoretical scientific knowledge" that was by then widely shared. The only "secrets" to be kept, urged veterans of the wartime weapons projects when testifying before Congress, were technical details and manufacturing processes.[26]

Throughout this first phase of discussions—much of it elicited by overt political debate regarding the proper role and scope of the new Atomic Energy Commission (AEC)—the emphasis remained on controlling matériel and industrial capacity.

---

20), 9. Historian Gregg Herken argued in 1980 that the only real "atomic secret" was General Groves's secret negotiations during the war to secure a monopoly on the world's supply of uranium ore: Herken, *The winning weapon: The atomic bomb in the Cold War, 1945–1950* (New York, 1980), chapt. 5. Groves similarly testified in Nov 1945 before the Senate Special Committee on Atomic Energy that the greatest challenge had been garnering sufficient fissionable material: Quist (ref. 17), chapt. 4, p. 28.

25. Quoted in Harry Hall, "Congressional attitudes toward science and scientists: A study of legislative reactions to atomic energy and the political participation of scientists" (Ph.D. dissertation, University of Chicago, 1961), on 96; see also 26, 33. A facsimile edition of Hall's dissertation was published by Arno Press in 1979.

26. Gustavson (ref. 21), 765; "The secret," *Time* (17 Sep 1945), 30; Hanson W. Baldwin, "Has Russia the atomic bomb? Probably not," *NYT*, 9 Nov 1947; Winifred Mallon, "Leaders condemn secrecy in science," *NYT*, 26 Oct 1945; E.R. [Eugene Rabinowitch], "Editorial: Freedom of scientific publication," *BAS, 2* (Dec 1946), 1, 32; Russell Porter, "Dr. Conant in plea for U.S. leadership," *NYT*, 20 Oct 1948. Truman similarly erected a division between "scientific information" on the one hand, and "manufacturing processes leading to the production of the atomic bomb itself": Truman (ref. 20), 8-10. This basic division appeared throughout Truman's other early statements about the bomb: cf. White House press release, "Statement by the President of the United States" (6 Aug 1945), in Kesaris (ref. 11), reel 1 part 6; the three-nation accord signed on 15 Nov 1945 with Britain and Canada, reprinted in Special Committee on Atomic Energy, *Essential information* (ref. 18), 3-4; and Henry Wallace's recollections of the 21 Sep 1945 cabinet meeting: Wallace, "The Maginot Line of secrecy," *BAS, 8* (Feb 1952), 36-37.

Driving the message home, several scientists, politicians, and journalists during this period spoke of a basic category error: bombs are not formulas. One week before the three-nation accord on atomic energy, signed on November 15, 1945, by President Truman and Prime Ministers Clement Attlee of the United Kingdom and MacKenzie King of Canada, former Prime Minister Winston Churchill delivered to the British Parliament a lesson on the sociology of knowledge. What the Americans sought not to disclose after the war, Churchill explained, [27]

> is the practical production methods, which they have developed at enormous expense and on a gigantic scale. This would not be an affair of scientists or diplomats sending over formulas. To be effective, any such disclosure would have to take the form of a considerable number of Soviet specialists, engineers and scientists visiting the United States arsenals, for that is what the manufacturing centers of the atomic bomb really are.
>
> They would have to visit them and would have to dwell there so they could have it all explained to them and the officials would then return to their own country with all the information they had obtained and with any further improvements which might have occurred to them.

No mere written exchanges here. Churchill evoked a notion of tacit knowledge to explain how working knowledge of atomic bombs could be transported from one country to another. His successor, Attlee, continued the lesson later that month, explaining that the ability to construct atomic bombs "cannot be given in a formula or a handbook or a blueprint. It can only be done by scientists and technicians being taken to the plant, everything being shown and explained to them in great detail."[28]

Back in the States, scientists tried to deliver a similar message to Congress and the public, testifying that atomic bombs "are not matters that can be stolen and transmitted in the form of information."[29] Smyth told reporters from *Life* magazine that "[t]here is no 'secret' of the atomic bomb in the sense of a mysterious formula that can be written on a slip of paper and carried in the sole of a shoe or the handle of a hunting knife." The message got through to some mainstream journalists. Responding to allegations in July 1947 that some servicemen at wartime Los Alamos had stolen classified information about the bomb project (by pilfering a few documents to keep as souvenirs), the *New York Times* replied that "we do not believe that the atomic 'secret' is something that can be written down in a page or two, like a recipe for corn bread. We suspect that some tons of blueprints might convey it, and that it will never be carried overseas in the heel of a shoe."[30] It would take much more than such pencil-and-paper information

27. Churchill's 7 Nov 1945 address to Parliament as quoted in Reuben (ref. 22), 10-11.
28. Attlee quoted in Reuben (ref. 22), 12.
29. Lloyd Berkner quoted "sworn testimony before Congress in 1945 by [unnamed] authorities on the matter," in Berkner, "Is secrecy effective?," *BAS, 11* (Feb 1955), 62-63, 68, on 63.
30. Smyth quoted in Press Schwartz (ref. 17), chapt. 2, p. 12 n. 33; "Mystery of the stolen atom," *NYT*, 11 Jul 1947; see also E.R. [Eugene Rabinowitch] (ref. 17), 33, 68; and Baldwin (ref. 26), E3.

to design and manufacture nuclear weapons; the "secrets" were not textual in nature but based on artisanal know-how and industrial production. Such was the dominant message during 1945-48.

The "bombs are not formulas" campaign was not without its detractors during this early period. Competing visions of atomic secrets emerged in the course of (often quite bitter) political debate over domestic and international proposals for the control of atomic energy. Allegations of atomic espionage—along with the assumption that various "secrets" existed that could be stolen, with great benefit to foreign nations—were used as political instruments as early as 1946. As Gregg Herken has shown, for example, news of a Canadian spy ring that had operated during the war was leaked by General Groves to certain hand-picked journalists who were friendly to the general's goal of erecting military rather than civilian control over the postwar atom.[31] Nor did Groves only enlist journalists while pursuing this tactic. In March 1946, in the midst of debate over competing domestic atomic energy bills, Senator Bourke Hickenlooper read into the *Congressional record* a letter from Groves claiming that British physicist Alan Nunn May (who had worked in Canada during the war and later confessed to sharing material with the Soviets) "has a general knowledge of the construction of the atomic bomb." Groves's letter continued, suggesting that May "understands the principles of design and construction" of nuclear reactors (which could be used to produce fissionable materials such as plutonium). The *New York Times* reported that Hickenlooper's sudden introduction of Groves's letter caught rival Senator Brien McMahon, backing a bill for civilian control of atomic energy, by surprise, coming as it did at a crucial juncture of the debate. Groves's and Hickenlooper's allegations were seen as adding great weight—at just the right moment—to Senator Arthur Vandenberg's proposed amendment to McMahon's bill, calling for a military advisory board to the civilian Atomic Energy Commission.[32] Claims about "general knowledge of the construction of the atomic bomb"—knowledge of "principles" that could be written down and given to Soviet agents—helped tip the balance, ensuring passage of Vandenberg's promilitary amendment. Never mind that the official report on the alleged Canadian spy ring concluded three months later that "no one in Canada could have revealed how to make an atomic bomb. There was no one in Canada who had that information"—as Groves, overseer of the entire wartime project, had well known all along.[33] Innuendo about secrets in the form of "general knowledge" and "principles" that could be stolen had come in handy.

31. Herken (ref. 24), chapt. 6. See also Reuben (ref. 22), chapts. 2-4, on the Canadian spy ring.
32. Indeed, Vandenberg had begun the Senate discussion that led to Hickenlooper's disclosure only after McMahon had made a passionate radio address the night before arguing against such military incursions into the civilian agency: Anthony Leviero, "Groves bares leak in U.S. atom plant," *NYT*, 20 Mar 1946; Groves's complete letter to Hickenlooper, dated 12 Mar 1946, is reprinted in JCAE, *Soviet atomic espionage* (Washington, D.C., 1951), 51-53. See also Hall (ref. 25), 230-231.
33. In his confession, May claimed that the report he wrote for the Soviets was brief and general,

If such political jockeying around matters atomic began in the Senate's Special Committee on Atomic Energy, it soon became the basic operating procedure for HUAC. As Jessica Wang has demonstrated, HUAC's forays followed a predictable political timetable. After orchestrating a series of press leaks throughout spring 1947, HUAC delivered its first explicit salvo on the "atomic secrets" front on March 1, 1948, when it announced that theoretical physicist Edward Condon, then director of the National Bureau of Standards, constituted "one of the weakest links in our atomic security"—an official report delivered just days before debate over appropriations for HUAC was set to begin in Congress.[34] The first attack by HUAC on theorists involved few details of the nature of their work; Condon *qua* theorist was not really under discussion. The theorist's outspoken support for the McMahon bill throughout 1946, combined with the fact that he had been hired to direct the bureau by then Secretary of Commerce Henry Wallace—and hence offered the Republican-dominated HUAC an opportunity to strike at Truman and his most left-leaning cabinet member—were probably sufficient to single Condon out as a target.[35]

The list of HUAC's "weak links" was quickly expanded and the committee conducted highly publicized hearings later that year into supposed Communist activities and atomic espionage during the wartime Manhattan Project. This time the nature of theoretical physics and the trustworthiness of theorists played a larger role. Releasing their new report in September 1948, in time for maximum election-year impact (the *New York Times* reported Truman's "no comment" on the report while campaigning in Texas), HUAC claimed that several Manhattan Project scientists had passed along "vital information" on the bomb to the Soviets.[36] Following charges against the chemists Clarence Hiskey, John Chapin, and Martin Kamen, HUAC included a titillating accusation against a person identified only as "Scientist X," who had worked at the University of California's Radiation Laboratory during the war, one of the contracting laboratories of the Manhattan Project. This person, HUAC charged, had visited

---

containing no more information than was soon released in the Smyth Report. The Report of the (Canadian) Royal Commission, appointed to investigate the spy-ring allegations, was published in June 1946, and was quoted in JCAE (ref. 32), 53. House member J. Parnell Thomas (R., N.J.) likewise invoked fear over losing (unspecified) "atomic secrets" in his efforts to block the McMahon bill; see Wang, *American science* (ref. 1), 45-46.

34. Wang, "Science, security" (ref. 1), 246, 252-253; see also Robert Carr, *The House Committee on Un-American Activities, 1945–50* (Ithaca, NY, 1952), 134.

35. Several scientists at the time attributed HUAC's attacks on Condon to their differing political views on domestic atomic energy legislation; see, e.g., "Scientists defend themselves on loyalty charges," *BAS, 9* (Feb 1953), 2-3; and Carr (ref. 34), 131-153.

36. "Atomic spy report will shock public, official declares," *NYT*, 26 Sep 1948; William White, "Indictment of five is urged in report on atomic spying," *NYT*, 28 Sep 1948; HUAC, *Excerpts from hearings regarding investigation of Communist activities in connection with the atom bomb* (Washington, D.C., 1948); and "Text of report by House Committee on Un-American Activities relating to atomic espionage," *NYT*, 28 Sep 1948, 22-23. See also "The atom spy report," *NYT*, 29 Sep 1948, criticizing HUAC for unsubstantiated charges drummed up for political gain.

the house of Steve Nelson, a Bay Area Communist Party member, at 1:30 a.m. one morning in March 1943. Once there, Scientist X allegedly[37]

> read to Nelson a complicated formula, which Nelson copied down. Scientist X gave as his reason for asking Nelson to copy it down that the formula was in the handwriting of some other person, and that he, Scientist X, had to return the formula to the University of California radiation laboratories in the morning.

HUAC's charge displayed a remarkable fetish for the written word—a medieval worship of the manuscript page and the authorial power of an individual's penmanship. Atomic bombs were not children of the Industrial Revolution after all, cried HUAC; bombs' "vital" innards were the stuff of manuscript culture, captured in symbols on a page.

The committee's sensational charges take on a still more bizarre cast when compared to the declassified transcript of the conversation, as captured by officers of the Military Intelligence Division. The speaker identified as "Steve" had actually asked his interlocutor for copies of an article that had "already been published." "I could certainly get reprints of it," the other person replied, although he added quickly that "the leaflet itself will give them [the Soviets] no knowledge" with which they could actually build anything. The conversation turned to whether or not the Soviet Union had "the means and raw materials" to construct atomic bombs, over and above any explicit information that might be conveyed to them. Near the end of the transcript, the officers noted that Steve's conversation partner "dictated and STEVE wrote down at this point approximately 150 to 200 words, largely indistinguishable, but believed to be from the conversation a basic formula of some type"; "the word 'spectograph' [*sic*] was mentioned." What was *not* mentioned anywhere in the transcript—nearly every page of which includes several interruptions because the speakers' words were "unintelligible," and thus "a few words were missed"—was anything about other people's handwriting or top-secret manuscripts that had to be returned to the laboratory. What HUAC later construed as "a complicated formula" was, in the first instance, only *assumed* to be "a basic formula" based on a conversation that had been largely "unintelligible." All the while, Steve's interlocutor consistently expressed doubts that textual materials (let alone formulas) would be of any use at all.[38]

Of course, nothing beyond HUAC's version was available to the public at the time, and HUAC's florid (and largely manufactured) account received immediate, feverish attention. *Time* magazine covered HUAC's report under the boldface subheading, "Hot formula."[39] "Atomic secrets" were back, cast now as explicit, textual information rather than tacit "know-how" or industrial capacity. HUAC milked the story relent-

37. The *New York Times* reprinted HUAC's entire 20,000-word report in "Text of report" (ref. 36), quotation on 23.
38. The transcript may be found in Record Group 77, Entry 8, Box 100, National Archives, Washington, D.C. My thanks to Shawn Mullet for bringing the transcript to my attention.
39. "The atomic spy hunt," *Time* (4 Oct 1948), 22-23.

lessly over the next year, sending out nearly identical press releases (dressed as "fresh news") and garnering at least eight front-page headlines in the *New York Times* for what remained essentially the same story. The "trial by newspaper" began to work: many soon proclaimed that bombs might be formulas after all.[40]

The secrets-as-formula formula thus emerged a full year before Truman announced the Soviets' detonation of their own atomic bomb, and seventeen months before the theorist Klaus Fuchs confessed to spying for the Soviets during the war. These later events solidified the "hot formula" message, but did not create it. One week after Truman's announcement about the Soviet "Joe I" bomb, HUAC identified the mysterious "Scientist X" as theoretical physicist Joseph Weinberg. Their voluminous, three-volume follow-up report on "Communist infiltration of the Radiation Laboratory" implicated Weinberg along with four other young theorists, all of whom had worked at the Manhattan Project site during the war, and all of whom were now accused of having been Communists. Politicians and the public grasped at this "explanation" for the Soviet bomb: theorists had stolen secret formulas during the war and shepherded them to the Soviets; only because of theorists' treachery did Stalin now possess nuclear weapons.[41] The association between theorists and atomic secrets grew stronger still following the revelation, in early February 1950, that theoretical physicist Klaus Fuchs, who had worked at Oak Ridge and Los Alamos during the war as a member of the British team, had confessed to spying for the Soviet Union. Fuchs's announcement set off a domino effect of accusations and confessions during spring and summer 1950, ultimately implicating Harry Gold, David Greenglass, Julius and Ethel Rosenberg, and Morton Sobell as members of an American "spy ring" working during the war to ferret "atomic secrets"—in the form of reports, formulas, blueprints, and graphs—to the Soviets.[42] The swirl of comment by politicians and journalists surrounding each of these events cemented the shift in assumptions about what constituted "atomic

40. On HUAC's repetition of their press releases about "Scientist X" and his clandestine formula, and the *New York Times*' front-page coverage, see Walter Gellhorn, *Security, loyalty, and science* (Ithaca, NY, 1950), 118-120. On HUAC's "trial by newspaper" (referring to their similar treatment of the theorist Condon), see J.T. Klapper and C.Y. Glock, "Trial by newspaper," *Scientific American, 180* (Feb 1949), 16; and Carr (ref. 34), chapt. 10.

41. "Named 'Scientist X,' he denies charge," *NYT*, 1 Oct 1949; "Soviet atom gains laid to U.S. laxity," *NYT*, 26 Sep 1949. The charge that scientists' espionage had single-handedly delivered Stalin the bomb received repeated tellings throughout the mid-1950s, subsiding only after the surprise launch of Sputnik in 1957—a scientific and technical achievement that the Soviets could not possibly have "stolen" from the Americans. Cf. "Vast aid to Soviet laid to atom spies," *NYT*, 9 Apr 1951; Edward Shils, "Conspiratorial hallucinations," *BAS, 10* (Feb 1954), 51-54; and Harry Schwartz, "The real threat of Moscow's missile," *NYT magazine* (15 Sep 1957), 11, 78-80. On the case against the five Rad Lab theorists, see also HUAC, *Hearings regarding Communist infiltration of Radiation Laboratory and atomic bomb project at the University of California, Berkeley, Calif.* (Washington, D.C., 1949), 3 vols; Carl Beck, *Contempt of Congress* (New Orleans, 1959), 65-70; Schrecker (ref. 1), 126-48; Olwell (ref. 2); and Mullet (ref. 2).

42. See esp. JCAE (ref. 32); Robert C. Williams, *Klaus Fuchs: Atom spy* (Cambridge, 1987); and Ronald Radosh and Joyce Milton, *The Rosenberg file*, 2nd edn. (New Haven, 1997 [1983]).

secrets" and what types of people were most likely to hold and transmit such secrets. The secrets now became solidly theoretical, and the secret-keepers seemed more than ever to be theoretical physicists.

During 1948–55, the formula linking textual secrets with theorist-keepers calcified into a robust cultural artifact. Five distinct but related candidates for "the atomic secret" were proposed during this later phase, all centered around textual information rather than tacit knowledge: (1) "complicated formulas," akin to HUAC's charge against "Scientist X"; (2) information about the nation's nuclear stockpile—the types and numbers of nuclear weapons being made; (3) the size and shape of the bomb; (4) the implosion mechanism; and (5) more general "principles" and "theories" of bomb design. All of these types of information could be written down on paper and smuggled out of the country; each "secret" candidate gained force from unfolding political events.

The charge by HUAC about "complicated formulas" soon received many echoes; two days after Truman's announcement about the Soviet atomic bomb, science journalist Waldemar Kaempffert declared, "No doubt Soviet agents in this country pieced together scraps of information" from American physicists, from which the bomb had been built. A year and a half later, the Joint Committee on Atomic Energy alleged that David Greenglass—an Army machinist at Los Alamos during World War II and the government's chief witness against his sister and brother-in-law, Ethel and Julius Rosenberg—had "met with a Russian official in New York and was asked to supply a mathematical formula concerning high-explosive lenses used in the Nagasaki-type weapon." Greenglass was not the only person charged with peddling secret formulas; J. Edgar Hoover, imperious director of the FBI, explained in *Reader's digest* in May 1951 how the "crime of the century" had unfolded: the chemist-turned-courier Harry Gold "received formulas and various other technical data about atomic research" from Klaus Fuchs, and then passed them on to Russian agents. Indeed, Hoover charged, between Gold and Fuchs, "the basic secrets of nuclear fission had been stolen"—as if there remained any "basic secrets of nuclear fission" to be stolen from wartime Los Alamos! (It was precisely the lack of such "secrets" that had allowed Smyth to focus so squarely on basic nuclear physics in his report.) In October 1952 a new film entitled *The thief* depicted a nuclear physicist-turned-traitor who surreptitiously photographed his colleagues' "top secret scientific papers" to pass them along to foreign agents.[43] Even the medium of film began to profess the power of texts.

43. Waldemar Kaempffert, "Soviet bomb speeds up battle of scientists," *NYT*, 25 Sep 1949; JCAE (ref. 32), 7; J. Edgar Hoover, "The crime of the century: The case of the A-bomb spies," *Reader's digest, 58* (May 1951), 149-168, on 155, 158; "Spy melodrama at the Roxy," *NYT*, 16 Oct 1952, 1. The idea of "hot formulas" spread still further. For example, J. Edgar Hoover alleged in a follow-up article that Russian "undercover agents...stole formulas for petroleum products, lacquers, textiles, industrial chemicals": Hoover, "Red spy masters in America," *Reader's digest, 61* (Aug 1952), 83-87, on 83. Similarly, a chemist was accused of being part of the Canadian spy ring of 1946; his alleged crime (of which he was exonerated) was passing along the "secret" formula for a new chemical explosive, RDX—even though the chemical formula for the mate-

Beyond "hot formulas," other kinds of textual information came under scrutiny. On the day that President Truman delivered his controversial decision for "crash course" development of a hydrogen bomb, Senator Brien McMahon delivered a speech in Detroit urging that the most important atomic secrets concerned the nation's nuclear stockpile. How could democracy function, McMahon asked, if neither the Joint Committee on Atomic Energy (of which he was chair) nor the public at large had access to basic information about "how many bombs and atomic weapons we possess and how fast we are producing them"? Several years later, the editors of *U.S. news & world report* concurred with McMahon that the real "secrets" had to do with stockpiles and strike capabilities—even if they disagreed with McMahon on whether or not such "secrets" should be released.[44] Others claimed that the real secrets concerned the outward appearance and basic dimensions of atomic bombs, likewise textual information that could easily fall into the wrong hands. Soon after being elevated to chair of the Atomic Energy Commission, for example, Gordon Dean told the editors of *U.S. news & world report* that the real secrets of the bomb concerned "the size, the weight, the shape and general characteristics of the gadget," while journalist Alan Moorehead explained part of Klaus Fuchs's treachery: "He gave the size of the bomb—a vital point....He gave his own calculations of the actual dimensions of the parts." Two of McMahon's fellow senators on the joint committee argued more generally that "one of the strong points about the atomic weapon is the fact that it creates mystery."[45] To Senators Tydings and Connally, "mystery"—engendered by textual "secrets" regarding stockpile and bomb-size—was deemed *politically* vital to the entire atomic weapons program, all replication and proliferation issues aside.

During and after the Rosenbergs' trial in spring 1951, most discussion of "atomic secrets" turned on information supposedly gleaned by the machinist Greenglass concerning the implosion mechanism for detonating plutonium bombs. At the start of the trial, *Life* magazine described the nature of Greenglass's thievery: he had stolen

---

rial had been published in the open chemical literature in 1904. The same charge was later made against the American chemist Alfred Dean Slack; even though he likewise protested that the chemical formula for RDX had been published for decades, he was sentenced to ten years in prison, and all of his requests for appeal were denied. See Reuben (ref. 22), 53, 300-312. Thus even in indictments of chemists during this period, their main crime was seen as *textual*, trading in illicit formulas.

44. Brien McMahon, "Atomic energy policy: Paradoxes which secrecy creates," *VSD, 15* (15 Feb 1949), 263-265, on 263 (also published as McMahon, "Should we reveal the size of our atomic stockpile?," *BAS, 5* [Mar 1949], 66-68); "The strange case of Dr. Oppenheimer," *US news & world report* (23 Apr 1954), 20-22, 24, on 24.

45. "Can atom secrets be kept?," *U.S. news & world report* (24 Nov 1950), 36-40, on 36; Alan Moorehead, "Traitor Klaus Fuchs: He gave Stalin the A-bomb, Part Two," *Saturday evening post* (31 May 1952), 32-33, 70-74, on 72; Senators Tydings and Connally quoted in "Secrecy debated in Joint Congressional Committee," *BAS, 5* (Mar 1949), 95. See also Waldemar Kaempffert, "Science in review," *NYT*, 6 Mar 1955.

"this nation's most closely guarded and worst kept secret: the mechanism of the atomic bomb." "Having wangled information out of loquacious scientists," *Life*'s coverage continued, Greenglass "was able to show Julius [Rosenberg] a rough design of the ultrasecret detonating lens producing the implosion that triggers the bomb into action." Fuchs, too, was accused of giving away "a description of an implosion lens" as part of the sheaf of papers he passed off to Harry Gold. Ironically, the Atomic Energy Commission had allowed David Greenglass to testify about the implosion lenses (even showing diagrams) during the Rosenbergs' trial—to a partially open court whose gallery had been cleared of all listeners *except* journalists!—precisely because it had deemed this information safe enough for release. Gordon Dean, AEC chief, conferring nearly every day with members of the prosecution team, had actually approved releasing this information as a kind of *de facto* declassification.[46] So much for "this nation's most closely guarded and worst kept secret."

More generally, in the wake of these allegations and court proceedings, journalists and politicians spoke of general "principles" and "theories" of bomb design that had been "stolen," all by means of textual transfer. The Joint Committee on Atomic Energy, for example, explained in its April 1951 report on *Soviet atomic espionage* that "it is little appreciated that Fuchs" was "the great betrayer of the theory" underlying uranium isotope separation at Oak Ridge. Based on "Fuchs's grasp of the theoretical principles involved," they reported, "he would be able to reconstitute our whole program [at Oak Ridge] from only scattered pieces of information"—rebuilding an entire city of industrial production from a few scraps of paper. (In contrast, the Chrysler Corporation in 1947 had tried to popularize how critical various large-scale industrial processes had been to the operation at Oak Ridge, such as certain metal-plating techniques to keep equipment from corroding in the presence of the highly noxious uranium hexaflouride gas—all techniques that built on Chrysler's in-house stock of tacit knowledge and large-scale capacity.[47]) Alan Moorehead, in his four-part series on Fuchs that ran in the *Saturday evening post* in late spring 1952, closed his report by arguing that Fuchs "thought the worst thing he had done was to give information

46. "Spy's version of the A-bomb," *Life* (26 Mar 1951), 51-52, on 51; "The case of the world's greatest secret," *Life* (16 Apr 1951), 53-56, on 53. On Gordon Dean's behind-the-scenes work with the prosecutors, see Roger Anders, "The Rosenberg case revisited: The Greenglass testimony and the protection of atomic secrets," *American historical review, 83* (Apr 1978), 388-400; and Roger Anders, ed., *Forging the atomic shield: Excerpts from the office diary of Gordon E. Dean* (Chapel Hill, NC, 1987), chapt. 3. Dean seems to have assumed that the testimony about implosion lenses was safe to release because the Soviets' "Joe I" bomb had used a similar design, and hence was no "secret" to them. But even if the Soviets knew, presumably the AEC would not have wanted such "secrets"—if they really were significant and dangerous—to fall into other countries' hands. The implication thus remains that such text-based information really was not as dangerous as the prosecution and Judge Irving Kaufman alleged during the Rosenberg-Sobell trial and sentencing.
47. Wesley Stout, *Secret* (Detroit, 1947).

about the principle of the design of the plutonium bomb."[48] Now when physicists protested that such claims relied on a category mistake, confusing equations with the tacit knowledge and industrial infrastructure required to design and produce nuclear weapons, their tone had become defensive. No longer educating congressional leaders, they were merely offering "heretical afterthoughts."[49]

Two sets of HUAC hearings frame most starkly the shift in discussions about "the atomic secret" from raw materials and industrial capacity to text-based information. In June and July 1948—a few months before releasing their "Scientist X" and "hot formula" report—the committee heard testimony about shipments of uranium metal that had gone to the Soviet Union under wartime lend-lease arrangements. A year and a half later, deep into the second phase of "secrets" discussions, HUAC held new hearings, this time on the allegation that suitcases of Manhattan Project *documents* had been shipped to the Soviets under lend-lease.[50] The "atomic secret" had come a long way since 1945. The second part of the Cold War formula—a notion of atomic secrets as "principles," "theories," and "hot formulas" that could be scribbled down on scraps of paper and passed off to foreign agents—thus emerged late in 1948 and solidified during the early 1950s.

Egghead Revolutionaries

The final portion of the Cold War equation—casting theorists as inherently left-leaning, indeed as especially susceptible to Communist influence—likewise took shape during 1948–49, rather than immediately after the bombings of August 1945. Only in this later period did many come to assume that theorists as a whole were different from other groups of scientists (let alone academics or federal employees). Generalizations about theorists eventually came to be found among working scientists, journalists, politicians, and even federal judges. Some began to argue that theorists' special mental states or patterns of thought marked them as different (indeed, some claimed, as dangerous). Others pointed to theorists' unbalanced educations, their ethnic backgrounds, and even their physiognomies as signs of their untrustworthiness. Theorists became the new "eggheads"—that term used increasingly to disparage intellectuals—and many feared they needed to be watched.[51] As discussion of "atomic secrets" centered

48. JCAE (ref. 32), 1, 22; Alan Moorehead, "Traitor Klaus Fuchs: He gave Stalin the A-bomb, Part Four," *Saturday evening post* (14 June 1952), 34, 170-174, on 170.
49. See, e.g., Eugene Rabinowitch, "Atomic spy trials: Heretical afterthoughts," *BAS, 7* (May 1951), 139-142, 157; "Atomic secrecy deemed overdone," *NYT*, 17 Mar 1954 (quoting James Beckerly, director of the AEC Classification Office); Ralph Lapp, "We meet the Soviet scientists: Many misconceptions about Soviet science," *VSD, 22* (15 Dec 1955), 156-160, esp. 157.
50. HUAC, *Investigation of uranium purchases by the Soviet Union* (Washington, D.C., 1948); HUAC, *Hearings regarding shipment of atomic material to the Soviet Union during World War II* (Washington, D.C., 1949-50). See also JCAE (ref. 32), 184-92; and Carr (ref. 34), 186-193.
51. On the rise of "egghead" as a term of derision against intellectuals, see Arthur Schlesinger

more and more narrowly around theoretical formulas and information, the purported keepers of such secrets were seen in a more ominous light.

Theorists were not singled out at first. In the eighteen months after Truman enacted his immense loyalty-security program in spring 1947, for example, Congress appropriated nearly $18 million to pay the FBI to screen *all* federal employees and applicants for federal jobs, from the Agriculture Department through the State Department. Nearly two and a half million people were examined by the end of 1948, triggering fewer than eight thousand full field investigations, at the end of which only two hundred or so "disloyal employees" had been found (although no evidence of espionage surfaced). By the end of 1951, the number of federal employees and applicants screened for loyalty had risen to more than four million (nearly 17,000 of whom had been subjected to full FBI investigations), yielding a grand total of 343 individuals who were denied employment as a result. The early loyalty net had been thrown wide.[52]

The employees and applicants to the newly established Atomic Energy Commission—and all individuals working with the fast-growing list of AEC contractors and subcontractors—fell under the new loyalty-security surveillance. Most required formal security clearance (whether or not they worked with classified data) in addition to loyalty screening; by one estimate, the AEC investigated four hundred thousand individuals between 1947 and 1952. The average cost of these FBI investigations ran between one hundred and two hundred dollars each, and in the early days averaged two full months per screening. During 1947 and 1948, the AEC commissioners devoted fully one-third of their formal meeting time to personnel security matters; soon the number of AEC employees working full time on secrecy and security rivaled the entire population of Los Alamos.[53] From the start, all kinds of scientists, engineers, and technicians were affected. During 1947 and 1948, the Federation of American Scientists reported early cases of security-clearance problems among chemists, biologists, physiologists, electrical engineers, and both experimental and theoretical physicists; no specialist group seemed to have been singled out for special scrutiny. The following year, the *New York Times* reported that somewhere between twenty thousand and fifty thousand scientists, engineers, and technicians were still waiting for their security clearances, working at reduced capacity, if at all, in the meantime.[54]

Jr., "The highbrow in American politics," *Partisan review, 20* (1953), 156-165; and Richard Hofstadter, *Anti-intellectualism in American life* (New York, 1964), 4.

52. Robert Cushman, "Freedom versus security," *BAS, 5* (Mar 1949), 69-72, on 71; Leonard White, "The loyalty program of the United States government," *BAS, 7* (Dec 1951), 363-366, 382, on 366. On Truman's loyalty-security program more generally, see also Gellhorn (ref. 40), chapts. 4-7; Schrecker (ref. 1), chapt. 4; Wang, *American science* (ref. 1), chapt. 3; and Ellen Schrecker, *Many are the crimes: McCarthyism in America* (New York, 1998), chapt. 6.

53. Wang, "Science, security" (ref. 1), 257; "Can atom secrets be kept?," *U.S. news & world report* (24 Nov 1950), 36-40, on 38; "Joint Committee's reports on AEC investigations," *BAS, 5* (Dec 1949), 330-336, on 335. See also Peter Westwick, "Secret science: A classified community in the national laboratories," *Minerva, 38* (2000), 363-391.

54. Byron Miller and Harrison Brown, "Loyalty procedures of the A.E.C.: A report and recom-

An important shift occurred between April 1948 and early 1949 in the AEC's official regulations, however; the change in rules unfolded just as HUAC made repeated headlines with its allegations of atomic espionage against various theorists. Now the AEC made an explicit distinction between two types of case. "Category A" referred to specific disloyal acts committed or strongly suspected of an individual; "category B" referred to the more nebulous region of *thought* rather than action. "Sympathetic interest" in various "political ideologies"—rather than party membership or actual activities—now qualified as sufficient grounds for denying clearance and triggering dismissal.[55] Once the criterion for dismissal became curiosity and thinking in certain ways rather than proclivities for action, the balance of suspicion—at least in congressional committees and the popular press—fell more narrowly upon that group most easily cast as professional thinkers: theoretical physicists. HUAC's "trial by newspaper" had pressured the Atomic Energy Commission to change its rules; the new regulations reflected broader changes in assumptions about what and who constituted the ultimate security risks.[56]

Sometimes scientists themselves began to assume that theorists as a whole shared more leftist political affiliations than any comparable group of scientists or academics. The University of California enacted a new "loyalty oath" in spring 1949, for example, and over the next two years Berkeley's physics department lost *all* of its theorists, either through resignation or firing for refusal to sign the anti-Communist oath. Although the department also lost two of its most esteemed and successful experimentalists, department chair Raymond Birge linked specific political proclivities only to the category of theorists. With the disagreeable oath in place, lamented Birge, the department would never be able to "induce a single first-class theoretical physicist in this country to accept a position at Berkeley"; he displayed no such concerns about finding "first-class" replacements for his lost experimentalists.[57] The former executive secretary of the Federation of American Scientists presented a similar taxonomy of scientific special-

---

mendations," *BAS, 4* (Feb 1948), 45-48; FAS Committee on Secrecy and Clearance, "Loyalty clearance procedures in research laboratories," *BAS, 4* (Apr 1948), 111-114; FAS Committee on Secrecy and Clearance, "Some individual cases of clearance procedures," *BAS, 4* (Sep 1948), 281-285; see also the special issue of *BAS, 4* (Jul 1948); *New York Times* report as quoted in Gellhorn (ref. 40), 193-194. See also Wang, *American science* (ref. 1), chapts. 2-3, 5.

55. "AEC criteria for security clearance," *BAS, 5* (Feb 1949), 62-63 (quotes on 63), which reprinted the newly released AEC "official policy statement." This superseded the AEC's "Interim procedure" of April 1948.

56. On HUAC's pressure, see Wang, *American science* (ref. 1), chapt. 5. The AEC's relationship with the JCAE was also fraying at this point, in part because the JCAE sought to fend off HUAC's incursions into its atomic-energy territory: Hall (ref. 25), 403-404.

57. Raymond Birge to Baldwin M. Woods, Chairman, Committee on Academic Freedom (Berkeley), 12 Sep 1950, Raymond Thayer Birge Correspondence and Papers, call number 73/79c, TBL. On the loyalty oath at the University of California, see David Gardner, *The California oath controversy* (Berkeley, 1967); and Schrecker (ref. 1), 126-31; on its effects on Berkeley's physics department, see Kaiser (ref. 1), esp. 245-247.

ties and political dispositions in June 1951. Unlike chemists, biologists, or engineers, explained Richard Meier, "the physicist by nature is politically radical":[58]

> His mind is schooled in the proposition that progress is made by discarding various assumptions and premises and thereby making it possible to create a more powerful theory upon a simpler underpinning. The physicist, more than any scientist, deals with abstractions which make nonsense out of observations based on the commonplace; he is educated in doubt and can disregard evidence which to the ordinary observer is both convincing and conclusive. Thus many physicists chose a vague leftist political philosophy, partly as the only relatively rational set of value premises which was offered at the time ('36 to '40) in the world of ideas.

Though he attributed these political leanings to physicists as a whole, he did so on the basis of their *theoretical* methods, emphasizing "abstractions" and styles of thinking rather than, say, skills in manipulating apparatus or designing experiments.

More and more the associations between scientists and political leanings became a topic of general fascination and comment. Psychologist Anne Roe and science journalist Waldemar Kaempffert offered contrasting views to readers of the *New York Times magazine* as to whether or not a unique "scientific mind" existed, and whether or not that "mind" had special affinities for particular political philosophies. Their debate played out between 1949 and 1954. Although Kaempffert denied that any special "scientific mind" existed (whereas Roe argued for it), he did argue that scientists' mode of thought encouraged them to "turn a Marxian proposal this way and that, so view it in all its aspects, and sometimes accept it. The communistic argument appeals to younger scientists." Five years later, Kaempffert elaborated on this thesis, narrowing its application to "thinkers" in biology or physics—theorists—who sometimes became enamoured of Communist arguments. These theorists ultimately failed to realize, asserted Kaempffert, that "it is much easier to establish the physical and chemical constitution of a star than it is to invent a social system that will bring happiness to mankind, abolish poverty and end war."[59] To such commentators, scientific "thinkers" were more inclined toward leftist political philosophies, and Communism in particular, because they mistakenly tried to tackle social and political problems the same way they tackled theoretical physics.

The assumption that theorists as a whole were the most susceptible of all scientists to Communist influence received repeated tellings, often in less benign reports than Kaempffert's and Roe's musings. Self-proclaimed experts held forth on a new and dangerous threat to the nation: "ideological espionage." According to a lengthy *New York Times magazine* article in May 1949, it was the number of "'idealists' or ideo-

58. Richard Meier, "The origins of the scientific species," *BAS, 7* (June 1951), 169-173, on 170.
59. Waldemar Kaempffert, "The atom and the scientific mind," *NYT magazine* (9 Oct 1949), 15, 28-35, on 30-32; Waldemar Kaempffert, "X-ray of the scientific mind," *NYT magazine* (25 Apr 1954), 7, 54-66, on 55; cf. Anne Roe, "What makes the scientific mind scientific," *NYT magazine* (1 Feb 1953), 10, 22.

logical traitors, the educated, sophisticated thinkers" who were inclined to act against the country on principle rather than for money or fame that increased so alarmingly in the postwar period—not the number of mercenaries or other opportunistic traitors. The latest psychoanalytic theories, the article explained, held that intellectuals committed espionage in part because they unconsciously likened the nation to their parents (who else?), against whom they sought revenge. One psychiatrist went so far as to write down a special equation for the new phenomenon:[60]

> C = T + S - R.
> Crime equals the criminal Tendencies ("which we all have to some degree") plus the given Situation—but minus the mental Resistance which normal people develop over the years.

Be on guard, the experts now warned: mentally unbalanced intellectuals were poised to wreak havoc.

Several people began to apply the notion of "ideological espionage," fashioned for critiques of intellectuals in general, to theoretical physicists in particular. Journalist Rebecca West argued that Klaus Fuchs's "mental processes"—including what Fuchs himself called his "controlled schizophrenia," compartmentalizing his thinking so he could continue his espionage even as he worked closely with American and British colleagues—pointed far beyond Fuchs himself. Fuchs stood in, West argued, for a new ideal type, "the traitor scientist." This new breed tried—with disastrous results—to apply their "rare and exalted gifts" at mathematical and theoretical abstraction to political thinking. Although the results of their political philosophizing might read "like the ramblings of an exceptionally silly boy of 16," the theorists' new relation to atomic secrets, and their peculiar thought processes, made them the most serious threat of the nuclear age. Fuchs's own attorney reasoned along virtually identical lines when arguing for leniency in sentencing, explaining that (all) theorists' overly rationalized and insufficiently "flexible" minds often arrived at poor political judgments.[61] Manhattan Project veteran and *Bulletin of the atomic scientists* editor Eugene Rabinowitch complained in exasperation at the new tendency to generalize from the single case of Fuchs to all theorists everywhere: why hadn't journalists like West made similar leaps to the in-built mental processes and political proclivities of all lawyers in the wake of the Alger Hiss case? Yet West's purple prose soon found many imitators, including the Joint Committee on Atomic Energy. Its April 1951 report on *Soviet atomic espionage*

60. Gertrude Samuels, "American traitors: A study in motives," *NYT magazine* (22 May 1949), 17, 63-66, on 17, 66. On "ideological espionage," see also Schrecker, *Many are the crimes* (ref. 52), chapt. 5; and Mullet (ref. 2).

61. Rebecca West, "The terrifying import of the Fuchs case," *NYT magazine* (4 Mar 1951), 6, 29-34, on 31-32; the appeal from Fuchs's attorney, Derek Curtis-Bennett, is quoted in Moorehead (ref. 48), 173. The JCAE quoted extensively from West's article in their April 1951 report: JCAE (ref. 32), 17-19, 21, 29-30, and reprinted the article in full on 200-204.

argued in precisely the same way that the "warped mentalities" of theorist-spies created "an almost diseased yearning to remold the world after the image of their own work in physical science." Theorists were disproportionately driven to Communism, this official government report concluded, because they tried to force sociopolitical ideas into the hyperrationalist straitjacket of their beloved nuclear theory.[62]

More than just mental states and thought processes were at issue. Part of the problem, West and the Joint Committee concluded, derived from theorists' unbalanced education: too much theoretical physics and mathematics had come at the expense of the liberal arts.[63] As late as December 1956 an influential federal judge took up the same issue. Judge Alexander Holtzoff of the United States District Court for Washington, D.C., reinstated a physics graduate student's contempt of Congress conviction for failing to name names during a HUAC investigation back in 1954. Although the physics student's conviction had been overturned on a technicality in November 1955, the Federal Court of Appeals reinstated the charge in July 1956, which is how Bernhard Deutch wound up in Judge Holtzoff's courtroom that December. Not content simply to sentence Deutch to a ninety-day prison term (which he did), Holtzoff seized the opportunity to deliver a broader sermon about young physicists and Communism. "The younger generation of pure scientists" had become a "fertile field for Communist propaganda," he declared; the *New York Times* carried his decision under the headline, "Pure scientists called red prey." Holtzoff made it clear that he did not mean just any scientists; he "was not referring to scientists in engineering, chemistry, and similar fields." Rather, he explained that "From evidence admitted in other cases that have come before the court"—Holtzoff had presided over the cases of all five of the theorists whom HUAC had accused of being Communists and of "infiltrating" Berkeley's Radiation Laboratory—"the court has gleaned the inference that the younger generation of pure scientists specifically engaged in research in physics has succumbed to Communistic propaganda." Like Rebecca West and the Joint Committee, Holtzoff attributed the problem to recent educational changes, which deprived young physics students of "a proper cultural background"; the new generation displayed "an abysmal ignorance" of such fields as history and economics.[64]

62. Eugene Rabinowitch, "Editorial: Scientists and loyalty," *BAS, 7* (Dec 1951), 354-355; JCAE (ref. 32), 12. David Katcher, former editor of *Physics today*, similarly wrote to the *New York Times* to complain about West's hasty generalization from Fuchs to all physicists: Katcher, "Science to blame?," letter to the editor, *NYT*, 18 Mar 1951. The JCAE's report in fact showed little coherence on the question of theorists' inherent loyalty, arguing two pages earlier that "The fact that three of the four spies are physicists does not indicate that men of science, as a group, are more vulnerable to communism than other groups. It suggests, instead, that physicists alone fully understood many of the secrets which counted most and that physicists therefore attracted the concentrated attention of Soviet intelligence directors." JCAE (ref. 32), 10. Such government reports might have displayed this lack of coherence because they were written by several different staffers.

63. West (ref. 61), 6, 31; JCAE (ref. 32), 12.

64. "Pure scientists called red prey," *NYT*, 14 Dec 1956; "Judge criticized on physicist view,"

Judge Holtzoff's accusation that "pure physicists" lacked a "proper cultural background" signaled still another layer in the unfolding image of the theorist. Nine of the thirteen theorists who fell afoul of protracted anti-Communist scrutiny were of Jewish background. Indeed, it had become common even among physicists to note the high proportion of Jews working in theoretical physics (rather than in other branches of the discipline); some physicists in the early 1950s still made note of job candidates' "Jewish features." Overt examples of anti-Semitism riddle the secret FBI and Military Intelligence Division reports on physicists during World War II.[65] Several conservative members of Congress repeatedly emphasized during 1945–46 the foreign- and Jewish-sounding names of theoretical physicists who had worked on the Manhattan Project.[66] After the war, however, as the magnitude of the Holocaust became better known, blatant and explicit anti-Semitism became less common, at least in popular discourse. Bigotry hardly disappeared; rather, it often reemerged in the guise of virulent anti-Communism. Many sociologists, historians, and social critics—people like Arthur Schlesinger Jr., Edward Shils, David Riesman, and Richard Hofstadter—observed during the early years of the Cold War that much anti-Communist sentiment

---

NYT, 17 Dec 1956. The second article included more quotations from Holtzoff's original decision, along with news of a formal letter of complaint from the president of the FAS to Holtzoff for his remarks. Soon physics graduate students at Columbia and Yale protested the judge's unfair characterization as well: "Judge is challenged: Students reply to charge of 'fertile ground' for reds," *NYT*, 18 Dec 1956; and Robert Gluckstern, McAllister Hull, Jr., Joseph Shapiro, and James Turner, "Loyalty of physicists," letter to the editor, *NYT*, 29 Dec 1956. On the background of Deutch's case, see "4 held for trial in House inquiry," *NYT*, 4 Dec 1954; [No title], *NYT*, 5 Nov 1955, on 12; "Court reinstates contempt charge," *NYT*, 27 Jul 1956. Deutch appealed Holtzoff's conviction, and his case eventually made it to the U.S. Supreme Court: "United States Supreme Court," *NYT*, 23 Mar 1961; although unfortunately no later article seems to report the outcome of Deutch's appeal. Meanwhile, Holtzoff continued a controversial career, being formally reprimanded by the Federal Court of Appeals for "presiding over criminal trials as if he were prosecuting the case"; Holtzoff also had more of his decisions reversed by the U.S. Supreme Court than any other District Judge during the 25 years that he was on the bench. See "Judge Holtzoff, 81, will retire Dec. 30," *NYT*, 25 Nov 1967; and "Judge Alexander Holtzoff, 82, of Washington District Court," *NYT*, 7 Sep 1969.

65. See, e.g., Raymond Birge to William Roth, 31 May 1955, Folder 5:143; Francis Jenkins's notes on job candidates for the Berkeley physics department, winter 1950, Folder 5:125, Department of Physics records, University of California, Berkeley, call number CU-68, TBL; and David Kaiser, "The postwar suburbanization of American physics," *American quarterly, 56* (Dec 2004), 851-888, on 863-865. On anti-Semitism within wartime intelligence files, see Rebecca Press Schwartz, "The military-industrial-scientific complex meets the international Jewish conspiracy: Antisemitism and anticommunism on the Manhattan Project," paper presented at the 2002 annual meeting of the History of Science Society. Of course, many commentators at the time and since have blamed anti-Semitism for the severity of the Rosenbergs' death sentence; see, e.g., Marjorie Garber and Rebecca Walkowitz, eds., *Secret agents: The Rosenberg case, McCarthyism, and fifties America* (New York, 1995).

66. Hall (ref. 25), 183-189, 193, 199-200.

drew its force from sublimated anti-Semitism. Older claims about Jews' supposed "cosmopolitan" leanings (not to mention "internationalist conspiracies") fit the new stereotype of the urban Communist equally well.[67] Indeed, as late as 1957 the retired intelligence chief for General Douglas MacArthur could denounce Soviet espionage undertaken by "East European immigrants, an ethnic group that has furnished the largest percentage of agitators and Communists who can combine the benefits of American citizenship with the betrayal of their refuge and their neighbors."[68] Major General Charles Willoughby's listeners at the 27th Annual Convention of the American Coalition of Patriotic Societies would have had little difficulty parsing just who this "ethnic group" of "East European immigrants" was. Theorists were Jews; Jews were Communists—still one more reason to treat this group as a special threat.

Beyond the basic bigotry often associated with McCarthyism, the theorists' case was tied even more broadly to longer-standing American anxieties over intellectuals. At times the charges and counter-charges of anti-intellectualism were made explicitly.[69] More often, however, anti-intellectualism surfaced in more subtle ways. Even physiognomy could invite suspicion: during this period, intellectuals were routinely cast as "effeminate" in appearance, their overdeveloped brains paired with emasculated bodies.[70] The trope worked to the detriment of theoretical physicists. In story after story during the early 1950s, theorists appeared as tall, thin (even lanky), with delicate features and retiring manners, whereas all other scientists, engineers, and technicians were described as short, overweight, even "burly," but certainly not effeminate. Reporting Julius Rosenberg's arrest in July 1950, for example, *Time* magazine called the engineer "a puffy, spectacled native New Yorker"; his wife Ethel, explained the *New York Times*, was a "plump, plain-faced girl." Rosenberg's alleged coconspirator, the electrical engineer Morton Sobell, was "short" and "chunky" with "unruly hair," reported *Newsweek* upon Sobell's arrest. During the Rosenberg-Sobell trial, the *New York Times* described Sobell as "swarthy-faced, with a receding shock of heavy black hair," while the machinist-turned-government witness, David Greenglass, was "a burly young man," reported *Life* magazine. The chemist-courier Harry Gold was "a little man, insignificant, average-looking," with "broad build [and] round face"; in short, he was "stocky," explained FBI director J. Edgar Hoover in *Reader's digest*. The *New York Times* agreed, describing Gold as "pudgy-faced." The chemist Alfred Dean Slack,

67. Schlesinger (ref. 51), 164; Shils (ref. 41), 53; David Riesman and Nathan Glazer, "The intellectuals and the discontented classes," *Partisan review, 22* (1955), 47-72, on 57-58; Hofstadter (ref. 51), 4, 42. See also David Hollinger, *Science, Jews, and secular culture: Studies in mid-twentieth-century American intellectual history* (Princeton, 1999).

68. Maj. Gen. Charles Willoughby, "Soviet espionage: The American Communist Party," *VSD, 23* (15 Mar 1957), 344-347, on 346.

69. See, e.g., the letter of protest against Judge Holtzoff's "red prey" comments signed by eighty-nine graduate students at Columbia, as quoted in "Judge is challenged" (ref. 64).

70. See, e.g., Louis Bromfield as quoted in Schlesinger (ref. 51), 159; also quoted in Hofstadter (ref. 51), 9; on intellectuals as "effeminate," see also ibid., 19.

convicted on charges of wartime espionage for ferreting out information about new chemical explosives via Gold to the Soviets, was likewise "of medium height [and] a little too heavy," reported *U.S. news & world report*.[71] Chemists, engineers, technicians: they might be disloyal, but they were certainly not intellectuals; they simply didn't have the body for it.

How did the theorists compare? Klaus Fuchs was "of feeble physique," explained Rebecca West; Hoover agreed, describing Fuchs as "thin, sallow-complexioned, with stooping shoulders, balding head and weak brown eyes behind thick lenses." He was a "frail scientist" with "delicate fingers," Hoover continued. Indeed, physiognomy should have revealed ignominy all along, declared journalist Alan Moorehead in the *Saturday evening post*:[72]

> Even in 1946, before anyone knew about Fuchs' real character, it was possible to notice superficial similarities between him and [Alan Nunn] May. Both were serious and shy and self-effacing. Both had the same bulging forehead, the receding hair, the horn-rimmed glasses, the set mouth and the slightly weak chin.

How could the British mission or Groves's security forces have missed Fuchs, since he had all the facial features of a devious, calculating spy? Nor was this body description limited to Fuchs. Other nuclear physicists at the time were described in the press as "lanky" and "scholarly-looking," or tall and "loose-jointed."[73] None seemed to have struck journalists as "burly," "stocky," "plump," "chunky," or "swarthy-faced"; intellectuals never were. By the early 1950s, the final element in the Cold War formula was thus complete: theorists were special, different from other types of scientists, engineers, or academics. They were especially prone to Communism, the worst type of intellectual in an anti-intellectual age.

71. "Espionage: No. 4," *Time* (31 Jul 1950), 12-13, on 12; Joseph Kraft, "Rosenberg spy case," *NYT*, 11 Jan 1953; "Atom arrest No. 8," *Newsweek* (28 Aug 1950), 30; "Spy defendants classmates at C.C.N.Y.," *NYT*, 30 Mar 1951; "The case of the world's greatest secret" (ref. 46), 53; Hoover, "The crime of the century" (ref. 43), 158-160; A.H. Raskin, "Story of the Rosenbergs: Two links in atomic conspiracy," *NYT*, 21 June 1953; "Inside story of a native American who turned spy," *U.S. news & world report* (24 Nov 1950), as reprinted in JCAE (ref. 32), 214-216, on 214. On Slack's case, see also Reuben (ref. 22), chapt. 15.
72. West (ref. 61), 29; Hoover, "Crime of the century" (ref. 43), 150, 155, 161; Alan Moorehead, "Traitor Klaus Fuchs: He gave Stalin the A-bomb," *Saturday evening post* (24 May 1952), 22-23, 96-103, on 23.
73. "Smyth and Libby men in contrast," *NYT*, 16 Sep 1954; Sandra Littlewood and Skip Garretson, "Espionage in the Rad Lab—Naw!," *Daily Californian*, 11 Dec 1953. As late as 1959, a *New York Times* reporter seemed stunned to find a nuclear physicist and veteran of the Manhattan Project who did not fit the prevailing image, explaining that "there is little of the so-called typical professor in his physical appearance": "A man's man in science: John Harry Williams," *NYT*, 17 Jul 1959, 2.

Proof in the Pudding?

In his study *Anti-intellectualism in American life* (1964), Richard Hofstadter noted that intellectualism in this country has often been "pitted against practicality, since theory is held to be opposed to practice, and the 'purely' theoretical mind is so much disesteemed." Trouble brews especially during those periods when "ordinary folk"—including politicians—develop *ambivalence* about formal knowledge and those who possess it: ambivalence driven by a mixture of "respect and awe with suspicion and resentment."[74] After World War II theoretical physicists, as supposed keepers of the "atomic secret," seemed to have acquired awe-inspiring power, and they became all the more suspicious for it.

Yet it didn't have to be that way. No portion of the Cold War formula—theorists built the bomb; bombs were made from formulas; and theorists were Communist dupes—was beyond doubt. Some of these elements were demonstrably false, even if the demonstrations were hampered by postwar security gag-orders. Whatever the origins of this train of assumptions (likely some admixture of the Smyth Report and Oppenheimer's escalating troubles, fired by debates over atomic energy legislation and cemented by news of the Soviets' bomb and Klaus Fuchs's confession), these allegations and representations marked an important shift in broader assumptions about science and scientists. Before the war, theoretical physicists—when garnering media attention at all—had usually been dismissed as ultimately amusing at best, and certainly not worth troubling over. The *New York Times* declared in 1923, for example, on the topic of "Einsteinism" and the new physics, that one should "just ignore it...as of no concern to us"; a few years later the newspaper advised its readers to file relativity and quantum mechanics under "things you needn't worry about just yet." By the mid-1930s, casual dismissals of theorists in the national media had become common: they were likened to medieval theologians debating how many angels fit on the head of a pin; theorists' equations, readers of the *New York Times* were told, gave no answers to life's truly important questions.[75] Yet within a few years after the war, the stuff of pencil and paper began to trump depictions of American technical ingenuity; the Thomas Edison types, hands-on tinkerers who had self-consciously projected an antitheoretical image, no longer claimed center stage.[76] Struggling to make sense of these fast-moving events, journalists and politicians poured the new Cold War wine into older anti-intellectual bottles, grasping at ready-to-hand descriptions of "effeminate" and devious intellectuals and quickly pressing them into action. If some of the characterizations were stale,

74. Hofstadter (ref. 51), 46, 19, 21.
75. "Topics of the Times," *NYT*, 6 Feb 1923; Simeon Strunsky, "About books, *More or less*: Excessively up to date," *NYT*, 29 Apr 1928; "By-products: In the matter of Einstein, tea-kettles, destiny, &c.," *NYT*, 22 Mar 1931; "New forces for atoms," *NYT*, 7 Jul 1936; and "Improving on relativity," *NYT*, 15 Mar 1939.
76. On Edison's prewar persona, see Thomas Hughes, *American genesis: A century of invention and technological enthusiasm, 1870–1970* (New York, 1989).

however, their target was new. The scientists requiring closest scrutiny were not the builders but the thinkers. "Science" now meant formulas, not know-how.

Some might claim that in the light of what we know now, many of these Cold War precautions about theorists and atomic secrets were ultimately justified. After all, Alan Nunn May and Klaus Fuchs admitted to supplying the Soviets with written information about the atomic bomb project, and theorist Ted Hall all but admitted his own wartime espionage more recently. Even the assistant director of the Soviet nuclear weapons project, Yuli Khariton, admitted repeatedly in the mid-1990s that his group received clandestine atomic information from the American project, and that these stolen reports helped to determine the form of "Joe I."[77] A few cautions are in order. First, many of the claims about the effectiveness of wartime espionage for the Soviet nuclear project were made by retired KGB agents looking to share the honor for having made the Soviet Union's first bomb. Espionage certainly occurred, but even with the purloined texts the Soviet group still had to redo most experiments, adapting stolen blueprints to the availability of different types of materials, and in general improvising at least as much as following explicit, text-based procedures.[78] The stolen information could have helped the Soviet group avoid certain blind alleys, and thus save time and resources (although in fact the Soviet leadership decided, much as General Groves had done, to march forward along several competing directions at once—even though the Soviets knew which of these processes had worked best for the American project). The major limiting factor of the Soviet effort, it is now clear, was its dire shortage of uranium—raw materials trumped textual information for several years.[79] Moreover, even with designs, reports, and data—whether received surreptitiously by the Soviets or legitimately by the British—*every single country* that has succeeded in making its own nuclear weapon has taken *longer* than the wartime Manhattan Project. Each attempted replication has had to build up both the industrial capacity and the tacit knowledge for its use from scratch. Texts simply can't deliver these. To this day, bombs have never been reified equations.[80]

77. On Ted Hall, see Joseph Albright and Marcia Kunstel, *Bombshell: The secret story of America's unknown atomic spy conspiracy* (New York, 1997); on Yuli Khariton's discussions of atomic espionage, see Yuli Khariton and Yuri Smirnov, "The Khariton version," *BAS, 49* (May 1993), 20-31; and Yuli Khariton, Viktor Adamskii, and Yuri Smirnov, "The way it was," *BAS, 52* (Nov/Dec 1996), 53-59. On the use of espionage in the Soviet nuclear weapons project, see also David Holloway, *Stalin and the bomb* (New Haven, 1994), 90-97, 102-108, 296-297, 310-312; and Alexei Kojevnikov, *Stalin's great science: The times and adventures of Soviet physicists* (London, 2004), chapt. 6.

78. Roald Sagdeev, "Russian scientists save American secrets," *BAS, 49* (May 1993), 32-36; Sergei Leskov, "Dividing the glory of the fathers," *BAS, 49* (May 1993), 37-39; Holloway (ref. 77), 199; and Arkady Brish, "We copied the charge design, not the bomb itself," *Literaturnaya Gazeta, 36* (7 Sep 1994), 10, available in translation at http://www.fas.org/news/russia/1994.

79. Kojevnikov (ref. 77), chapt. 6.

80. See esp. MacKenzie and Spinardi (ref. 7); McNamara (ref. 7); and Gusterson (ref. 7).

Obviously, security measures are essential; even the postwar atomic scientists' movement argued that certain technical details, materials, and processes needed to be tightly controlled. What I find fascinating about the early Cold War period is the image of theory—and of theorists—that drove know-how and tacit knowledge from view, especially since these had long been staples of a triumphalist American self-image. Nor was the postwar valorization of theory (with attendant demonization of theorists) a temporary phenomenon. In the mid-1970s, the Princeton undergraduate physics major who drew up his own plutonium bomb design from publicly available sources was repeatedly asked where he was keeping his bomb and what he planned to do with it—the distinction between blueprint and arsenal consistently evaded people.[81]

More troubling is the recent case of Wen Ho Lee, the Chinese-American scientist at Los Alamos who was accused of nuclear espionage in the late 1990s. Lee worked with computer codes in part of the lab's X-division (an outgrowth of the wartime Theoretical Physics or T-division). The Departments of Energy and Justice, along with the FBI, alleged that Lee had committed the single gravest case of nuclear espionage since the Rosenbergs, likening the text-based codes that Lee handled to the "crown jewels" of bomb making—"the nation's most prized nuclear secrets," as a Justice Department panel wrote.[82] As Hugh Gusterson has remarked recently, government officials and the national media acted as if Lee had "emailed the U.S. nuclear arsenal to a foreign power," betraying the familiar slippage from text to artifact.[83] Several lab experts had to explain that bombs are no more made from million-line computer codes than they are from secret formulas. The material that Lee downloaded had not even been marked as "Top Secret Restricted Data"; in fact, it had never even been classified at all. Racial profiling, meanwhile, marred the case from the beginning, as counterintelligence officials consistently asserted that a Taiwanese-born American citizen was more likely than anyone else to aid Communist China. Lee spent close to one year in solitary confinement before fifty-eight of the fifty-nine felony charges against him were finally dismissed.[84]

Even today, as news of Pakistani nuclear scientist A.Q. Khan's illegal activities come to light, Cold War assumptions shine through. Khan has been accused of selling equipment, including specialized, high-precision centrifuge parts that can be used to

81. John Aristotle Phillips and David Michaelis, *Mushroom: The story of the A-bomb kid* (New York, 1978).
82. See esp. Dan Stober and Ian Hoffman, *A convenient spy: Wen Ho Lee and the politics of nuclear espionage* (New York, 2001). A redacted version of the classified May 2000 Justice Department review of the mishandling of the Lee case is available at http://www.usdoj.gov/ag/readingroom/bellows.htm, quotations from chapt. 1, pp. 6, 12.
83. Gusterson (ref. 7), 94.
84. Stober and Hoffman (ref. 82). The material that Lee downloaded had been marked "Protect as Restricted Data" (PARD), which was an informal measure devised to handle the large volumes of paper print-outs from the early days of computing at the national weapons labs; it referred to material that had never been reviewed for classification, and was quite distinct from even low-level classifications such as "Restricted data" (293-294).

enrich fissionable uranium, as well as having sent his own trained personnel to various places to set up and use the equipment. So it seems that Khan trafficked in matériel and the tacit knowledge to use it, precisely the most dangerous ingredients feeding nuclear proliferation. And yet at least some press reports have dismissed Khan as "merely" a metallurgist; after all, he's no theorist.[85] All these lesson-filled years later, we still live with the legacy of the early Cold War.

85. Bronwyn Curran, "A.Q. Khan, godfather of the 'Islamic bomb': Hero or proliferator?," Agence France Presse report (25 Jan 2004), available via LexisNexis; William Broad et al., "A tale of nuclear proliferation: How Pakistani built his network," *NYT*, 12 Feb 2004.

CHAPTER TEN

Something Resembling Justice
John Francis Neylan and the AEC Personnel Hearings at Berkeley, 1948-49

Stephanie Young

When a Personnel Security Board of the Atomic Energy Commission inquired into J. Robert Oppenheimer's fitness for a security clearance in 1954, it enacted a drama already played out in a still undisclosed number of similar hearings focusing on lesser scientists and technical personnel throughout the early Cold War. Recently available archival materials reveal the dynamics of these quasi-trials, enabling us to see as never before the contours of the terrain onto which Oppenheimer walked when he agreed to meet with the board appointed to interrogate him. These materials, which concern the hearings of three Berkeley scientists in 1948 and 1949, shortly after Oppenheimer departed for Princeton, shed light on the administrative culture in which scientists and the political authorities of that era dealt with their misgivings about each other.

Despite attempts by Atomic Energy Commission Chairman David Lilienthal to design a security screening process for Commission personnel consistent with the spirit of American judicial institutions, the Berkeley hearings are an example of the degree to which these proceedings were in fact shaped not by well-defined standards and procedures but by powerful personalities. The security program as outlined by the Commission left open wide avenues for senior scientists and administrators to exert substantial influence. Such opportunities to shape the outcome of hearings

I would like to thank David Hollinger and Gregg Herken for their enthusiastic support and guidance of this project, Cathy Carson for her thoughtful comments, and David Farrell of the Bancroft Library for helping me get access to Neylan's restricted papers.

The following abbreviations are used: AEC, Atomic Energy Commission; CM-EL, Herbert Childs, Materials assembled for a biography of Ernest Lawrence, CU-369, TBL; EOLP, Ernest Orlando Lawrence Papers, BANC MSS 84/129 c, TBL; FBI, Federal Bureau of Investigation; HUAC, United States Congress, House Committee on Un-American Activities; JCAE, United States Congress, Joint Committee on Atomic Energy; JFNP, John Francis Neylan Papers, BANC MSS C-B 881, TBL; JRO, J. Robert Oppenheimer; JROP, J. Robert Oppenheimer Papers, Manuscript Division, Library of Congress, Washington, D.C.; PSB, Personnel Security Board; TBL, The Bancroft Library, University of California, Berkeley.

were amplified by the seating of local citizens in judgment over regional AEC cases, increasing the likelihood of preexisting personal relationships between board members and parties interested in the outcome of cases. In Berkeley such a preexisting conflict of interest manifested itself in the person of John F. Neylan, chairman of the local security board and intimate friend to scientist and laboratory administrator Ernest O. Lawrence. If by virtue of this personal relationship Lawrence had special access to the Chairman's ear and confidence, Wendell Latimer, a second senior Berkeley scientist appearing on behalf of a subject of inquiry, enjoyed no such connection. The Berkeley personnel security example is thus deeply revealing both about AEC procedures allowing clearance recommendations to be shaped by personal influence and about the different manners in which senior scientists wielded such procedures in an attempt to protect younger protégés. These three cases, then, present a picture of the climate at Berkeley shortly after the most famous victim of the AEC personnel security apparatus, Robert Oppenheimer, moved east, but also, in the cast of characters and in the informal rules and structures of authority defining these proceedings, foreshadow Oppenheimer's own troubles to come.

Context and Institutions

While instruments of the federal apparatus conducted personnel security investigations of employees seeking access to classified atomic material during World War II, the scope and frequency of such proceedings in all fields of sensitive work picked up significantly after the war. During the war the United States Army's Manhattan Engineer District controlled all aspects of atomic energy, and personnel security questions fell under the jurisdiction of Military Intelligence. In accordance with wartime priorities, the Army granted top clearances as needed, and suspect employees on sensitive projects were booted without much discussion. The military needed manpower, and was thus willing to take some risks, but did not have the patience to give the benefit of the doubt to workers whose loyalty was questionable.

After the war Congress sought to develop more formal and permanent institutions consistent with peacetime for dealing with personnel security decisions. In July 1946 President Truman signed into law Senator Brien McMahon's bill granting civilian control over atomic energy to the Atomic Energy Commission, effective January 1, 1947.[1] On the question of personnel security, the McMahon bill made it the AEC's responsibility to evaluate Federal Bureau of Investigation reports into the "character, associations, and loyalty" of all new AEC applicants.[2] Through the criteria of "character, associations, and loyalty," the Atomic Energy Act charged the Commission with devel-

1. For an excellent discussion of Truman's role in the dialogue surrounding the establishment of the AEC in the context of postwar antistatist culture see Michael J. Hogan, *A cross of iron: Harry S. Truman and the origins of the national security state, 1945-1954* (Cambridge, 1998), 234-252.
2. U.S. Public Law 585, 79th Cong., 2nd sess., 1 Aug 1946, sec. 10.

oping a personnel security program to determine whether employment of the individual on atomic energy projects would "endanger the common defense or security."[3]

Truman's Executive Order 9835 of 1947 established the federal employee loyalty program, making sympathetic association with totalitarian, fascist, Communist, and subversive organizations grounds for dismissal for all federal employees.[4] The fundamental assumptions and procedural flaws of Truman's program would be reproduced by the AEC in developing its own personnel security apparatus. Under the loyalty program federal workers, regardless of access to sensitive information, could lose their jobs solely based on political beliefs and associations. This program assumed that any vaguely defined "sympathetic association," past or present, with a group decreed Communist, fascist, or subversive by the Attorney General made the subject unfit for employment in any federal job. Procedurally, Executive Order 9835 stacked the deck against the employee by allowing judgments to be based on often unsubstantiated information from FBI files by accusers who could not be confronted.[5]

The numerous personnel security arenas of postwar America, including military security programs, the House Committee on Un-American Activities (conventionally called HUAC), and the Atomic Energy Commission, each integrated the precedents set by Truman into their own personnel security programs.[6] In each of these government programs, the approach to personnel security was based on the assumption that employment on sensitive projects was a privilege, not a right, and thus any claims on due process rights did not apply. While the framers of personnel security programs argued that any protections extended to employees were measures of significant goodwill, for the employee, investigations were a grand survey of intimate relations and personal politics with accusations capable of stretching from the absurd to the very serious, but always with potentially grave consequences. A negative finding likely meant that the individual was shut out of classified and often even unclassified research, as many institutions participating in classified work of any kind had an informal policy of only

3. Ibid., sec. 10b.

4. Executive Order 9835 is reprinted in Athan G. Theoharis, *The Truman presidency: The origins of the imperial presidency and the national security state* (New York, 1979), 257-261. Certainly the loyalty-security program was not the first incident in which the federal government declared political views and associations of personnel relevant to employment. The Hatch Act (1939) barred members of communist or fascist organizations from federal employment, and the Smith Act (1940) forbade the teaching, advocacy of, or association with groups advocating the forceful overthrow of the American government. Truman's loyalty-security program, however, would surpass preceding programs in scale and scope. Jessica Wang, *American science in an age of anxiety: Scientists, anticommunism, and the Cold War* (Chapel Hill, NC, 1999), 85.

5. Hogan (ref. 1), 255. Ellen Schrecker has argued that the federal loyalty-security program more than any other factor shaped domestic Cold War policy, establishing anticommunism as the national ideology. Ellen Schrecker, *No ivory tower: McCarthyism and the universities* (Oxford, 1986), 4-5.

6. For a discussion of scientists wrangling with HUAC see, for example, S.S. Schweber, *In the shadow of the bomb: Oppenheimer, Bethe and the moral responsibility of the scientist* (Princeton, 2000); cf. Wang (ref. 4); and Schrecker (ref. 5).

hiring individuals with top clearances, even for unclassified work.[7] A denial also carried a stigma that hampered future employment and strained the personal and professional associations so vital to professional success.[8]

Although the original intent of civilian control deemphasized secrecy in favor of more open scientific discourse, much of this early intent was diluted in the final McMahon bill. Civilian control activists such as the Federation of American Scientists were willing to accept the bill's security and secrecy restrictions as preferable to the more burdensome procedures arising under military control.[9] To military control proponents, however, these grudgingly adopted measures did not go nearly far enough. This compromise, like most, left everyone dissatisfied, and many in Congress soon challenged the Commission's ability to effectively maintain security. In the first year of AEC control, no fewer than six bills were pending in Congress advocating the return of atomic energy to the army.[10]

It was in this atmosphere of political pressure that AEC chairman David Lilienthal sought to develop procedures rigorous enough to assuage his Republican critics, while maintaining his unique commitment to individual rights. Personnel security proceedings deeply troubled Lilienthal, who proved eager to develop an AEC program compatible with civil liberties. He was concerned that personnel security proceedings resulting in a denial of clearance saddled AEC employees with a burden similar to a police record, but without the benefit of legal safeguards.[11] The methods of FBI investigations disgusted him, as did attempts to draw conclusions from them. He wrote of his own feelings of helplessness and violation when subjected to FBI scrutiny during his confirmation hearing; once a commissioner, he described being a part of the evaluation process even worse than being subjected to it.[12] Five months into his tenure, Lilienthal expressed in his diary his anxieties about AEC personnel security. In reviewing AEC cases, Lilienthal lamented: [13]

7. Wang (ref. 4), 95. The AEC defined three levels of clearance. P and S clearances allowed employees and frequent visitors onto AEC sites but prohibited access to restricted data. The highest clearance level, Q, did allow for access to restricted data and areas; it is this latter type, Q, that will be relevant for the purposes of this paper. Richard G. Hewlett and Francis Duncan, *A history of the United States Atomic Energy Commission, Atomic shield, 1947-1952,* (University Park, PA, 1969), vol. 2, 25.
8. Wang (ref. 4), 108.
9. Wang (ref. 4), 25. Ellen Schrecker (ref. 5) observed university administrations initiating or facilitating anticommunist purges as a means of protecting their institutions from external attacks. Wang argues that leaders of atomic scientists' organizations made similar fundamental concessions as a means of protecting civilian control of atomic energy.
10. Hewlett and Duncan (ref. 7), 91.
11. David E. Lilienthal, *The journals of David E. Lilienthal* (New York, 1964), 7 June 1947, vol. 2, 189.
12. Ibid., 190. At his confirmation hearing Congressional conservatives attacked Lilienthal's politics and ancestry as evidence that he could not be trusted to uphold national security and American values. Hogan (ref. 1), 252-253.
13. Lilienthal (ref. 11), 5 Aug 1947, 233.

> [T]here are really no cases of "disloyalty." These are usually cases of people who have a mother or a brother or wife who is, or is reputed to be, a Communist, or the equivalent, and the "evidence" to confirm these conclusions is only some FBI agent's rendition of what someone has said, or a conclusion from some very flimsy, thin stuff indeed. And so we sweat and agonize about the injustice to these people by such a travesty as our examination of these files must be.

Lilienthal wrote that the political scrutiny under which the Commission operated made deciding personnel security clearances particularly painful. The agitation for returning personnel security to the firm hand of the military meant that the commissioners had to take flimsy FBI evidence seriously "rather than 'take chances'"; the Commission thus had to "decide very important matters affecting people and events on a basis of caution, not justice."[14] Lilienthal was caught between his conscience and a Congressional hard place: damage reputations and careers of innocent employees by denying security clearance on the basis of substandard evidence, or grant clearances and endanger civilian control.

In his attempt to develop a personnel security program more consistent with civil liberties, Lilienthal enlisted distinguished jurists to review AEC procedures and offer recommendations. In June 1947 reviewers concluded that while "substantial justice" was done by current personnel security measures, the AEC could improve procedures by prescribing precise, written security standards, allowing for appellate review, and permitting employees to respond in writing to charges or be granted a hearing to respond to derogatory information.[15] The AEC adopted all of these suggestions, and Lilienthal even succeeded in establishing such a board, with former Supreme Court justice Owen J. Roberts serving as chairman.[16] The Roberts board was to both issue recommendations on specific cases and make general recommendations with respect to AEC personnel security standards and practices.[17] Lilienthal greeted this as a great triumph, which "may save this Commission and may contribute an important by-product to American institutions."[18]

In coping with the tremendous overload of new personnel security cases, the AEC would in its early years adopt a bureaucratic division of labor that one critic likened to a Rube Goldberg cartoon.[19] As the Commission needed to decentralize operations in order to evaluate the FBI files of approximately 70,000 individuals seeking security clearance annually, it established regional AEC offices that gave FBI reports a preliminary review. The scope of this task was tremendous and, in 1948 the AEC reported to

14. Ibid., 25 May 1947, 180.
15. Hewlett and Duncan (ref. 7), 92.
16. Hogan (ref. 1), 257.
17. AEC, *Fourth semiannual report* (Washington, D.C., 1948), 51.
18. Lilienthal (ref. 11), 10 Dec 1947, 267.
19. Senator William Knowland in JCAE, *Investigation into the United States atomic energy project,* 81st Cong., 1st sess., 26 May-11 Jul 1949, on 8 June 1949, 192.

Congress that the personnel security program accounted for a significant portion of the Commission's workload and administrative expense.[20]

If, as in the vast majority of cases, no material in the file raised doubts to the individual's clearance status, the local office could grant clearance. However, if the local AEC offices identified "substantial derogatory information" in the report making access to restricted material questionable, the AEC area manager sent its negative recommendation to Washington for final review.[21] Until 1950, if an individual seeking to begin employment with the AEC was denied clearance, it was the end of the road; he enjoyed no recourse for appeal beyond Washington. An AEC employee holding a wartime clearance, however, did enjoy further consideration from the Commission before it could be revoked. In these cases, the AEC advised the employee in writing that he had the right to request a hearing before a local Personnel Security Board.

The Interim Procedure

On April 15, 1948, the AEC integrated these recommendations into a written policy for conducting personnel security hearings termed the "Interim Procedure."[22] This procedure was part of an AEC effort to streamline personnel security through decentralization by delegating clearance decisionmaking to a three-person board assembled by the AEC field manager at one of ten regional field offices. The document defined standards for the rights of employees and outlined appropriate behavior for board members deciding personnel security cases. The employee was guaranteed a timely hearing, at which he was allowed to be represented by counsel of his own choosing, to present supporting evidence through witnesses or documents, and to challenge any member of the board for cause.[23] While the Army, Navy, and other defense agencies had previously allowed personnel to appeal negative clearance finding to a board, the conduct of these boards was not governed by standardized goals or methods. The AEC program as outlined in the Interim Procedure was exceptional in the extent to which it sought to develop standards and extend safeguards to employees.[24] At a time when other federal employees were rarely given any indication why they were denied positions, the AEC offered employees the opportunity to hear the derogatory information against them, and even the chance to respond to charges in person at a hearing.

The local AEC field manager was responsible for assembling the three-person board of locally prominent citizens—not necessarily Commission employees or contractors. Board members were called on to avoid "the attitude of the prosecutor," and behave in a manner reflective of the purely investigatory nature of PSB hearings. The AEC dictated that board members "always bear in mind and make clear to all concerned in

20. AEC (ref. 17), 51.
21. JCAE (ref. 19), 8 June 1949, 194.
22. "Interim Procedure," 15 Apr 1948, JFNP, box 175, folder 9.
23. Ibid.
24. Hewlett and Duncan (ref. 7), 333.

the proceeding that this is an inquiry," lest the participants confuse the hearing with a criminal prosecution. In developing a recommendation, the board could consider the FBI report, the evidence presented by the employee, witnesses testifying before the board, and the effect of denying clearance on the overall health of the atomic energy project. In evaluating this information, the board was offered only the guideline that recommendation "should be guided by the same considerations that would guide them in making a sound decision in the administration of their own lives."[25] This language left the evaluation of already problematic evidence wide open to interpretation, impressions, and personal influence.

While the Interim Procedure extended protections to individuals, the burden of proof remained squarely on the shoulders of the employee. The hearing was not a time for the AEC to prove its case, but an opportunity for the employee to present evidence sufficient to convince the board of the worthiness of his character, associations, and loyalty. Furthermore, as this was just an inquiry, the AEC explained, the individual would not be protected by some normal rules of evidence. Hearsay evidence, for example, was admissible and "accorded such weight as the circumstances warrant."[26] Indeed, the FBI file, consisting almost entirely of hearsay evidence, typically formed the backbone of the AEC's case. For the employee, the damaging effect of the admissibility of FBI files was greatly exacerbated by the fact that the files' content would not be disclosed to the employee. The AEC recognized that keeping FBI files secret would severely limit the employee's ability to challenge accusations appearing within them, but accepted it as necessary to protect informants. To further protect the confidentiality of sources, PSB inquiries were to be secret proceedings, their methods and findings closed to public scrutiny.[27]

These developments did not pass without criticism. Just over a month after the Atomic Energy Commission adopted the Interim Procedure, James Stewart, chairman of the Association of Oak Ridge Engineers and Scientists, wrote a letter of protest to AEC General Manager Carroll Wilson. The historian Jessica Wang has discussed Stewart's letter to Wilson as an example of scientists seeking to effect change by negotiating within the personnel security system, but the exchange also reveals a fundamental reason why the standards governing adverse findings by a PSB were so easy to meet.[28] Stewart demanded that the AEC accompany the statement of charges with an "explanation given as to why it is expected that this may lead to future disloyalty."[29] He also urged the Commission to disallow information from witnesses unavailable for cross-examination and affidavits such as FBI reports shielded from

25. "Interim procedure" (ref. 22).
26. Ibid.
27. Joseph Volpe, the AEC's deputy counsel who had a hand in drafting the Interim Procedure, described AEC personnel security hearing procedures in 1949 as "not much more than a kangaroo court." Gregg Herken, *Brotherhood of the bomb: The tangled lives and loyalties of Robert Oppenheimer, Ernest Lawrence, and Edward Teller* (New York, 2002), 269.
28. Wang (ref. 4), 173-174.
29. Carroll Wilson to James Stewart, 16 June 1948, JFNP, box 175, brown folder.

scrutiny. Wilson was unmoved. He responded, "It is improper to infer that these hearings are confined solely to the question of a person's loyalty." In addition to loyalty, Wilson explained, "experience has shown that both character and associations are of the utmost importance in determining the qualifications of an individual to protect restricted data." Thus, the AEC, acting in accordance with the requirements of the Atomic Energy Act, could revoke clearance for either character or associations criteria as *cause*, not just as it implied disloyalty. Indeed, Wilson wrote, most of the cases thus far appearing before local boards revolved around charges not at all related to questions of loyalty.[30]

To Congressional critics of civilian control, however, the safeguards the AEC extended to employees did not go unnoticed. In the weeks immediately following adoption of the Interim Procedure, and amid worsening relations with the AEC, Congress's Joint Committee on Atomic Energy made it known to Lilienthal that AEC personnel security measures were not tough enough. Lilienthal learned late in April of 1948 that JCAE chairman Bourke Hickenlooper was incensed that the AEC had been dawdling over the cases of men the chairman thought should have been fired immediately. Lilienthal blazed in his journal: [31]

> Hickenlooper was terrified of Parnell Thomas [HUAC chairman] and his criticism of the Joint Committee, and was prepared to outdo them if necessary...for my part I knew enough about the institutions of freedom and how they are undermined that I would be God damned if I would start lynching these poor devils just because Hickenlooper or anyone else didn't have the backbone to insist on decency in these things.

Lilienthal's frustration with JCAE criticism of the Commission and wariness of HUAC illustrates the tremendous political pressure Congress was under to be tough on communism, as well as the extent to which such a climate made political leadership by a Lilienthal liberal exceptionally harrowing.

The conflict in Washington over how best to balance individual liberties with security concerns would play out in local offices across the country. In Berkeley, a little over a month after Hickenlooper marked Commission procedures as dangerously lenient, the AEC set into motion personnel security proceedings against three Manhattan Project scientists. All three men would request hearings before a local security board. Serving as chairman at all three hearings was powerful University of California Regent John Francis Neylan, whose appointment to the board and role in the venture sheds light on the extent to which avenues for interjecting personal influence shaped the character and outcome of the personnel security hearing.

30. Hewlett and Duncan also report that most AEC security cases involved questions of character and associations, not loyalty. Hewlett and Duncan (ref. 7), 333.

31. Lilienthal (ref. 11), 24 Apr 1948, 326. Herken (ref. 27) argues that Joint Committee hearings investigating the AEC were part of a continuing effort to undermine Lilienthal's leadership.

John Francis Neylan: "Repugnant to My Sense of Justice"

Due to secrecy limitations, transcripts of AEC personnel security hearings are notoriously hard to come by. Other than Oppenheimer's sensational hearing in 1954, the few available accounts of proceedings have come to historians' attention more often through the personal papers of participants than through AEC disclosures. It is precisely these circumstances under which the personnel security board in Berkeley came to light. The restricted and underutilized papers of Berkeley personnel security board chairman, John Francis Neylan, and an only recently available interview conducted with him in 1960, reveal in striking detail the informal rules and structures of authority governing AEC personnel security procedure.[32]

As the inadvertent narrator of the story of Berkeley AEC personnel security, John Neylan deserves a short introduction. Neylan's friend Ernest Lawrence had specifically requested he sit on the local security board, a recommendation honored by area manager Al Pollman.[33] In 1948, Neylan, at sixty-three, was a prominent San Francisco lawyer and chairman of the University of California Board of Regents. Born in New York in 1885, Neylan moved west to work first as a newspaper journalist in Tombstone, Arizona, before moving to San Francisco in 1909 to cover Hiram Johnson's first gubernatorial campaign.[34] While in San Francisco, Neylan earned a law degree in night school and soon became one of the most successful lawyers in California. Among his most notable clients in this period was William Randolph Hearst, who made Neylan publisher of Hearst's *Call* and chief editor of five other West Coast newspapers. In 1925 Neylan was tapped as general counsel for the massive Hearst organization.[35]

In 1928 Governor C.C. Young appointed Neylan to the University of California Board of Regents. As a freshman regent, Neylan first had occasion to meet the University's charismatic young star Ernest O. Lawrence, and his band of cyclotroneers.[36] Neylan remembered years later being struck that between Lawrence and his "boys," "there weren't two in the place who had to shave more than twice a week. It was just a bunch of children."[37] Fondly, Neylan recalled how Lawrence's face "lighted up" as he discussed accelerating and smashing particles; Neylan was "tremendously impressed"

32. Neylan's papers are JFNP; the interview is John Neylan, interview with Herbert Childs, tape recording, 6 May 1960, CM-EL, box 6, folder 16. The interview, conducted by historian Herbert Childs as part of his biography of Lawrence (Herbert Childs, *An American genius: The life of Ernest Orlando Lawrence* [New York, 1968]), was carried out in May 1960. Neylan was 75 and would be dead in August. Despite his age, Neylan appears mentally sharp; his memory of the details of the Berkeley hearings was strong, and he dominated the conversation, at which Donald Cooksey was also present. While the 1960 interview is obviously of limited utility in reconstructing Neylan's thought process in 1948, I have attempted to limit use to those recollections demonstrably consistent with other historical evidence.
33. John Neylan to Ernest Lawrence, 14 Aug 1950, EOLP, carton 13, folder 44.
34. "John Francis Neylan, civic leader, dies," *San Francisco chronicle,* 20 Aug 1960.
35. Gray Brechin, *Imperial San Francisco: Urban power, earthly ruin* (Berkeley, 1999), 230.
36. Neylan interview (ref. 32).
37. Ibid.

by Lawrence and "attracted to him."[38] In the years following this meeting, the affection between the men would only grow.[39] In his decades as a Regent, Neylan facilitated Lawrence's career by protecting his interests with the University and counseling him on policy matters; Lawrence would call him his "father confessor."[40]

Although a supporter of labor and liberals through the 1920s (even managing Upton Sinclair's unsuccessful gubernatorial campaign on the socialist ticket), by the 1930s no trace of Neylan's left-leaning sympathies remained.[41] He despised Franklin Roosevelt and the Soviet Union and with the outbreak of war in 1939 was a staunch isolationist.[42] After the reality of atomic energy burst on the global scene in August 1945, Neylan's isolationism extended into an enthusiasm for maintaining atomic secrets. He opposed his friend Bernard Baruch's ill-fated plan for international control of atomic energy in 1946 as dangerous to American security.[43] In an interview with the *San Francisco chronicle* Neylan argued that "the fundamental and fatal weakness of the Baruch plan relative to atomic energy is that we could trade the world's most valuable secret for worthless promises. If nobody is going to use the secret, what sense is there in sharing it with others and subjecting them to temptation? It is about time we were cured of our child like faith in great and generous gestures."[44] Writing to Baruch a month later, Neylan expressed his concern that in negotiating with Russia, "We are buying peace in our time as fatuously as Chamberlain did."[45] With subsequent battles over atomic secrets, Neylan always advocated a tight-fisted policy. In 1949 Neylan wrote to his friend, the AEC commissioner Lewis Strauss, "There is just no common denominator between us and the Russians. We think in terms of human life and human rights. These are the cheapest commodities in Russia."[46] Parenthetically, Strauss would

38. Ibid.
39. Brechin describes Lawrence's friendship with Neylan as it developed in the 1930s in the atmosphere of the elite Bohemian Club of San Francisco power brokers to which both men belonged. Brechin (ref. 35), 312.
40. Neylan interview (ref. 32); In 1960 in correspondence with regard to the interview, Herbert Childs wrote to Neylan, "I find at almost every turn that Ernest Lawrence depended upon you more than anyone I have yet to hear of, for advice." Childs to Neylan, 9 May 1960, CM-EL, box 2, folder 19.
41. Herken (ref. 27), 144. A hint as to how Neylan viewed his youthful dalliance with leftist causes appeared in his comment at a 1949 PSB hearing that the Atomic Scientists Educational Association was "still afflicted with the mental measles that we all had in the sophomore year." "Proceedings before Personnel Security Board, Atomic Energy Commission: In the matter of Melvin Calvin," 23 Jun 1949, JFNP, box 171, 84.
42. "John Francis Neylan" (ref. 34); Brechin (ref. 35), 312.
43. As others have noted, Baruch was not the originator of the Acheson-Lilienthal Report, and is indeed often credited with sabotaging international control through a brash 1946 presentation to the United Nations.
44. Neylan, statement of 17 June 1946, *San Francisco chronicle,* undated clipping, JFNP, box 171, folder 1.
45. Neylan to Bernard Baruch, 10 Jul 1946, JFNP, box 171, folder 1.
46. Neylan to Strauss, 1 Aug 1949, JFNP, box 171, folder 2.

prove to be Robert Oppenheimer's principal adversary in 1954. Strauss, like Neylan, was a proponent of security through secrecy policies. The two men were well suited politically and temperamentally; Neylan would describe Strauss as a great American. Neylan's move to the political right and attendant distrust of the Soviet Union gave way to an often fanatical anti-Communism.[47]

Neylan described his service on the Berkeley personnel security board in part as a patriotic duty to guard against Communist influence. Neylan argued, "[W]e must face the fact realistically and unequivocally that Communism is not a political movement but a criminal conspiracy aimed at world conquest," and faced with this reality, "only the gullible or the deliberately blind can fail to recognize that Communism is a far greater danger to civilization than Nazism or fascism ever was."[48] He saw his participation in AEC personnel security hearings as his contribution to protecting atomic secrets and American institutions. As he informed one subject of inquiry, "in the present condition of the world the United States happens to be in possession of scientific information and instruments of warfare that are of great destructive power." In this context, Neylan explained, "it is deemed extremely important to protect that against disclosure."[49]

In addition to protecting national security, Neylan depicted his chairmanship of PSB hearings as a way to defend the individual rights of employees. In 1960, Neylan described the alternative to AEC procedure, firing questionable employees outright, as "repugnant to my sense of justice." Neylan held the work of FBI agents in tremendous regard, explaining "every report of theirs on a man gave you the details on what he was accused of but they also gave you the estimate of him by his neighbors so you had a balanced picture."[50] Neylan opined that the FBI report provided a balanced analysis of an individual and a reasonable substitute for ordinary rules of evidence. Shortly after his board first convened, Neylan reiterated his active commitment to civil liberties and his support for the Interim Procedure as a framework "outlined by someone thoroughly familiar with the Anglo Saxon tradition of individual rights."[51] Neylan explained to one subject of inquiry that his AEC board had the dual purpose of protecting national security *and* civil liberties. It hoped to determine from available evidence, [52]

> whether, on the one hand, the U.S. is secure in having you have access to such information as is available in the Rad Lab, and, on the other hand, we are charged with the responsibility of seeing that your rights as an American citizen are protected to the uttermost and that you are not unjustly dealt with. That is the whole objective of the Board.

47. Herken (ref. 27), 144.
48. Neylan to Flaherty, 28 Aug 1948, JFNP, box 171, folder 2.
49. "Proceedings before Personnel Security Board, Atomic Energy Commission: In the matter of Carl Robert Hurley," 4 Aug 1948, JFNP, box 175, brown folder, 6.
50. Neylan interview (ref. 32).
51. Neylan to Flaherty, 28 Aug 1948, JFNP, box 171, folder 2.
52. "In the matter of Melvin Calvin" (ref. 41), 85.

Neylan's commitment to the defense of both the security of America and the preservation of American freedoms appears, from all available evidence, to have been genuine.

Joining Neylan on the Berkeley board in the summer of 1948 were two war heroes: Admiral Chester Nimitz and Major General Kenyon Joyce. As the Interim Procedure called for the board to be constituted of "locally prominent citizens," it is perhaps not surprising that the Berkeley board consisted of such historically significant players. A week before hearing their first case, Neylan informed Nimitz that "in anticipation of the hearing, we have been furnished with confidential files and with an outline of procedure," and suggested that "it will be essential for us to review in advance the matter of procedure," as "the Commission is pioneering in the setting up of the Boards and will have to rely in large degree on the judgment of the members to avoid serious misconceptions."[53] In the months after adoption of the Interim Procedure, boards such as Neylan's were among the first to interpret the Procedure as they saw fit, forging a personnel security policy with tremendous ramifications for employees.[54]

Carl Robert Hurley

First to face Neylan's security board was Carl Robert Hurley, a Berkeley graduate student in the College of Chemistry and Manhattan Project alumnus working under Wendell Latimer. Years later, Neylan remembered, "his real name was Hurulin," but "he was using the name Hurley and so I was summoned...to serve as chairman of a security board."[55] Hurley was raised in Michigan by American-born parents and studied at the Michigan College of Mining and Technology beginning in 1932. After college, Hurley worked in industry as a chemist first for General Motors and then for the Udylite Corporation of Detroit. In 1942 the Navy employed Hurley on military research projects for the Bureau of Aeronautics. In 1945 he began work on the Manhattan Project at Columbia University, where he stayed for one year before being recruited to work at Los Alamos. Hurley continued work as a chemist at Los Alamos until fall

53. Neylan to Nimitz, 22 Jul 1948, JFNP, box 175, brown folder.
54. The Berkeley hearings in August were early, but not the first such AEC proceedings. In a 1948 report to Congress, the AEC reported that "as of July 15, 1948 there have been fifteen hearings conducted before local personnel security boards, including one which occurred prior to the issuance of Interim Procedure of April 15th. While most of these cases are still in process, six have been finally disposed of; and of these six, security clearance has been granted in five instances and denied in one." AEC (ref. 17), 53.
55. In both the 1948 hearing and his 1960 recollection, Neylan called attention to Hurley's name. Neylan hinted that both the fact that his family name was Hurulin (a Finnish name Hurley said his father informally abandoned as too difficult to pronounce) and that he never committed to either Carl or Robert as a first name made Hurley's names look suspiciously like aliases. "In the matter of Carl Robert Hurley" (ref. 49), 53-55.

1947, when he left to pursue a Ph.D. at the University of California with Professor Wendell Latimer. At Berkeley Latimer offered Hurley a position working on an AEC project for pay, to be the topic of an eventual dissertation.[56] Although this project required access to restricted information, Hurley already held top-level Q clearance from his work during the war.

Any false sense of security, however, disappeared in June when area manager Pollman notified Hurley in a "security statement" that the Commission found substantial derogatory information in his FBI file and intended to challenge his continued employment on AEC projects. The statement consisted of charges on eight counts and included assertions of neither disloyalty nor Communist Party membership. Rather, the Atomic Energy Commission accused Hurley of associating with radicals, sympathizing with the Soviet Union and the Communist form of government, and participating in Communist front activities and organizations.[57]

Hurley denied outright most of the charges in his written response two weeks later. He protested, "It is not a fact that I have ever declared myself a Communist sympathizer or have openly or otherwise advocated the Russian form of government as being superior to that of the United States" and requested "to be advised when and to whom I am supposed to have made the declaration specified in order I may reply more specifically." Of course, Hurley's request to know the FBI origins of derogatory information against him was contrary to the rules of evidence as outlined in the Interim Procedure. His complaint that he could not defend himself against unknown charges made by unknown persons would be repeated often as a criticism of PSB policies. Appealing to his past service to the United States government, Hurley pleaded to the AEC:[58]

> All during the war, I worked harder and longer than was demanded of me, as anybody for whom I worked can attest, for the preservation of the American way of life as defined by our constitution, customs, and traditions. I will continue to do so in the future. I firmly believe that my conduct over the years proves my loyalty. I wish to emphatically deny that there can be any valid questions concerning my character, associations or loyalty, nor any justification based on disloyalty, for discontinuing my security clearance for my present employment as a chemist.

Neither Hurley's war service, nor his rejection of "valid questions" concerning character, associations, and loyalty confronted the AEC requirements as outlined in the Interim Procedure. War service was insufficient to guarantee future clearance, and only board members could speak to the validity of charges.

Hurley's case came before Neylan's board on August 4. Neylan later remembered, "We had this young Hurulin and he was the smart alec type. In anticipating the thing I

56. "In the matter of Carl Robert Hurley" (ref. 49), 20.
57. Pollman to Hurley, 24 Jun 1948, JFNP, box 175, brown folder.
58. Hurley to Pollman, 6 Jul 1948, JFNP, box 175, brown folder.

prepared like I'd prepare a law case, and figured how I'd protect the rights of the person who was under investigation in the absence of the guarantees of being confronted."[59] In the proceedings, Neylan focused on Hurley's personal associations, his participation in "front" organizations, and his political views. Hurley's wife Lucille had been active, for instance, in a college organization called the Peace Foundation, which organized speakers on the campus of the University of Wisconsin.[60] A confidential informant's report appearing in Hurley's FBI file declared this student group to be "dominated by members of the Young Communist League in attendance at the University of Wisconsin at that time, as evidenced by the fact that James K. Hudson, Chairman of the Young Communist League during the years 1938, 1939, and 1940 was also chairman of the University of Wisconsin chapter of the Peace Federation [sic]."[61] In 1960 Neylan would continue down this slippery slope and describe Hurley's wife as "a member of the Young Communist League or something on the campus at Wisconsin."[62]

In addition to charges of associating with radicals, Neylan also prodded Hurley on his political sympathies. Hurley was charged with sympathizing with the Soviet Union during the war, being in possession of radical literature such as Marx's *Das Kapital* and local Communist newspapers, and occasionally engaging colleagues in politically charged lunchtime discussions. Although one informant assured the FBI that Hurley was "never a member of any of the Communist party's front groups or any groups that favored communism, he continually had in his possession literature of a communist nature." Hurley admitted owning *Das Kapital*, but maintained that it never held his interest. Similarly, one former colleague described Hurley as "of high moral character and an entirely honest and capable employee," but admitted to being disturbed by Hurley's propensity a decade earlier to argue politics and take the side of Russia. This informant said Hurley's politics raised "doubt as to Hurley's 100% loyalty to the United States" and expressed his belief "that Hurley should not be considered for a position in which he would have access to restricted information," as "there are undoubtedly deserving Americans about whom there is no doubt as to their loyalty to this country."[63] This informant's comments make clear the extent to which the employee carried the burden of proof in these proceedings. The Interim Procedure made Hurley responsible for countering evidence he could not be privy to, in an attempt to assure the board of the impossible standard of his "100% loyalty."

On the question of Hurley's association with certain so-called front organizations, Neylan was most concerned about Hurley's membership in the United Federal

59. Neylan interview (ref. 32).
60. "In the matter of Carl Robert Hurley" (ref. 49), 28.
61. Carl Robert Hurley FBI file, JFNP, Box 175, brown folder. To historians, the FBI file illustrates the range of accusations deemed relevant to determining eligibility for employment; to Hurley, who could see neither the accusations nor the parties accusing, the file points to the near impossibility of developing an effective response.
62. Neylan interview (ref. 32). Neylan's disinterest in making distinctions between Communists and vaguely communist associations in 1960 is consistent with his behavior in 1948.
63. Hurley FBI file (ref. 61).

Workers union and the Washington Bookshop Association. Hurley claimed that he joined the UFW in 1945 as a way to meet women at social events and denied any real interest in union politics. When Neylan asked if UFW members were "somewhat 'pink,'" Hurley shrugged that "they were certainly to the left of the Kiwanis Club."[64] Through his membership in the local union, Hurley came to associate with another establishment which made the Attorney General's list of subversive organizations: the Washington Bookshop Association, which offered discounts to cooperative members. Hurley admitted that before the Attorney General made such a proclamation, he purchased two or three folk music albums from this bookshop between 1943 and 1944.[65] He downplayed the significance of his association with this and other "front" organizations and claimed no knowledge of others. In a final plea to Neylan, Hurley expressed his frustration with questions pertaining to his activities in years past: "I can't remember everything I have done in the last ten years and how it will effect [sic] me in the future."[66] Neylan was likely unmoved; indeed the sarcasm marking Hurley's often dismissive testimony likely did not inspire Neylan to give his fuzzy memory the benefit of the doubt.

In addition to the character and associations charges dominating the substance of the hearing, one FBI informant also floated direct questions about Hurley's loyalty. This anonymous critic claimed Hurley "stated everything Moscow does is right and everything the United States does is wrong in effect." Such a claim illustrates the potentially devastating impact hearsay evidence from anonymous accusers could carry. This informant continued that, in his opinion, "there is absolutely no doubt that Hurley's loyalty to the United States is secondary to his interest in the Communist Party" and had "no hesitation in stating that Hurley would have no qualms in passing on confidential information to Russian agents."[67] In making the very serious claim that Hurley would be amenable to spying for the Soviet Union, this accuser was neither required to justify his assessment, nor to make himself available to Hurley for response. The Interim Procedure made such evidence admissible, and worse, gave no firm guidelines for evaluation.

Hurley next invited his adviser to appear on his behalf. This was a potentially powerful move. But Wendell Latimer's decision to use his testimony to grandstand against AEC procedures likely served only to aggravate Hurley's situation. Latimer did not share Neylan's faith in the balanced picture presented by FBI reports. He protested that their investigations were an unreliable means of evaluating employees, and he emphasized that persecuting employees would have a negative impact on the United States atomic energy program. Disdainful of security intrusions, Latimer said, "Hardly a day passes that there aren't two or three FBI men in my office investigating these boys," and objected that the men "making these investigations are not in a position

64. "In the matter of Carl Robert Hurley" (ref. 49), 50.
65. Ibid., 49.
66. Ibid., 156.
67. Hurley FBI file (ref. 61).

to form opinions."[68] Rather than relying on the judgment of lab administration to inform employment decisions, Latimer protested that the FBI drew conclusions on piecemeal information from anonymous or confidential sources. To which Neylan snapped back, "I don't think that you want to be in the position of disparaging the FBI as an institution."[69] In addition to his objection to FBI procedure, Latimer argued that investigations destroyed lab morale and discouraged scientists from working on AEC projects. As dean of the College of Chemistry, Latimer felt responsible for "maintaining morale among our men, and of placing men on these programs," and thus was "beginning to feel very bitter toward the type of case which has been repeatedly brought up."[70] Latimer's concern that draconian security measures damaged AEC recruitment was an oft-repeated criticism of personnel security programs, though no such exodus from AEC projects appears to have occurred in reality.[71] However, in a statement submitted to the AEC on Hurley's behalf, Latimer threatened that if inquiries such as this one continued, he would no longer recommend that his students work for the AEC.[72] An indication of how Latimer's rhetoric and threats sat with Neylan appeared in his comment in 1960: "Latimer showed up during the hearing, made a stump speech about this boy, and made a damn fool of himself."[73]

One week later, Neylan reported the unanimous findings of the board that Hurley was "lacking in frankness, and in many instances, evasive in relation to substantive issues." They "found it impossible to reconcile his professed academic interest in radical theories with a naïve unawareness of the aims and activities of radicals." Neylan recommended Hurley be fired—not a justifiable risk in the atomic energy program.[74] Neylan laughed in a 1960 interview that Hurley "took a look at our white hairs" and thought he need not take "these three nice old goats" seriously.[75] Years later, Neylan recalled that Hurley "just toyed with us," and "I thought to myself, well young man, you're just about tying the knot around your neck as securely as anybody could possibly tie it, and the more you fed him, the more he said things that were obviously untrue."[76]

For the purposes of exploring how Neylan's AEC security board functioned, it is less important to reconstruct Hurley's past political commitments than it is to understand how the board evaluated evidence. While most informants placed Hurley's commitment to radical politics at a level consistent with the foolishness of youth, Hurley did not

68. "In the matter of Carl Robert Hurley" (ref. 49), 93, 99.
69. Ibid., 99.
70. Ibid.
71. Wang (ref. 4), 180.
72. Wendell Latimer, "Statement on the effect which denial of security clearance would have upon the atomic energy program," 5 Aug 1948, JFNP, box 175, brown folder.
73. Neylan interview (ref. 32).
74. "Recommendation in re. Carl Robert Hurley," 11 Aug 1948, JFNP, box 175, brown folder.
75. Neylan interview (ref. 32).
76. Ibid.

concede any such youthful indiscretions to John Neylan. Neylan no doubt interpreted this discrepancy as evasive—a conclusion entirely justifiable by the standards dictated by the Interim Procedure. It is important to note, however, that Neylan could have attributed Hurley's testimony with equal justice to the fuzzy memory of decades past, or indeed to a different interpretation of what constituted significant radical character or associations. Furthermore, Neylan's decision to weight Hurley's evasion to levels of commitment to a labor union or a cooperative bookstore as sufficient to deny clearance was in itself a choice not dictated by the Interim Procedure.

In fact, upon receiving Neylan's recommendation, AEC officials did not immediately find the derogatory evidence against Hurley as convincing as had Neylan. The Commission's Chicago manager, Alfonso Tammaro, and overseer of Berkeley operations, wrote to Neylan on August 30 requesting an explanation for the board's negative findings in Hurley's case. Equipped with the board's recommendations and the hearing transcripts for the Hurley and Robert Serber cases, but lacking access to the FBI file, Tammaro found the board's documentation of its thought process unsatisfying and requested "a statement from the Board briefly summarizing its findings and reasoning with respect to the alleged derogatory information concerning Mr. Hurley...and any other circumstances the board deems relevant to its conclusion."[77] Tammaro needed this information as he was responsible for making a final recommendation on personnel security decisions to General Manager Carroll Wilson who in turn actually granted or denied clearance. This exchange points to at least two shortfalls of Lilienthal's carefully crafted personnel security hearing. First, the FBI file of unsubstantiated claims and unaccountable informants, and not the hearing at which the employee had (however limited) opportunity to respond to charges, was paramount in clearance decisions. For example, Tammaro noted that in Hurley's case "the transcript tends to build up the employee's case," and thus "the basis for the denial of clearance must be found in the Federal Bureau of Investigation report."[78] Second, the checks and balances anticipated through use of a series of recommendations and independent evaluations broke down as Tammaro, rather than issuing independent oversight, fell back on Neylan's reasoning in issuing his own recommendation and as Tammaro added, preparing the Commission for appeal.[79] The AEC would eventually defer to the local board's assessment and remove Hurley from employment. In a strange aftermath, Neylan discovered "a couple of months later...by God," that Hurley was back "in the chemistry department; Latimer had hired him back."[80] He was not back for long; Neylan immediately fired him for a second time.

77. Tammaro to Neylan, 30 Aug 1948, JFNP, box 172, folder 9. For more on Tammaro, see Hewlett and Duncan (ref. 7), 102.
78. Tammaro to Neylan, 30 Aug 1948, JNFP, box 172, folder 9.
79. Ibid.
80. Neylan interview (ref. 32). Childs takes up Neylan's chairmanship of the AEC PSB briefly as part of a discussion of postwar personnel security at the Berkeley Rad Lab. Childs accepts without supporting evidence Neylan's evaluation of his own role in the "loyalty hearings" as

Although Hurley appealed the decision to the Personnel Security Board in Washington led by Owen J. Roberts, the Roberts Board upheld the Berkeley findings.[81] Hurley's unsuccessful appeal to the Roberts Board illustrates the dilemmas of allowing appeals only within the AEC structure. Although the AEC attempted to separate levels of its personnel security apparatus, political pressures meant that the safeguards of separation of powers were not truly independent. When Tammaro wrote to Neylan asking for more information on Hurley, Tammaro was seeking to reconcile his findings that Hurley's testimony did not justify denying clearance by asking Neylan to convince him. The mechanics of AEC personnel security boards encouraged conservative behavior at any appellate level. For both Tammaro and the Roberts board, overturning the Berkeley recommendation would make them vulnerable to criticism in case Hurley proved to be a real problem. Deferring to someone else's judgment was always safe.

Given the informal rules governing personnel security, Hurley and Latimer could not have handled Neylan's board worse. Whether Hurley's blanket denials of radicalism arose from an attempt to hide a leftist past or were more benign, this is less important than the fact that the Interim Procedure left the decision of how heavily to weight such evidence to Neylan's discretion. The tremendous flexibility that the AEC gave to boards in evaluating information and making recommendations meant that personal appeals and impressions took on a great deal of import. Hurley's dismissal of Neylan's charges certainly did not endear him to Neylan's board; Latimer's testimony only cemented Hurley's fate. Then, by speaking not on the specifics of Hurley's case but rather by attacking the FBI and AEC procedure in general, Latimer chose the wrong audience to protest PSB proceedings. He did not serve to bolster his student's character, associations, or loyalty and likely served only to further inflame the board.

"The Philosophy of the Law, Merits of Procedure and Other Extraneous Matters"

Latimer was furious with the proceedings against his graduate student. A week after the hearing, he voiced his grievances directly to David Lilienthal. The ensuing exchange sheds light on the procedural shortcomings of the AEC personnel security apparatus. In Latimer's letter, copied to Neylan, Latimer wrote that he "wish[ed] to register a strong protest against the procedure in the conduct of these investigations."[82] As he had at the hearing, Latimer expressed concern that the board acted as both prosecutor and

---

a measured and fair protection of employee rights. Childs makes veiled reference to Hurley's case, stating that "one of the accused, of whose guilt there could be no doubt, Neylan found had been immediately rehired by Dean Latimer, who had testified at the man's hearing. He was fired again." Childs gives no indication that he has availed himself of documentation other than Neylan's own personal assessment, and he does not clarify of what Hurley is "no doubt" guilty. Childs (ref. 32), 406.

81. Neylan to Harold Fidler, 8 Aug 1950, EOLP, carton 13, folder 44.

82. Latimer to Lilienthal, 10 Aug 1948, JFNP, box 172, folder 9.

judge and based its recommendation on the partial picture of a man provided by the FBI. Indeed, Latimer protested that the board seemed much more interested in the contents of the FBI security file inaccessible to Hurley than in hearing the opinions of "responsible scientists who expressed the greatest confidence in his loyalty." In an attempt to show how susceptible security boards were to drawing conclusions from isolated information, Latimer naïvely offered to Lilienthal his own case: "I twice have had cocktails with a Russian Consul. In 1945 I permitted the Friends of the Soviet Union to use my name to sponsor a lecture...I had a secretary who was a communist. I have communicated with Russian scientists, and a number of whom [sic] I once regarded as friends. These facts" he suggested, "could be presented to a reviewing board to give a very unfair picture of my loyalty."[83] Latimer pointed out that if someone chose to bring charges against him, he would be essentially defenseless, as the "[c]harges are based upon testimony which is not given under oath, indeed the charges may be entirely anonymous. As far as I see, anyone who wishes to attack me may bring the most serious charges merely by contacting the FBI."[84] Furthermore, Latimer reiterated to Lilienthal that AEC security measures made recruiting scientists for atomic energy projects difficult and thus actively threatened national security.

Latimer may have believed he would get a sympathetic hearing from the AEC chief. His opponent would not concede so easily, however. Neylan retorted to Latimer via Lilienthal two days later. Neylan grumbled that unwarranted criticisms from Latimer positively hampered the board's ability to, in his words, "discharge our thankless and distasteful task and retire to the obscurity from which we were drawn."[85] Waving off Latimer's criticisms, Neylan explained to Lilienthal that "[d]uring his testimony, Dr. Latimer was quite emotional and disposed to discuss the philosophy of the law, the merits of procedure and other extraneous matters."[86] Neylan felt confident that there was, in fact, "nothing original in Latimer's criticism" and that procedural questions had been thoroughly examined by the Commission and his board prior to Hurley's case. Neylan was also, predictably, very concerned about the personal information Latimer offered pertaining to his own Russian connections. Indeed, Neylan pointed to Latimer's subversive associations as evidence of his poor judgment. That a Communist would be found working as Latimer's secretary, Neylan noted to Lilienthal, led him "respectfully [to] suggest the good doctor is not fully aware of the capabilities of potential enemies."[87]

The matter would not rest there. The same day Neylan wrote this letter to Lilienthal, he also revealed to Nimitz and Joyce that he had taken it upon himself to investigate Latimer's disclosures further. He reported to Nimitz, "For your confidential information, I have learned today that the Security Department uncovered a card-carrying

83. Ibid.
84. Ibid.
85. Neylan to Lilienthal, 12 Aug 1948, JFNP, box 172, folder 9.
86. Ibid.
87. Ibid.

Communist in the [chemistry] department six months ago and also moved against a lady in the department who may be prosecuted shortly."[88] Neylan attributed this Communist incursion to Latimer's naiveté in personnel security issues and cited it as further evidence of the necessity of maintaining security investigations and PSB hearings. The very next day, tossing aside the pretense of addressing letters to Lilienthal, Latimer wrote directly to Neylan. "Your remarks," he expostulated, "as to my competence to select personnel because I once had a Communist secretary I think illustrates the danger which lurks in conclusions from isolated statements. I am sure you will be pleased to learn that the lady was sent to me by our personnel office and that she was discharged when the facts were known."[89] Clearly Latimer did not intend to simply let the matter drop. Joyce in fact wondered a few days later "if some kindly but pointed advice from Ernest Lawrence would not be beneficial in getting Latimer on a more conservative and less emotional track"; Neylan replied to Joyce that he had already spoken with Lawrence, who assured him that Latimer was just overworked.[90] Joyce's suggestion to involve Lawrence in the conflict hints at the extent to which Lawrence was able to serve as a buffer between scientists and the national security state, a trusted and influential voice, able to mediate between individual scientists and the anti-Communist establishment.

Meanwhile, as Lilienthal vacationed on Martha's Vineyard, Atomic Energy Commissioner Sumner Pike piped up as the voice of the AEC. He responded diplomatically, lauding the strong sense of public duty that moved Neylan to sit on the AEC board, and thanking Latimer for his views. Pike expressed gratitude to the men for airing out their differences to the Commission, as the dialogue "will prove helpful to the AEC in carrying out its personnel security program in that it will promote a better understanding of the objectives of this program by those who participate in it and are affected by it."[91] Both parties, then, were thanked for their contribution to the development of a still evolving personnel security program. Pike's letter of appeasement, however, would be insufficient to settle their differences; two years later Robert Hurley would again be caught in another, and this time very public, dispute between Latimer and Neylan.[92]

88. Neylan to Nimitz, 12 Aug 1948, JFNP, box 172, folder 9.
89. Latimer to Neylan, 13 Aug 1948, EOLP, carton 13, folder 44.
90. Joyce to Neylan, 16 Aug 1948; Neylan to Joyce, 17 Aug 1948, JFNP, box 172, folder 9.
91. Pike to Neylan, 20 Aug 1948, JFNP, box 172, folder 9.
92. In May 1950, at the height of California loyalty oath furor, then Dean of the College of Chemistry Joel Hildebrand criticized the Regents' handling of the controversy. In a public meeting of the University of California Board of Regents, Neylan lashed back at Hildebrand for employing in his department a man held to be a bad security risk. Neylan publicly identified the man as Carl Robert Hurley and cited unfair and inflammatory details about his case as part of a personal attack against Hildebrand. Neylan boomed, "And this gentleman, this Mr. Dean Hildebrand is the man who is telling the public that the Regents are incompetent and inefficient and that the faculty would not stand for a loyalty oath!" Neylan's speech dredged up Neylan's feud with Latimer anew—this time to be played out in the press. Latimer publicly

As a postscript, given Latimer's outspoken criticism of the methods of personnel security hearings in 1948, it is perhaps surprising that he testified *against* Robert Oppenheimer in 1954 in a substantively identical proceeding. As a strong proponent of a crash H-bomb effort, Latimer was critical of Oppenheimer's resistance to the project.[93] His 1954 testimony against Oppenheimer contained innuendo, but it shied away from questions of motivations or loyalty. Latimer stated, indeed, "A man's motives are just something you can't discuss, but all his reactions were such as to give me considerable worry about his judgment as a security risk."[94] What offended Latimer was Oppenheimer's judgment in intentionally hindering the H-bomb effort through manipulation of others, not his association with Communists or participation in liberal causes. Poor judgment, even well intentioned, that was then acted upon in such a way as to have deleterious impact on the mandate of the AEC was much more threatening to Latimer's sense of security than were the questionable associations of Robert Hurley. Furthermore, Latimer was likely heartened that the charges against Oppenheimer were voiced not only through anonymous whispers to FBI men but through the open testimony of Oppenheimer's colleagues and acquaintances, himself included. While it is unclear what Latimer would have thought of Oppenheimer's problems had he known just how much subterranean FBI whispering was going on, it seems likely that in the end Latimer trusted his own judgment more than anyone else's—whether those other judgments were anonymous or not.

Robert Serber

On August 5, 1948, one day after Hurley's hearing, and still a week before Latimer penned his grievances to Lilienthal, Robert Serber became the second subject of Neylan's personnel security board. Serber, a theoretical physicist, first came to Berkeley to work as a National Research Fellow under the direction of Robert Oppenheimer in the summer of 1934. After a two-year fellowship, Serber stayed on at Berkeley as Oppenheimer's research assistant.[95] At Berkeley, Serber, like most of Oppenheimer's students, grew close to his mentor. From 1934 up until the start of the war, Serber spent many summers on Robert and Frank Oppenheimer's ranch in New Mexico.[96]

---

criticized Neylan and AEC procedures, the American Civil Liberties Union criticized the AEC for allowing Hurley's case to be made public, and Neylan lashed back that Latimer started it all by gossiping about Hurley's case around campus. *San Francisco chronicle,* 27 May 1950.

93. Herken (ref. 27), 201.

94. Richard Polenberg, ed., *In the matter of J. Robert Oppenheimer: The security clearance hearing* (Ithaca, NY, 2002), 241.

95. Robert Serber, *Peace and war: Reminiscences of a life on the frontiers of science* (New York, 1998), 25.

96. "Proceedings before Personnel Security Board, Atomic Energy Commission: In the matter of Robert Serber," 5 Aug 1948, JFNP, box 175, brown folder, 53.

Indeed, at Serber's 1948 hearing he would say of Robert Oppenheimer: "For many years, he was my best friend."[97] In 1938, with glowing recommendations from Oppenheimer and Ernest Lawrence, Serber accepted a faculty position at the University of Illinois. Serber's peacetime role as theoretical physicist was redirected in 1942 when Oppenheimer recruited him to work on the Manhattan Project.[98] During the war both Serber and his wife Charlotte, employed as the lab librarian, held top security clearances at Los Alamos. After the war, Ernest Lawrence successfully lured Serber away from Illinois to a position as professor at Berkeley and head of the theoretical physics division at the Radiation Laboratory—a position recently vacated by Robert Oppenheimer.[99]

As Barton Bernstein has noted, Serber's exceptionally radical family and friends had been the subject of security scrutiny for years before Neylan's inquiry.[100] Perhaps the first sign of security problems brewing for the Serbers after the war came early in 1947, when the AEC denied Charlotte a position as Rad Lab librarian. Charlotte was the daughter of left-leaning Russian émigrés and often shared their political inclinations. During the Serbers' earlier Berkeley years, Charlotte, together with Oppenheimer's communist lover Jean Tatlock, had organized a local committee dedicated to providing medical supplies to Spanish loyalists.[101] Years later, John Neylan described Charlotte's family as "high up in the Consuls [sic] of the Communist thing."[102] At Los Alamos, however, Charlotte Serber earned the highest accolades as laboratory librarian from director Robert Oppenheimer.[103] Oppenheimer wrote her a letter after the war thanking her for her work—a letter Charlotte would offer three years later to Neylan's security board as a positive reflection on her own character and loyalty, and thus upon her husband's associations. When Charlotte was rejected for a similar position at the Rad Lab, the lab protested to the AEC that her unique experience at Los Alamos made her the ideal candidate for similar employment at Berkeley. The AEC security division, however, likely weighed her prior radical associations more heavily than personal recommendations and refused her the clearance she needed.[104] John Neylan in 1960 would blame Serber's problems on the fact that Serber was "a nice guy who had a gabby wife."[105]

97. Ibid.
98. Serber (ref. 95), 53.
99. Ibid., 144.
100. Barton Bernstein, "Interpreting the elusive Robert Serber: What Serber says and what Serber does not explicitly say," review of Serber (ref. 94), *Studies in the history and philosophy of modern physics, 32* (2001), 443-486.
101. Herken (ref. 27), 64.
102. Neylan interview (ref. 32).
103. JRO to Charlotte Serber, 2 Nov 1945, Alice Kimball Smith and Charles Weiner, eds., *Robert Oppenheimer: Letters and recollections* (Stanford, 1980), 313.
104. Daniels to Cooksey, 10 Jan 1947, EOLP, carton 24, box 16, folder 16; Herken (ref. 27), 190.
105. Neylan interview (ref. 32).

In the summer of 1948, Robert Bacher, an AEC commissioner and old friend, visited Serber in Berkeley and warned him that the Commission intended to subject him to a personnel security hearing. Serber recalled many years later that Bacher "assured [him] that they had no doubts about me, but said that they were forced into such actions to defend themselves from the (pre-McCarthy) hysteria in Washington."[106] As expected, late in July 1948, Pollman, the Berkeley area manager, sent Serber a security statement and invited his response to nine charges. In looking over the list of charges, Serber noted that "none of the charges seemed very serious," as "the only thing they had directly against Charlotte was that she had been secretary of a medical aid committee for Spanish loyalists during the Spanish Civil War, and during the Second World War she had been active in the British and Russian war relief committees. All the rest was guilt by association."[107] Unbeknownst to Serber, the Interim Procedure dictated that association with radical persons or organizations was indeed a serious charge and grounds for a denial of clearance.

Ernest Lawrence, likely very concerned about the potential loss of the new backbone of Berkeley's theoretical physics program, wrote a glowing letter of support for Serber to area manager Al Pollman. Lawrence declared that he and everyone at the Lab had "the highest regard for Professor Robert Serber and are deeply concerned that some questions have been raised as to his clearance status." Indeed, with Oppenheimer's departure, "the whole responsibility for theoretical physics here was thrust upon Serber's shoulders." Lawrence emphasized Serber's contribution to the Manhattan Project and his tremendous responsibility at Berkeley. He further addressed the specific concerns of personnel security boards, assuring Pollman that "quite apart from his great scientific talents he has even more personal qualities of character, loyalty, and integrity."[108] Lawrence extended his lobbying efforts on Serber's behalf to Neylan himself, who commented in 1960 that before the hearing "Ernest said to me...I hope you'll give him the benefit of every doubt because he's a fine fellow."[109]

The hearing took place August 5 in Neylan's San Francisco office. As he had done the previous day for Robert Hurley, Neylan assured Serber that in AEC proceedings the rights of the individual were to be protected: "members of this Board not only subscribe to that in theory, but their disposition is to protect every right of yours."[110] Joining Neylan on the PSB were Admiral Nimitz and Major General Joyce, who again remained essentially silent throughout. As indicated in his statement of charges, Serber's personal associations took center stage in Neylan's office. The only mild assertion of disloyalty Neylan raised regarded a scientific indiscretion to Frank Oppenheimer.

Neylan thus pressed Serber on a number of his family members and friends. He established that Charlotte Serber's family was "somewhat radically inclined," and that

106. Serber (ref. 95), 162.
107. Ibid., 163.
108. Lawrence to Pollman, 5 Aug 1948, EOLP, carton 24, box 16, folder 16.
109. Neylan interview (ref. 32).
110. "In the matter of Robert Serber" (ref. 96), 5.

Charlotte herself often shared this propensity for liberal politics. Two further associates of interest to the Atomic Energy Commission were left-leaning Berkeley physicists and fellow Oppenheimer students Joe Weinberg and David Bohm. Weinberg and Bohm would suffer HUAC persecution and political dismissals from university posts in subsequent years, and their association with Oppenheimer would prove similarly troublesome to him in 1954. While Serber knew both men in a professional capacity he downplayed any social or political connection.

Although the "Chevalier Affair" had not yet taken on the notoriety it would during Oppenheimer's own personnel security hearing, already by 1948 newspapers had reported a version of the story that the Berkeley professor of French literature Haakon Chevalier had approached Oppenheimer during the war with the aim of garnering Oppenheimer's receptivity to transferring scientific secrets to the Russians. At his own security hearing, Serber described Chevalier as "one of the moving spirits of the left wing faculty element." But, he continued, "I didn't take those views very seriously. About that time it was sort of fashionable in the literary world to express fairly radical views."[111] Besides, Serber testified, Oppenheimer had personally assured him that the approach was much more innocent than as represented in the media. Skeptical, Neylan in 1948 posed the Chevalier affair as a hypothetical situation for Serber's consideration. "If somebody sympathetic to Russia," he asked, "were to approach you and ask you to get some secret information from Ernest Lawrence, and retail it back to a Russian agent, would you put that matter up to Lawrence?"[112] Perhaps misunderstanding the question, or to avoid criticizing Chevalier, Serber responded, "Yes, I think so."[113] Aghast, Lawrence interrupted: "What?" What happened next is telling. While Serber stumbled to clarify his position—that he would only inform Lawrence of the incident, not pose the question as an invitation to espionage—Lawrence and the board guided him through rough water. Lawrence clarified that Serber "said he would stall because he would want to have a breathing spell" before notifying security personnel, while Neylan himself offered, "In fairness to yourself, you wouldn't for a minute permit anybody to decide for you whether or not you would violate security."[114] A flustered Serber assured him absolutely not. With this, Neylan seemed satisfied. But had he ever been dissatisfied? The softballs Neylan tossed to Serber, and his leading questions in getting Serber out of a sticky admission, indicate that Neylan really had very little intention of firing him. However, Lawrence's worried interjection to clarify the record also indicates that he did not necessarily consider Serber's exoneration a foregone conclusion.

The only significant inquiry Neylan pursued that was not based entirely on Serber's associations concerned Serber's supposed transmission of classified information to Robert's brother Frank. After the war, Luis Alvarez, Frank Oppenheimer, and Serber

111. Ibid., 94.
112. Ibid., 101.
113. Ibid., 102.
114. Ibid., 103, 104.

worked together in Berkeley on a new linear particle accelerator design.[115] When Ernest Lawrence banned Frank Oppenheimer from the Rad Lab in February 1947, Frank took a faculty position at the University of Minnesota.[116] In September of that year, Alvarez and Serber confirmed that the linear accelerator worked properly, and each wrote to Frank in Minnesota to congratulate him and to notify him that his name would appear on the group's upcoming publication. This narrative seemed troublesome to Neylan, who believed that at the time of Serber's communication with Frank, the linear accelerator design was still classified—a claim Serber vehemently denied. During the PSB's lunch break, Serber confirmed that by July 1946 the accelerator design was entirely declassified for publication as a story in the *newspaper*: the subject of Serber's supposed leak had been removed from under security restrictions and brought into the public domain more than a year prior to his communication with Frank Oppenheimer.

In 1960, however, John Neylan recalled the question of the transmission of the linear accelerator design to Frank as the issue most critical to the board. Notably, in this interview Neylan remembered the incident all wrong. He told Herbert Childs that Serber wrote to Frank just three days before the declassification of the design, and forgot to mention that Frank was a very active member of the design group (indeed, *more* active than Serber), saying instead that Serber described the accelerator in a letter to Frank so he could present the findings at a meeting which neither Serber nor Alvarez could attend. This account conveniently makes the Commission's investigation of Serber on loyalty grounds sound much more reasonable than it in reality was. However, it is unlikely that this incident, so easily cleared up as entirely innocent, could really have been the basis for the AEC investigation. More likely, Serber's subversive associations set off flags with an AEC eager to crack down on personnel security. Serber's correspondence with Frank Oppenheimer, a man of known left leanings, called more attention to the incident than did the timing of the disclosure. Neylan's selective memory hints at the extent to which guilt by association charges that enjoyed so much legitimacy in the mid 1950s had lost steam by 1960.

Lawrence himself soon appeared on Serber's behalf. Answering in a frankness that could not be imitated by men less intimate with John Neylan, Lawrence boosted Serber's positive attributes and downplayed the significance of Serber's personal associations. During the war, Lawrence stated, Serber "was right at the center of developing the bomb," and of delivering it as well, as he was supposed to be on board the photography plane flying with the Nagasaki mission.[117] Lawrence further opined that Charlotte's "parents, or her relatives being radicals, perhaps Communists, to me is beside the point. I mean in the matter of this clearance."[118] Without questioning AEC standards or procedures, he reiterated Charlotte's superb contributions to Los Alamos and his

115. Ibid., 59.
116. Herken (ref. 27), 192.
117. "In the matter of Robert Serber" (ref. 96), 114.
118. Ibid., 119.

regret that the Rad Lab was unable to secure her for similar work. Lawrence thus acknowledged but downplayed the Serbers' subversive associations, offering: "I have known that Robert Oppenheimer had radical friends in years gone by. In fact, I have had radical friends in my time."[119]

In the end, Neylan's board cleared Robert Serber as a justifiable risk in the Atomic Energy Commission Program. They declared him "frank and forthright," finding that "in communicating with Frank Oppenheimer, Mr. Serber did not violate any security restrictions." Neylan declared Serber's failure to criticize Russia "consistent with a retiring nature" and expressed his confidence that "the radical tendencies of his wife and her relatives would not induce him to fail in his duty."[120] As was his negative finding for Hurley, Neylan's positive recommendation for Serber was entirely justifiable within the AEC's new security regime. Again, however, it is important to note that an opposite finding would also have been legitimate under the auspices of the Interim Procedure. Neylan's decision to weight lightly Serber's extensive and well-documented radical, even Communist associations, as well as the extent to which Serber himself was committed to leftist causes, was a choice.

For Lawrence's part, his successful appearance before Neylan's board was in marked contrast with Latimer's debacle. Lawrence's written statement in support of Serber expressed concern that *Serber* was the subject of inquiry. By contrast, Latimer's expressed indignation at AEC personnel security procedures as a whole. Lawrence emphasized Serber's positive contributions to the war effort and postwar AEC projects, and acknowledged but downplayed radical associations. He understood his audience better than did Latimer; his suggestion that Serber wanted to accompany the Nagasaki mission impressed the two military men on board.[121] In all this, it is clear that Lawrence's eager clarifications and qualifying comments imply that he felt that his role in protecting the clearances of his staff did not end when he secured John Neylan as chair of the Board.

Melvin Calvin

The techniques Lawrence used to defend Robert Serber illustrate the malleability of the Interim Procedure to personal influence, and Serber's case would not be an isolated example of the successful deployment of such tactics. One year later, on June 23, 1949, a Berkeley personnel security board again convened with Neylan as chairman. Replacing Admiral Nimitz for this hearing was fellow University of California Regent Farnham Griffiths—an indication that Neylan had influence over his supporting cast. The target was Melvin Calvin, a Berkeley chemist and Manhattan Project veteran who would in a little over a decade win a Nobel Prize.

119. Ibid., 121.
120. "Recommendation in re. Robert Serber," JFNP, box 175, brown folder.
121. Serber (ref. 95), 164.

Calvin was born in St. Paul, Minnesota, to Russian-born parents and raised largely in the Detroit area. He earned a bachelor's degree from the Michigan School of Mines in 1931 and completed a doctorate at the University of Wisconsin by 1935. Among American scientists during this period, Calvin told Neylan, "it was more or less the style to take a year or two in Europe after obtaining the doctor's degree," and he did this beginning in the fall of 1935.[122] For the next two years, Calvin studied on a Rockefeller grant with the well-known English chemist Michael Polanyi before being hired by Gilbert Lewis to join the University of California chemistry department. In 1941 Lewis and Latimer attended a National Defense Research Council meeting and returned with defense projects for Berkeley chemists. Calvin worked on one such military assignment until about 1944, when Glenn Seaborg and Latimer recruited him as a consultant to the Manhattan Project in Chicago. While Calvin had no direct affiliation with the Rad Lab prior to the war, afterwards, Ernest Lawrence gave him support to develop a chemical biodynamics laboratory working in concert with Lawrence's brother John primarily on peacetime biological and medical applications of radioactive isotopes.[123] It was in this work in which Calvin was employed on June 14, 1949, when he received the "security statement" of charges from Tammaro.

On June 16, Neylan was confirmed as chair of Melvin Calvin's security board. That same day he met with Berkeley AEC security officials to discuss Calvin's upcoming hearing, and the following morning Neylan began organizing for what he termed "Colim's" security hearing. So security-conscious was John Neylan that he even used a pseudonym for the subject of inquiry in his personal to-do list. He telephoned General Joyce, Griffiths, Lawrence, and, as noted in his memo book, decided to hold the hearing June 23.[124]

Mirroring his efforts the previous summer, Lawrence wrote a glowing letter of support for Calvin in language echoing the one he had sent regarding Serber. Lawrence again declared that he and "all his colleagues in the laboratory have the highest regard for Professor Melvin Calvin and are deeply concerned that some questions have been raised as to his clearance status." This time Lawrence also noted that upon hearing of Calvin's problems, he had held his own informal hearing into these matters, which only "strengthened and reaffirmed" his high opinion of Calvin's character, associations, and loyalty—a pre-emptive move which confirmed Lawrence's cooperation with AEC security measures. And as he had for Serber, Lawrence added that "quite apart from his great scientific talents, I have always regarded him also as a man of fine personal qualities of character, loyalty and integrity."[125]

As in the previous cases, when the hearing commenced in John Neylan's San Francisco office in June 1949, Calvin's associations took center stage. Foremost in Neylan's

122. "In the matter of Melvin Calvin" (ref. 41), 13.

123. Ibid., 30. For more on Calvin's lab and extent to which Lawrence supported Calvin's science financially and institutionally, see Roger L. Geiger, *Research and relevant knowledge: American research universities since World War II* (New Brunswick, NJ, 2004), 77-78.

124. JFNP, box 179.

125. Lawrence to Fidler, 18 June 1949, EOLP, carton 3, folder 26.

inquiry were Melvin Calvin's personal and professional associations, as well as what the AEC interpreted as an abnormal affinity for Russia. Neylan prodded Calvin on his wife's college activities and the political leanings of his immediate family. Of particular interest to Neylan was the fact that the FBI had once noted a car registered to Calvin's mother close to the location of a Communist meeting in Los Angeles. Ernest Lawrence would again appear on behalf of his colleague and offer Neylan's board his strong assurance of Calvin's character, associations, and loyalty.[126]

The bulk of Calvin's derogatory information came from professional associations. Michael Polanyi, with whom Calvin had chosen to do his postdoctoral fellowship, had since defected from chemistry to a position as professor of economics. Calvin described Polanyi's interests as far to the right, the antithesis of communism—an explanation Neylan accepted. Of further concern to Neylan was Calvin's relationship with Martin Kamen, a fellow Radiation Lab chemist. Kamen had been fired in July of 1944 under suspicion of passing classified data to the Russians.[127] Asked if he ever discussed classified data with Kamen while Kamen still held a Q clearance, Calvin answered that he did not remember, but would not be surprised if he had. This time Neylan interjected before Lawrence needed to, clarifying "so that the record will be very clear on that, Doctor: Of course all scientists employed on the project who were cleared of necessity discussed restricted matters with one another?"[128] Again Neylan tossed Calvin softballs.

Also appearing in the statement of charges against Calvin was his association with George Eltenton. Eltenton was the third man in the Chevalier affair that had come up in Serber's hearing and would later have graver consequences for Robert Oppenheimer. In 1946 Eltenton admitted to the FBI that in 1943, at the urging of a Soviet diplomat, he had called on mutual friend Haakon Chevalier to act as an intermediary to Robert Oppenheimer in an attempt to pass secrets to Russia.[129] Oppenheimer was sufficiently disturbed by Eltenton's role that he turned Berkeley security officials on to him later that year. Calvin testified that Eltenton was a chemist at the Shell Development Corporation in Emeryville, a few miles from the University of California. They had met on a number of occasions since Eltenton "was a member of the Chemical Society and [Calvin] was a member of the Chemical Society. We used to go to meetings. I went to the meetings quite regularly. He worked in chemical kinetics, as I did."[130] As the handful of meetings Calvin could remember with Eltenton occurred in the same period as did Eltenton's approach to Chevalier, Neylan asked Calvin if he was ever similarly approached, and if he ever disclosed any restricted information to Eltenton.[131] Calvin emphatically rejected such a scenario, and Neylan let the matter drop.

126. Donald Cooksey also appeared and offered Calvin similar support.
127. "In the matter of Melvin Calvin" (ref. 41), 69.
128. Ibid., 67.
129. Herken (ref. 27), 91-92.
130. "In the matter of Melvin Calvin" (ref. 41), 38.
131. Ibid., 47.

Neylan and his board did not take long to clear Melvin Calvin. Likely eager to get to the Regents' meeting in Los Angeles the following day, Neylan sent the findings of the board to Berkeley area manager Harold Fidler immediately after the hearing concluded. At the Regents' meeting, a few hours after handing the AEC its recommendation, Neylan led the Board of Regents in adopting the controversial anti-Communist policy and loyalty oath for the University. Neylan's role as the non-signers' chief adversary would make him a controversial and nationally known figure, defining him in history as a symbol of the excesses of anti-Communist hysteria.[132] But on the eve of such a destiny, Neylan wrote, "it is the unanimous judgment of the Board that Dr. Calvin should be declared eligible to continued employment on Atomic Energy Commission work."[133]

Melvin Calvin's experience before Neylan's board in 1949 indicates that Lawrence had access to a strategy for protecting his staff from the AEC anti-Communist security apparatus, a strategy that he repeated identically in Serber and Calvin's cases. That his support for Serber, a personal friend, was consistent with his support for Calvin, a professional colleague, indicates that Lawrence was protecting his Radiation Laboratory, not granting personal favors. In 1960 Childs asked Donald Cooksey about Lawrence's role in personnel security clearances, in reply to which Cooksey noted "there were several Ernest helped very much by getting them not to fight it."[134] "Getting them not to fight it" is likely an apt description of Lawrence's role before Neylan's board. In playing by AEC rules Lawrence gave himself personnel security credibility, and in similarly preventing Rad Lab scientists from fighting the system all the way into unemployment, Lawrence maintained control. During the war, Lawrence had already played an active role in protecting the security clearances of his Rad Lab personnel. For example, Martin Kamen remembered that in 1941 Lawrence was furious with his and Robert Oppenheimer's continued participation in leftist causes, because such activities were making it more difficult for him to procure clearances.[135] Clearly Lawrence knew that as director of the Radiation Laboratory overseeing hundreds of employees, he held great capacity to influence personnel security decisions. That Lawrence would withdraw support from employees such as Kamen or even Frank Oppenheimer, whose personal choices or biographies made them difficult to control, also seems plausible. Lawrence's lobbying would be ineffectual without cooperation from his employees,

132. David P. Gardner, *The California loyalty oath controversy* (Berkeley, 1967), 3.
133. "Recommendations in re: Dr. Melvin Calvin," 23 June 1949, JFNP, box 171, folder 2.
134. Donald Cooksey, in Neylan interview (ref. 32). In his description of Neylan's chairmanship of the Berkeley boards, Herbert Childs makes veiled reference the Serber and Calvin cases, stating, "There were several for whom Ernest fought valiantly, who were not evasive, and for whom there was considerable doubt of conscious intent to betray; they were cleared." As he did in reference to Hurley's case, Childs represents Neylan's methods and findings as infallible. Childs (ref. 32), 406.
135. Martin Kamen, *Radiant science, dark politics* (Berkeley, 1985), 185.

and extending his support to uncooperative employees would jeopardize his own credibility in future campaigns.[136]

It is an oversimplification, however, to attribute Lawrence's successful campaigns on behalf of these men to a simple matter of taking over Neylan's responsibility for Serber or Calvin's future security indiscretions. It is true that board members only stood to lose if they granted employees the benefit of the doubt, and thus there existed in all clearance evaluations a tendency toward negative findings. This tendency could be halted if someone else offered to shoulder criticism in the case that suspect employees proved problematic. That Lawrence would fill the role of fall guy in such a case is true, but this fact is insufficient in understanding Lawrence's success. In standing up for Hurley, Latimer was offering to shoulder responsibility in the case that Hurley proved disloyal; Neylan had no problem turning down that offer. Lawrence could successfully shape security findings to the extent to which he maintained Neylan's confidence; Latimer was unsuccessful when he refused to court it.

Neylan often commented that he served on the Berkeley PSB largely at the personal urging of Ernest Lawrence. In the midst of particularly ugly relations with Wendell Latimer, Neylan wrote to Lawrence with feigned resentment for getting him into the "thankless drudgery of serving on security boards."[137] While it is impossible to deny the very real affection between Lawrence and Neylan, it is also unlikely that Lawrence was unaware that their friendship could be wielded in his own best interest.[138] However, the idea of Neylan sitting as a stooge on the Berkeley board protecting Lawrence's interests or doing the AEC's bidding is wholly implausible. Neylan was not a man inclined toward granting security clearances in remotely questionable cases. Rather, all available evidence points to his thorough commitment to Cold War anti-Communist ideology, the validity of associations as criteria for protecting classified information, the necessity of maintaining atomic secrets, and the belief that radicals posed a grave threat to America. In framing his defense of employees, Lawrence was not only taking responsibility for their future actions, but making it palatable to Neylan to grant clearances to *anyone* with a questionable background.

Something Resembling Justice

The personal side of personnel security that Lawrence exploited is exactly what David Lilienthal sought to banish when he struggled to develop an AEC program

136. For further examples of Cold War era administrators maintaining control and institutional autonomy by initiating internal purges of uncooperative faculty, see Schrecker (ref. 5).
137. Neylan to Lawrence, 14 Aug 1950, JFNP, box 171, folder 3.
138. Jennet Conant has explored another example of Lawrence's forging a personal relationship with a very powerful man – Walll Street tycoon and patron to science Alfred Loomis—offering tremendous benefits for Lawrence's professional aspirations. Jennet Conant, *Tuxedo Park: A Wall Street tycoon and the secret palace of science that changed the course of World War II* (New York, 2002).

consistent with civil liberties. Despite his efforts, in the years following adoption of the Interim Procedure, Lilienthal appears to have been unconvinced that the AEC had been successful on this front. While it is thus far unrecorded what, if anything, Lilienthal thought about Neylan's personnel security board, the AEC Chairman described a similar proceeding in his journal, contemporaneous with the Berkeley board, as "a sickening personnel security case...these things tear me up inside; the gossip and the horrid things in that file that for centuries we have been building defenses against."[139] In 1949 JCAE hearings into the AEC, Lilienthal's foe Bourke Hickenlooper, looking to highlight the shortfalls of its personnel security, pointed to one case in which the sheer number of anonymous informants speaking ill of the employee was clearly incongruous with the local security board's recommendation for clearance. Lilienthal eloquently responded with a clear indication that while he shared Hickenlooper's frustration with AEC measures, the two men were not speaking the same language: [140]

> [I]t is a new thing, strange, and to me a very troublesome thing, in American life, and I believe it will be to the members of this committee, even to approximate the idea that the making of charges, the raising of doubts, and the raising of suspicion are in themselves a basis for condemnation of an individual. We just do not do things that way;...Charges and justice are two quite different things....The procedures we have set up, with the advice of many individuals, just as the procedures in the case of the President's loyalty program, are efforts to bridge the gap between charges, suspicions, doubts, ex parte derogatory information, so-called, and something resembling justice under the very difficult conditions of these cases.

Despite Lilienthal's best efforts, he knew that a year after the Interim Procedure was in place, the AEC, within the Cold War national security state, could at best hope to approximate justice.

As the atomic scientist most associated with the excesses of AEC personnel security, and as a figure looming in the shadows of Neylan's board, it is notable that Robert Oppenheimer in the years after 1948 also weighed in on the validity of the Interim Procedure. At the same 1949 Joint Committee hearings at which Lilienthal clashed with Hickenlooper, Oppenheimer defended Commission personnel security procedures. While not enthusiastic about the realities of balancing security with free scientific discourse, Oppenheimer stamped his approval on the current measures. "My feeling," he said, "is that the Commission has sweated blood over that one and come up with about as good an answer as it is possible to get." On the issue of civil liberties, Oppenheimer testified, "I think the Commission has balanced very carefully in the few cases I know about, the requirements of security and the requirements of progress and humanity. It has not been easy."[141] It is very likely, in fact, that Robert Serber's case was

139. Lilienthal (ref. 11), 28 Sep 1948, 413.
140. JCAE (ref. 19), 6 June 1949, 162.
141. Ibid., 13 June 1949, 302.

one of the few personnel security cases with which Oppenheimer was acquainted. As General Advisory Committee chairman, Oppenheimer had reason to avoid publicly criticizing the AEC to an audience as hostile as Hickenlooper's JCAE.

Notably, despite his public support for Lilienthal and the AEC, Oppenheimer's concern about the realities of AEC personnel security ran much deeper than voiced before the Joint Committee. Through 1949-1950, Oppenheimer struggled to organize a study of looming and troublesome legal questions raised by AEC personnel security measures. He applauded the work of some scholars pointing out infringements on civil rights, but contended that such studies did not address the big legal questions. Previous studies, he suggested, [142]

> do not seem to me to come adequately to grips with the legal and constitutional essentials of the problem. They do not, for example, deal with the increasing activity of the Federal Bureau of Investigation...they also do not come to grips with the questions of what the objectives of our current security and loyalty programs are, nor raise, nor resolve the issue of whether those objectives are in themselves, and whether or not they could be achieved by methods less disruptive to our health and freedom.

Oppenheimer envisioned a grand study into the legitimacy and methods of the FBI, the composition of personnel security boards, the relationship of political views to loyalty, the effect of personnel security measures on employee morale, and the implications of invasive investigations for governmental powers. He shopped around for prestigious legal scholars to carry out his hypothetical study and even offered Institute for Advanced Study money to fund it. It is unclear whether this study Oppenheimer so fervently wanted ever came to fruition. His efforts however, are striking as an exceptionally thoughtful formulation of the fundamental problems with peculiarly Cold War notions of loyalty and security.[143] The discrepancy between Oppenheimer's public statements and private actions are illustrative of the tense relationship between the AEC and the Joint Committee, and his frustration with the slow progress an indication that his concerns were not universally of such high priority.

Despite the efforts of Chairman Lilienthal, the AEC procedure was obviously not without its critics, and in subsequent years scientists' organizations such as the Federation of American Scientists, the American Association for the Advancement of Science, and the National Academy of Sciences continued to quietly press for reforms. In December 1948 the AAAS's six-month-old Special Committee on Civil Liberties issued a scathing critique of the Interim Procedure. This panel of legal scholars criticized the lack of formalized standards for rendering decisions, the prevalence of vague charges, the characteristic overreliance on FBI evidence, the lack of an obligation to provide reasons for negative decisions, and the absence of procedural

142. JRO to Grenville Clark, 17 May 1949, JROP, box 233, folder Legal Studies.
143. See also Kai Bird and Martin J. Sherwin, *American Prometheus: The triumph and tragedy of J. Robert Oppenheimer* (New York, 2005), 405.

recourse for AEC applicants (non-current employees).[144] In response to such criticism, the AEC issued early in 1949 its "Criteria for Determining Eligibility or Ineligibility for Personnel Security Clearance."[145] The Criteria, while a straightforward-sounding document, did nothing to clarify the informal rules and structures of authority that plagued the Interim Procedure. It offered examples of potentially derogatory information, none of which were unexpected, but did not put any tighter bounds on the evaluation of such evidence. The influence of the new Criteria on Melvin Calvin's case, the only Berkeley case after their adoption, was likely zero. In September of 1950 the AEC revised its Interim Procedure for the last time to extend PSB rights to AEC applicants in addition to employees.[146]

Conclusion

That Lawrence was more successful than Latimer in protecting his protégés is not to imply that all scientists quietly cooperating within the security apparatus were similarly successful. Indeed, Jessica Wang has shown that scientists and administrators taking the safe route when navigating the Cold War security state often weakened their defenses against political attacks. But Lawrence's behavior before the Berkeley security board does not fit neatly into Wang's characterization of quiet diplomacy. Wang argues convincingly that many scientists walked a thin line between wanting to show support for the AEC and questioning the agency's increasingly burdensome personnel security tactics. The result was "quiet diplomacy." As big questions were left unstated, the scientific community passively legitimated the Cold War personnel security apparatus. One should not, however, overstate the extent to which such big picture concerns motivated individual scientists and administrators. There is little evidence to indicate that Lawrence approached Neylan's board with any political agenda beyond the immediate goal of protecting employees. Lawrence's postwar vision for atomic energy policy was by all accounts muddled. Originally offering support for the May-Johnson military control plan, Lawrence soon withdrew support upon learning that under such legislation his own routine wartime security indiscretions would have landed him in prison for ten years and earned him a hefty fine.[147] After withdrawing support for the May-Johnson bill, Lawrence refused to support any of the pending civilian control bills either. Lawrence was not a quiet diplomat out of concern for the political health of the Atomic Energy Commission. Nor was he, however, an unproblematic example of Cold War ideology personified. As evidenced by his own inability to maintain secrets, and his commitment to

144. Wang (ref. 4), 198.
145. "Criteria for determining eligibility or ineligibility for personnel security clearance," 5 Jan 1949, JFNP, box 171, folder 2.
146. "Security clearance procedures," 19 Sep 1950, JFNP, box 171, folder 2.
147. Herken (ref. 27), 156.

fewer, not more, personnel and physical security restrictions at the Berkeley Rad Lab, Lawrence was not a case study in the security-through-secrecy ideology marking Cold War conservatives. Lawrence's successful defenses of employees through well-orchestrated campaigns and apparent acquiescence to security measures are more revealing as an indication of the methods *he* found most fruitful, given his exceptional personal connections and sheer force of personality, than representative of generally promising protections against the Cold War national security state. While stories of personnel security cases often present flimsy standards and informal rules as ruinous for employees, it is notable that in Berkeley, AEC procedures allowed Lawrence to spin cases in favor of his staff. However, it is precisely this kind of focus on local and immediate concerns that Lawrence demonstrated, that legitimated and propagated the kind of anti-Communist policies which proved disastrous to scientists who did not share Lawrence's political and personal savvy.

Folding Robert Oppenheimer's hearing into the broader AEC personnel security story is obviously of only limited utility. Oppenheimer's case was exceptional in ways that have been studied exhaustively elsewhere.[148] Yet John Neylan's Berkeley board is remarkable to the extent it was a premonition of the 1954 case in both its procedures and its cast of characters. The objections raised by Latimer in the early years of AEC personnel security procedures would be echoed by Oppenheimer's camp. The heavy reliance on confidential FBI evidence, and the vague standards of character, associations, and loyalty, were not tricks cooked up by Lewis Strauss to attack Robert Oppenheimer; they were part of a dubious apparatus that had been wielded against AEC employees for years. The gaping holes in the Interim Procedure allowing for legitimate manifestations of personal bias that Lawrence exploited in favor of his staff, were similarly open to exploitation against Oppenheimer. When Strauss chose a personnel security hearing as the best means of discrediting Oppenheimer, he chose a model whose possibilities for manipulation had already been demonstrated. Then, in its cast of characters, the scientists who came before Neylan's board, were as troubled by their personal associations—with men such as Haakon Chevalier, George Eltenton, Martin Kamen, Joe Weinberg, David Bohm and Frank Oppenheimer—as would be Robert Oppenheimer himself. Neylan's proceedings against Hurley, Serber, and Calvin brought to the fore questions that could have just as easily been asked of Robert Oppenheimer. In a few years they would be. Indeed, that the AEC dredged up the same decades-old Berkeley associations as evidence against Oppenheimer in 1954 demonstrates the extent to which Oppenheimer enjoyed special treatment from the AEC for years.

148. Most recently, Bird and Sherwin (ref. 143) have highlighted ways in which the 1954 attack on Oppenheimer was exceptional. The AEC's use of a prosecutor (Roger Robb) and Oppenheimer's employment of a legal defense team indicate the unusual stakes on both sides. The FBI's steady feed on Oppenheimer and his meetings with the defense team no doubt gave the Commission a substantial advantage over the nearly month-long hearing, while Strauss's aggressive execution of the proceedings, likewise, had no parallel in Berkeley.

However unacceptable AEC personnel security proceedings such as Neylan's look to us today (and indeed as they looked to Lilienthal at the time), it is important to remember that when they were promulgated, they appeared a significant step toward protecting civil liberties. The AEC developed a personnel security apparatus going far beyond HUAC or the military in making procedures more consistent with individual liberties, and this fact often made the AEC vulnerable to Congressional attacks. Yet, despite David Lilienthal's best intentions, the AEC could not overcome the fundamental limitations of peculiarly Cold War notions of personnel security. The intense political pressure on the AEC in its early years weakened Lilienthal's intended system of checks and balances. The decentralization necessitated by the AEC's heavy caseload made personnel security even more local. As a result, opportunities for the interjection of personal influence, such as occurred in Berkeley, likely increased. Furthermore, the precedent set by the loyalty-security program set the standard for clearance unattainably high in some cases. This meant that AEC employees, despite Lilienthal's procedural reforms, faced the near impossibility of satisfying this Cold War expectation to prove 100% loyalty, when the standard for disloyalty came to include the academic interests and associations of endless years past. Personnel security procedures were not intended as a means of evaluating concrete actions, but of anticipating future indiscretions on the basis of what might be shoddy evidence and dubious assumptions. The informality of making projections about a person based on his character, associations, and loyalty made personnel security board hearings inherently improvised and personal affairs.

CHAPTER ELEVEN

Killing the Messenger
Robert Oppenheimer and Caltech's Project Vista

W. Patrick McCray

In June 1954, Lee DuBridge received a letter from an angry Pasadena citizen. It warned Caltech's president that the sympathies he expressed for J. Robert Oppenheimer when he testified before the AEC panel on his friend's behalf would not be tolerated by patriotic citizens. The country, the letter opined, was in "an ugly mood" and would make "short shrift" of people like DuBridge and Oppenheimer.[1] DuBridge shot back with uncharacteristic emotion that such sentiments were perhaps "the ugliest part" of the famous physicist's public humiliation.[2]

One of the topics Roger Robb had grilled DuBridge about was Oppenheimer's contribution to Project Vista. Vista was a classified "summer study" which Caltech hosted in 1951. More than one hundred scientists and other experts worked to determine how existing technologies—tactical nuclear weapons, in particular—could repel a massive

This article is adapted from "Project Vista, Caltech, and the dilemmas of Lee DuBridge," *Historical studies in the physical and biological sciences, 34:2* (2004), 339-370.

The following abbreviations are used: *BAS, Bulletin of the Atomic Scientists*; CBMP, Clark B. Millikan Papers, Archives, CIT; CCLP, Charles C. Lauritsen Papers, Archives, CIT; CIT, California Institute of Technology; HF, Historical Files, Archives, CIT; *HSPS, Historical studies in the physical and biological sciences*; HVP, Gen. Hoyt Vandenberg Papers, Manuscript Division, Library of Congress, Washington, D.C.; JRO, J. Robert Oppenheimer; *JRO Hearing*, United States Atomic Energy Commission, *In the matter of J. Robert Oppenheimer* (Cambridge, 1971); JROP, J. Robert Oppenheimer Papers, Manuscript Division, Library of Congress, Washington, D.C.; LADP, Lee A. DuBridge Papers, Archives, CIT; NARA, National Archives and Records Administration, Washington, D.C., and College Park, MD; NBL/AIP, Niels Bohr Library, American Institute of Physics, College Park, MD; RAMP, Robert A. Millikan Papers, Archives, CIT; RG, Record Group; RPFP, Richard P. Feynman Papers, Archives, CIT; TMM, Trustees' Meeting Minutes, President's Office, CIT; WAFP, William A. Fowler Papers, Archives, CIT; Vista Report, "A study of ground and air tactical warfare with especial reference to the defense of Western Europe."

1. Arthur Little to Lee A. DuBridge, 24 June 1954, LADP, 33.1.
2. Lee A. DuBridge to Arthur Little, 28 Jun 1954, LADP, 33.1.

Soviet invasion of Europe many perceived as likely if not imminent. A local Pasadena paper boasted that the project was even more secret than the Manhattan Project.

The early 1950s were the golden era of "summer studies," so called as they coincided more or less with the academic summer recess. Vista was an example of these specialized, goal-oriented examinations, typically done using newly emerging concepts such as systems engineering and operations research, of defense-related problems.[3] These projects generated new research opportunities (both classified and unclassified) and forged beneficial military and industrial connections that universities used to their advantage. They also helped create a "social reality in which working on military problems was an accepted norm," if not indeed a sign that one was a well-connected, top-notch researcher.[4] Summer studies often left behind a legacy of well-funded, university-managed follow-up projects or even laboratories where classified research continued to be pursued. This was a prize and responsibility Caltech chose not to pursue after Vista.[5]

For Caltech, an elite yet relatively small institution which prided itself on its basic research and close faculty-student interactions, the Vista study was a major commitment. Over a quarter of its full-time faculty took part; the number grows to over 120 if one counts scientists visiting from other institutions, military liaisons, and secretarial and security staff. The entire project consumed some 7,800 days of effort and disrupted campus teaching and research for months. In return for its participation, Caltech received $750,000 in compensation.

Vista was much larger and more expensive than previous summer studies. MIT, for example, carried out Project Hartwell between March and December 1950, which cost $124,000 and employed only 21 scientists.[6] Vista drew upon the talents and expertise of a wider range of experts as well. People from industry and the military played an active part as did scholars from fields other than the physical sciences.

From its inception, the leaders of Vista counseled the military not to be disappointed that their report did not suggest specific ways to make weapons more destructive, reliable, or efficient.[7] William A. Fowler, a nuclear physicist from Caltech and

3. J.R. Marvin and F.J. Weyl, "The summer study," *Naval research reviews, 19* (1966), 1-7, 24-28.
4. Silvan S. Schweber, "The mutual embrace of science and the military: ONR and the growth of physics in the United States after World War II," Everett Mendelsohn, Merritt Roe Smith, and Peter Weingart, eds., *Science, technology, and the military* (Dordrecht, 1988), 33.
5. For example, Project Michael led to the establishment of Hudson Laboratories at Columbia University, while MIT's participation in Project Charles resulted in the formation of Lincoln Laboratories; Marvin and Weyl (ref. 1).
6. Daniel J. Kevles, "K_1S_2: Korea, science, and the state," Peter Galison and Bruce Hevly, eds., *Big science: The growth of large-scale research* (Stanford, 1992), 312-333, on 316.
7. "Letter of transmittal" (p. vii) in the final report of Project Vista, entitled "A study of ground and air tactical warfare with especial reference to the defense of Western Europe," Feb 1952, HF, section Y (War Activities), box Y3. The Vista Report was published in two volumes; the first volume contains the monographs which comprise the actual report, while the second is a

Vista's Scientific Director, emphasized that the military should not expect new gadgets to emerge from Vista. Instead, Vista advised the military on how to use tools already at its disposal, whether nuclear-tipped rockets or sympathetic partisans armed with Molotov cocktails, to resist a Soviet incursion into Europe.

Oppenheimer's actual role in Vista was limited. As DuBridge told the AEC panel, Oppenheimer simply helped the group responsible for studying "atomic warfare" clarify its thoughts. Nonetheless, the final Vista report contained many policy suggestions which Oppenheimer had championed for some time. Furious at the report's suggestions to strengthen U.S. tactical air power and develop smaller and less powerful atomic weapons, the Air Force suppressed the Vista report in early 1952.

The story of Project Vista provides a fascinating look at the nexus of science, technology, and national security. In this paper I examine the history of Project Vista with an eye toward three topics. First, I consider how Project Vista came to Caltech. Second, I discuss Oppenheimer's contribution to Vista as part of the project's overall attempt—as the final Vista report stated—to bring the battle back to the battlefield. Finally, I explore some of the Vista's fallout, both for Caltech and Oppenheimer as well as for other scientists who contributed to defense-related research during the militarization of the Cold War.

Vista Comes to Caltech

North Korea's sudden invasion of its southern neighbor in the summer of 1950 stunned Americans. Many believed North Korea's incursion was the harbinger of an even greater geopolitical and military crisis—the invasion of western Europe by an emboldened Soviet army backed by swarms of tanks, artillery, and jet aircraft. For many in the government and those who followed world events closely, a Soviet invasion of Europe was not only possible but something NATO's outnumbered conventional forces appeared all but powerless to prevent.[8] It was this specific threat that Project Vista addressed.

In the fall of 1950, Charles C. Lauritsen, a Caltech nuclear physicist who was highly regarded as a military advisor, took an extended tour of Korea's battlefields. During World War II, Lauritsen had a prominent role in developing many key military

---

collection of technical appendices and supporting information. The Vista Report was classified until 1980, when a redacted version of it was made available for study. Since then, several documents pertaining to the report in the National Archives and Records Administration have been reclassified.

8. American politicians' and policy makers' fear of Soviet intentions was crystallized when President Truman approved NSC-68 in April 1950 after it was presented to the National Security Council. A profound influence on American thinking and strategy in the early stages of the Cold War, NSC-68 predicted that 1950-1953 would be the years of maximum peril. Ernest R. May, ed., *American Cold War strategy: Interpreting NSC-68* (New York, 1993).

technologies. He had helped persuade the military that rocket technology and proximity fuses could serve as practical weaponry and had participated in the Manhattan Project at Los Alamos. He had also helped establish the Naval Ordnance Test Station at China Lake, California. Lauritsen was accompanied on his trip by other members of the Weapons Systems Evaluation Group, a Defense Department advisory committee.[9] They arrived in Korea on the heels of General Douglas MacArthur's stunning amphibious assault at Inchon and were escorted by General James M. Gavin, an outspoken proponent of using tactical nuclear weapons against North Korean troops.[10]

Lauritsen briefed the Institute's trustees on his return from Korea. He explained that the U.S. military needed to greatly improve its tactical air support of ground troops.[11] Lauritsen also told the trustees about his recent participation in an MIT-managed summer study. Project Hartwell examined how the Navy could protect shipping from attacks by Soviet forces in the event of a major war and questioned the prevailing military view that all nuclear arms should be high-yield weapons delivered from big bombers.[12]

The idea that nuclear weapons could be used for purposes other than incinerating Soviet cities was gaining currency among some scientists and military leaders who were beginning to take the case for tactical nuclear weapons to the public. Prominent advisors like Robert Oppenheimer, Vannevar Bush, and James Conant recommended that the military diversify its atomic arsenal and develop tactical nuclear weapons, opinions that naturally found opposition from the Air Force's Strategic Air Command.[13] DuBridge, along with Oppenheimer, had opposed the hydrogen bomb while several Caltech scientists, including Lauritsen and William Fowler, shared reservations about the dangers of nuclear testing and America's emphasis on the large-scale nuclear bombing of Soviet cities.[14]

9. This group included Edward Bowles of MIT and William Shockley of Bell Labs.
10. Gavin's views and a recounting of Lauritsen's visit to Korea are in his book *War and peace in the space age* (New York, 1958).
11. "Undated presentation [likely November 1950]," CCLP, box 11, folder 10; "Minutes—November 6, 1950 meeting of the Trustees," TMM.
12. Project Hartwell is described in a number of articles including Daniel Kevles, "Cold War and hot physics: Science, security, and the American state," *HSPS, 20:2* (1990), 239-264, and Schweber (ref. 4).
13. For example, Vannevar Bush, "The weapons we need for freedom," *Reader's digest*, Jan 1951, 48-51. Oppenheimer gave a public speech in January 1951 to the New York Bar Association which criticized strategic bombing. It was later reprinted in the February 1951 issue of the *BAS*. In the same issue, there also appeared James. M. Gavin's article "The tactical use of the atomic bomb," *BAS, 7* (1951), 46-47, 50 which gave a military perspective. Excellent presentations of scientists' lobbying against strategic nuclear bombing and in favor of tactical weapons are found in Gregg Herken, *Counsels of war* (New York, 1987), chapt. 7; Matthew Evangelista, *Innovation and the arms race: How the United States and the Soviet Union develop new military technologies* (Ithaca, NY, 1988), 116-118.
14. William Fowler, for instance, gave public speeches presenting the case against building

In the wake of the Korean invasion, the revelation of Soviet nuclear capability, and Truman's decision to undertake a crash program to build the hydrogen bomb, the Air Force decided that it needed further input from university scientists. It was at this time that representatives of the Research and Development Division of the Air Force began to talk informally with DuBridge, Lauritsen, and Robert F. Bacher (the head of Caltech's Physics and Astronomy Division and a former AEC commissioner) about a possible summer study at Caltech.[15] The Air Force's initial request was that Caltech examine strategic air power and nuclear strategy.

While not enthusiastic about this topic or Caltech's involvement in general, DuBridge believed it was unlikely that the school would avoid conscription of some sort. In January 1951, he warned Caltech's faculty and trustees that escalating international crises made it "probable that...the Institute will be called upon to render services of many sorts to the nation...in this time of need."[16] Consequently, DuBridge encouraged Caltech's trustees to apply for security clearances.[17]

The Air Force's dogged courtship of Caltech for Vista can be traced to two people. Louis N. Ridenour, the Air Force's first chief scientist, earned his Ph.D. from Caltech in 1936. During the war, Ridenour worked on radar technology at MIT's Radiation Laboratory, which DuBridge directed.[18] Throughout 1950, Ridenour negotiated with scientists and administrators at MIT to establish new summer studies which would examine the defense of the United States against a Soviet bomber attack. This accomplished, Ridenour turned his attention to the problem of tactical air warfare, a neglected field he believed "desperately needed help."[19]

Caltech trustee John McCone supported Ridenour's lobbying efforts. Trained as a mechanical engineer at Berkeley, McCone had owned large shipping and construction firms during the war. After the war, McCone—by now a wealthy industrialist active in Republican politics—turned his attention toward government and civic service and remained an advocate for Southern California defense firms. In 1947, a year after McCone became a Caltech trustee, Thomas K. Finletter, a Washington lawyer and future Secretary of the Air Force, invited him to serve on a small panel established by

---

the hydrogen bomb, such as a 23 Mar 1950 talk to the Altadena-Pasadena Young Democrats, WAFP, box 142, folders 12 and 13.

15. Lauritsen had been considering the possibility of some type of study "to analyze the overall American position and to make specific proposals of the general course to follow including military plans" early in 1950, as noted in Lauritsen to Frederick Seitz, 15 Mar 1950, CCLP, box 3, folder 3. Actual meetings for what became Vista began even while Lauritsen was still in Korea; Ivan Getting to Robert Bacher, 26 Sep 1950, WAFP, box 3, folder 2.

16. Lee A. DuBridge, "The President's report," *Bulletin of the California Institute of Technology, 60:1* (1951), 1.

17. DuBridge memo to Trustees, 11 Sep 1950, LADP, box 130, folder 12. The clearances came through in July 1951 when Vista was underway.

18. Dwayne A. Day, *Lightning rod: A history of the Air Force's Chief Scientist's Office* (Washington, D.C., 2000).

19. Quote from Ridenour's colleague Ivan Getting, Day (ref. 18), 41.

President Truman to create a national policy for strengthening military and civilian aviation.[20] In June 1950, McCone briefly resigned from Caltech's Board of Trustees to become an undersecretary for the Air Force under Finletter.

McCone was well situated to steer Caltech toward participation in Project Vista. As a trustee, he understood how Caltech functioned as an organization and was acquainted with the particular strengths of its scientists. Privy to the Air Force's needs and concerns, his government service exposed him to broader issues associated with national security. Finally, McCone was a staunch anti-Communist who strongly believed in strengthening the United States military. Throughout late 1950 and into 1951, McCone made "urgent requests" for Caltech to undertake the project.[21]

Despite these entreaties, DuBridge remained skeptical. Ridenour concluded that the Army might have better luck convincing his colleague.[22] In mid-February, DuBridge met with Frank Pace, the Secretary of the Army, who requested that Caltech study how Army troops might receive better combat air support. Back in Washington less than a month later, DuBridge was surprised at the airport by Ridenour. The Air Force scientist alerted him that the Pentagon planned to formally ask Caltech to undertake "a study of ground-air tactical operations under an Army contract with close collaboration of the Air Force."[23] A few weeks later, DuBridge bowed to the inevitable and met with an entourage of Air Force and Army generals to finalize arrangements for what became Project Vista.

DuBridge now sought approval for Vista from Caltech's trustees. Despite the right-leaning views of most board members, several were initially leery of Caltech's participation in a major classified study.[24] Putting aside his own doubts, DuBridge marshaled persuasive arguments to convince the trustees this was the right decision. First, DuBridge pointed out Caltech's history of national service in times of emergency. On the basis of Caltech's experience in areas such as designing rockets for tactical warfare, the Institute, DuBridge claimed, had "been 'drafted' for this job." Moreover, given the pervading atmosphere of international crisis, he believed many Caltech faculty wanted to make a patriotic "contribution to the nation's defense effort."[25]

Institutional interests also motivated Caltech to accept Project Vista. It could help keep the Institute's staff together at a time when there was a strong temptation for faculty to be lured elsewhere. Meanwhile, money Caltech received from the secret study

20. The President's Air Policy Commission, *Survival in the air age: A report by the President's Air Policy Commission* (Washington, D.C., 1948).

21. DuBridge memo to Trustees (confidential), 5 Apr 1951, WAFP, box 116, folder 2.

22. Ridenour memo to General Gordon P. Saville, "Proposed study of tactical warfare" (confidential), 20 Feb 1951, NARA, RG 341, box 9, "360.3 Tactical Air Warfare."

23. DuBridge memo to Fowler, 27 Mar 1951, WAFP, box 116, folder 1.

24. Lee A. DuBridge, oral history interview with Finn Aaserud, 14 Feb 1986, NBL/AIP, 17-18.

25. DuBridge memo to Trustees (confidential), 5 Apr 1951 (ref. 21).

might, according to DuBridge, "mean the difference between a deficit and breaking even on the budget."[26]

DuBridge was also concerned that if schools like Caltech refused studies like Vista, the military would award these to entities like RAND with "the probable result that numerous nonprofit industrial research laboratories will be set up with government support." These labs, with their higher salaries and their "patriotic incentives," might out-compete university facilities, perhaps even causing "permanent damage" to the university infrastructure as a whole. The only choice available, as he explained it to the trustees, was for universities to administer such studies and fulfill national needs while not crippling their "fundamental research programs" at the same time.[27]

Not since World War II, DuBridge concluded, had so much "detailed consideration" been given by "so many individuals on the campus" to a government project. Moved by appeals to patriotism, financial prudence, and institutional preservation, the trustees approved Project Vista in April 1951.

Carrying Out Vista

As soon as Caltech had committed itself to Project Vista, organizational efforts for the summer study moved into high gear. DuBridge asked William Fowler, a forty-year-old nuclear physicist at Caltech, to be Vista's Scientific Director. A veteran of Caltech's wartime rocket work and a witness to the Trinity test, Fowler recognized that Vista would be controversial. Writing DuBridge from Washington, Fowler predicted that "the tactical use of nuclear weapons will be one of our biggest problems" and any recommendations the study put forward would "be meaningless" without the "views of experts like Oppenheimer."[28]

Recruiting the competent scientists Project Vista needed taxed the social and professional networks of Fowler and DuBridge. With several military-sponsored studies underway, it was difficult to find qualified scientists not already engaged in defense work or who were willing to put aside their personal research.[29] A more serious recruitment problem was that some scientists did not believe studies like Project Vista were useful.

26. Ibid. Scientific manpower was a topic DuBridge was familiar with not solely through his service on the National Security Resources Board's Scientific Manpower Advisory Committee in 1950 and 1951.
27. "Policies and procedures governing contracts undertaken at the request of governmental agencies," 15 Apr 1951. Part of meeting minutes, Executive Committee of Trustees, 17 Apr 1951, TMM.
28. Fowler to DuBridge, 31 Mar 1951; Fowler memo to DuBridge, 23 Apr 1951, WAFP, box 117, folder 2.
29. A fact noted in Office of Defense Mobilization—Science Advisory Committee, draft, "Scientists and mobilization: General remarks on development of a plan," 6 Sep 1951, LADP, box 186, folder 7.

This reluctance was evident when Fowler courted physicist Merle Tuve to lend his expertise on electronics to Vista. Fowler told Tuve he hoped Caltech's study might provide help for the "heretofore scientifically neglected doughboy." Unconvinced, Tuve criticized such "expert kibitzing projects," and urged that the scientific community not take on any more "arm chair analysis and advising."[30]

Nonetheless, by summer of 1951, a critical mass of physicists and other scientists convened in Pasadena for the study. The personnel roster identifying persons enrolled in Vista in some manner reads like a who's who list for mid-century science—Carl Anderson, Lloyd Berkner, Enrico Fermi, Leland Haworth, Edward Teller, and Jerrold Zacharias were all among the eighty-odd participants.[31]

Like Project Troy, an MIT-based inquiry into political warfare and propaganda techniques, Vista was a hybrid study. It drew upon the knowledge of experts from the physical sciences as well as the humanities.[32] Historians, psychologists, and economists participated. Vista's ranks went beyond the academy. Frank Capra, a Caltech graduate who created the famous *Why we fight* series of propaganda films, contributed to Vista. Capra was not the only person from Hollywood to take part, either. Cinematographer Alfred L. Gilks, who had shot the 1951 Gene Kelly film *An American in Paris*, led a small study group. They helped prepare a dozen short films that showed a different sort of choreography—weapons demonstrations and military exercises.

Management of Vista participants was accomplished by breaking them into several different study groups. Most of these were chaired by Caltech faculty; Robert Bacher, for example, led the "Special Weapons" group (SPECON in Vista parlance) that Oppenheimer advised as they examined the role for tactical atomic weapons.[33]

Project Vista was a response to the current crisis of foreign relations, historical circumstance, military capabilities, and geopolitics. During the execution of Vista and the other summer studies, what Eisenhower later called the "scientific-technological elite" acted as strategists. These activities, as Tuve had warned Fowler, were seen by some as beyond scientists' expertise. Just as importantly, Vista became caught up in a larger debate happening in military and in policy circles about how to divide and deliver America's nuclear arsenal. As the summer of 1951 drew to a close and they began to draft their report, Fowler, Lauritsen, DuBridge, and other leaders of Vista

30. Fowler to Tuve, 23 Apr 1951; Tuve to Fowler, 7 May 1951, WAFP, box 116, folder 10.

31. Of the many institutions that contributed scientists to Vista, the Cornell Aeronautical Laboratory was most generous and provided more than a dozen scientists, some of them on a full-time basis. At the time Vista got underway, CAL was engaged in developing the Lacrosse surface-to-surface guided missile, a complement to JPL's Corporal program.

32. Allan A. Needell, "Truth is our weapon: Project TROY, political warfare, and government/academic relations in the national security state," *Diplomatic history, 17:3* (1993), 399-420.

33. Bacher's colleagues on SPECON included Robert Christy (Caltech), Leland Haworth (Brookhaven), William Higinbotham (Brookhaven), Charles Lauritsen (Caltech), Robert Oppenheimer, Homer Stewart (Caltech), and Alan Thorndike (Brookhaven); Vista Report, 344. Edward Teller also contributed, according to Vista records, fifteen days of effort in July 1951: Vista personnel lists, WAFP, box 116, folder 7.

began to be more acutely aware of the powerful political currents circulating beneath Project Vista and around Robert Oppenheimer.

Vista Meets Opposition

Ultimately, it was Vista's recommendation to develop tactical nuclear weapons for the battlefield and the participation of Robert Oppenheimer that brought so much notoriety to it in the fall of 1951.[34] In language that alarmed the Air Force's Strategic Air Command, DuBridge and Fowler—both opponents of the hydrogen bomb—stated in Vista's introduction that the project had "found no great new weapons...we can get along with those we have."[35] The Vista report as a whole, and specifically the infamous Chapter Five ("Atomic Warfare") to which Oppenheimer contributed, was an academic ratification of views expressed publicly by well-known spokesmen like Oppenheimer and James Conant.[36] Their argument that the nation's relatively scarce fissionable materials would be better used to destroy an invading Soviet army rather than Russian cities was, as the *New York Times* reported, a "blow to the 'big bomber' school of thinking" that dominated Air Force thinking.[37]

In the early fall of 1951, Oppenheimer traveled back to Caltech, where he used to teach, to offer advice to Vista. By this time, he was a lightning rod for criticism from scientists and military leaders suspicious of his policy views on atomic warfare. During his stay in Pasadena, Oppenheimer helped Vista's "special weapons" panel organize their ideas. They drafted a report that outlined how the nation's nuclear stockpile should be divided between tactical weapons, strategic atomic bombs, and thermonuclear hydrogen bombs (the first prototype had been tested in the South Pacific the previous May) and how these weapons could be used in battle. Oppenheimer spent less time at the Vista del Arroyo Hotel than Edward Teller had earlier in the summer.[38] Nevertheless, his input to Vista's chapter on atomic warfare sparked the concern of high-ranking Air Force leaders.

34. The Vista Report and Oppenheimer's role in preparing it is commonly presented as evidence of his alleged bad judgment. See, for instance, John Major, *The Oppenheimer hearing* (New York, 1971), chapt. 5; also Herken (ref. 13), 66-70.
35. "Letter of transmittal," Vista Report, vi.
36. Conant's views, for example, on the defense of Europe are articulated in "The defense of Europe in the atomic age," 17 May 1951, talk to Chicago Association of Commerce and Council on Foreign Relations, JROP, box 27, file Conant, James B. (correspondence). See also James G. Hershberg, "'Over my dead body': James B. Conant and the hydrogen bomb," Mendelsohn, Smith, and Weingart (ref. 4), 379-430.
37. Hanson W. Baldwin, "Experts urge tactical air might; score stress on big atom bomber," *New York Times*, 5 June 1952.
38. Vista personnel lists (ref. 33). According to Air Force scientist David Griggs, Teller's contribution to Vista was to persuade the study that thermonuclear weapons deserved a more prominent role in their report. *JRO Hearing*, 762-763.

When news of Oppenheimer's participation filtered back to Washington, Air Force Secretary Thomas Finletter—already eager to keep Oppenheimer away from classified matters—was enraged. Acting on a suggestion from John McCone, Finletter dispatched the Air Force's new chief scientist, David Griggs, to hear an early presentation of Vista's findings.[39] Griggs was not sympathetic to the views on atomic strategy held by people like Oppenheimer and DuBridge.

After Griggs and his Air Force colleagues left, DuBridge and Lauritsen discussed their options. At this point, McCone intervened again and suggested that Caltech's president present the Vista report directly to General Eisenhower and other top leaders of NATO's forces in Europe.[40] Consequently, Oppenheimer arrived in Paris on December 4, 1951, accompanied by DuBridge and Lauritsen.[41]

Their trip was hardly secret, as accounts in the *New York Times* and the *Herald tribune* connected their visit to recent atomic tests in Nevada.[42] The scientists had lunch with Eisenhower and summarized Vista's activities for him.[43] The scientists next had a meeting with General Lauris Norstad, the commander of the United States Air Force in Europe. Oppenheimer did most of the talking. Norstad noted in a secret message back to Washington that the Princeton physicist exhibited "no wavering" in his opposition to strategic nuclear bombing. Nevertheless, Oppenheimer and his companions agreed to modify certain parts of the report the Air Force objected to. When they convened again the next day, Norstad reported that his "general impression of the report was [now] favorable."[44]

In February 1952, DuBridge traveled to Washington with Lauritsen and Fowler to brief high-ranking Pentagon brass. It was a tough audience.[45] Even Clark Millikan, a Caltech aeronautical engineer and frequent defense consultant, noted in his diary that he "got the shakes before the group."[46] Aware of a potentially hostile reaction, DuBridge reminded his audience that Vista did not originate at Caltech: "I will not

39. McCone's role is described by Griggs in *JRO Hearing*, 758.
40. McCone's involvement as a go-between between the Vista group and Finletter is described DuBridge memo to JRO, 3 Nov 1951, JROP, box 31, file DuBridge, Lee A.
41. Vista's trip to Europe is discussed in a number of places, including David C. Elliot, "Project Vista and nuclear weapons in Europe," *International security, 11:1* (1986), 163-183, on 174-176; and Evangelista (ref. 13), 136-140.
42. This was not a wild guess, as the Buster-Jangle test series was primarily directed toward testing low-yield atomic weapons that could have been suitable for tactical use.
43. It is interesting to note that Fowler and other Vista participants considered soliciting former German generals for their input on the best way to defend Europe against a Soviet invasion, advice Fowler noted they were uniquely positioned to give. Fowler memo to Horace Gilbert, 2 May 1951, WAFP, box 117, folder 14.
44. Norstad's comments are from a series of Red Line Messages between him and Vandenberg, 7 and 8 Dec 1951, HVP, box 86, security file.
45. Hostility described, for example, in Dale Corson to Jerrold Zacharias, 26 Aug 1954, LADP, box 187, folder 4.
46. Diary entry, 19 Feb 1952, CBMP, box 36.

say you rammed the idea down our throats—and yet I know of no more accurate way of describing what happened."[47]

DuBridge outlined Vista's basic assumptions and recommendations. He made it clear that the summer study had not set out "to invent new weapons" but rather to illustrate how NATO could be "smart...[and] make up for inferiority in numbers by superior intelligence and know-how...to take advantage of the best of American technology." Besides strengthening the Air Force's ability to support Army troops with conventional firepower, "tactical employment of our atomic weapons resources holds outstanding promise."[48]

Not all of Vista's recommendations were controversial. The report contained recommendations for a lightweight tank destroyer and explained how better reconnaissance and intelligence-gathering techniques, especially when coupled with resistance from local populations, could help stop a Soviet invasion. Themes of technological solutions achieved through scientific expertise and rational analysis were pervasive throughout the briefings that followed.

Nonetheless, the controversy accompanying the Vista Report regarding Oppenheimer's participation and its recommendations for tactical airpower and atomic warfare persisted. After the briefings, the Air Force moved to suppress the report. Copies of the Vista Report were recalled to Washington and Vista was largely forgotten until the Oppenheimer hearing brought renewed attention to it.

Oppenheimer and the Fallout from Vista

At the local level of Caltech, Vista's influence may be seen as the "dog that didn't bark." In August 1951, Fowler and DuBridge discussed how Caltech could expand upon the summer study. One possibility considered was the creation of a "war college" in Pasadena which would make a "continuing and detailed study of the European problem." As envisioned, a permanent organization—perhaps similar to MIT's Lincoln Laboratory—at Caltech would serve as a "national center" to direct research on both tactical and strategic problems. This seed never sprouted and Caltech never institutionalized Project Vista by establishing a "war college" in Pasadena. Nor did Caltech's leaders parlay their summer study into any permanent research facilities on campus devoted to classified research and publications by students and faculty. As DuBridge later recalled, "I think at the end of it...we hoped we wouldn't have to get into this again."[49]

47. "Vista briefing," 13 Feb 1952, WAFP, box 117, folder 11.
48. "Summary," Vista Report, xxi.
49. Judith R. Goodstein, "A conversation with Lee Alvin DuBridge, pt. II," *Physics in perspective*, *5*, (2003), 281-309, on 302.

MIT, in comparison, used its participation in summer studies to symbiotically expand the school's expertise and experience in research areas related to the military's needs and to substantially develop the school's infrastructure.[50] As Bill Leslie has shown, these labs performed classified research which MIT's faculty closely integrated into their educational and regular research programs. These studies had a substantial impact on the campus as a whole and made a significant contribution to MIT's educational mission in a major way that Caltech never exploited.

There is no evidence that DuBridge or Caltech's trustees expressed any interest in taking such a path. The controversy over Vista and the hostile reception the report received at the Pentagon certainly would not have encouraged DuBridge or the trustees to look favorably on any formal continuation of Caltech's summer study. Moreover, the already ambivalent relationship between Caltech and the Jet Propulsion Laboratory that the school managed suggests that the response to such an idea would not have been favorable. Adding to this apparent disinterest were the suspicions of Caltech's trustees, who were ambivalent if not outright opposed to supporting the Institute further with federal funds.[51]

At the national level, Vista helped expose a growing rift between scientists and their military sponsors and, indeed, within the research community itself. Compared to MIT's Project Hartwell (which the Navy embraced as the bible of undersea warfare) Vista received a hostile reception.[52] Military opposition to the Vista report was not solely due to its controversial recommendations. There was general resentment in the Pentagon of "theoretical" studies by supposedly uninformed and un-uniformed "long-haired scientists."[53] One Army general, when asked to critique the Vista Report, complained that the military really did not want "criticism of the work of the fair-haired boys."[54] Years after study had ended, DuBridge still bristled at the suggestion that Caltech had produced a flawed report, noting that it reflected the constant stream of military advice offered to himself, Oppenheimer, and others.[55]

The accusation that scientists who contributed to summer studies like Vista were operating outside their areas of expertise was starkly made by Roger Robb in 1954. The AEC's counsel exhorted Oppenheimer to tell "...why you felt it was your function as a scientist to express views on military strategy and tactics" when this "went

50. A point well made in Stuart Leslie, "Profit and loss: The military and MIT in the postwar era," *HSPS, 21:1* (1990), 59-85.
51. As Vista was winding down, the trustees told DuBridge of their concerns about "wholly classified contracts" which did not contribute to Caltech's educational mission. "Minutes—January 7, 1952 meeting of the Trustees", TMM. See also Reese Taylor to DuBridge, 14 Feb 1952, LADP, box 123, folder 3.
52. Marvin and Weyl (ref. 3), 7.
53. Baldwin (ref. 37).
54. Brigadier General Henry J.D. Meyer, memo, 25 Apr 1952, NARA, RG 381, box 18, file 2. Also Corson to Zacharias, 26 Aug 1954 (ref. 45).
55. Goodstein (ref. 49), 303.

beyond the scope of [a scientist's] function."[56] A scandal-mongering article written by an Air Force insider for *Fortune* magazine shrilled, "there was a serious question of scientists' trying to settle such grave national issues alone, inasmuch as they bear no responsibility" for a successful outcome.[57]

Even as Vista was germinating at Caltech, Philip Morse, an MIT physicist, asked his colleagues, "Must we always be gadgeteers?" Morse insisted that scientists should "contribute to the better use of the tools and weapons they have helped make."[58] Not all agreed with this suggestion. When John Wheeler was trying to recruit Richard Feynman from Caltech to work on thermonuclear research at Princeton, he argued that the job of physicists was not vetting strategy but doing "primordial" research. Oppenheimer must have been in Wheeler's mind when he deprecated the "hair-shirt philosophy of many nuclear physicists" who wanted to "tell the admirals and generals how to do tactical and strategic this-and-that."[59]

Scientists who participated in Vista and other summer studies continued to debate whether and how to contribute to national security. They certainly noticed how the Air Force had squelched the Vista report along with the subsequent pillorying of Oppenheimer for his policy views. These incidents provided some with cautionary tales. Dale Corson, a Vista participant and future president of Cornell University, later refused to take part in an MIT-sponsored summer study. Based on what he saw during the Vista briefings at the Pentagon, Corson came to believe the national security establishment was hostile toward unwelcome advice. Another Vista alumnus resisted recruitment for further defense studies, choosing instead to pursue "astronomy-as-usual" where he would be "freer to think my own thoughts."[60] Even scientists like Charles Lauritsen, who had long-standing ties to the military, questioned whether their advice on defense matters would still be welcome in the wake of the controversy over Vista and Oppenheimer's AEC hearing.[61]

Historians are generally agreed that it was Oppenheimer's recommendations on nuclear weapons and policies for their development and use that brought him such harsh condemnation in 1954. While not as critical (or as criticized) as his views on the hydrogen bomb, Oppenheimer's participation in Vista became another vulnerability for him. Opponents applied pressure on the charismatic physicist and his "disciples" with allegations that they had undermined America's national security. Such tactics

56. *JRO Hearing*, 959.

57. Anonymous [Charles V. Murphy], "The hidden struggle for the H-bomb," *Fortune* (May 1953), 230.

58. Philip Morse, "Must we always be gadgeteers?", *Physics today*, *3:12* (1950), 4-5.

59. Wheeler to Feynman, 29 Mar 1951, RPFP, box 3, folder 10.

60. Corson to Zacharias, 26 Aug 1954 (ref. 45); A.E. Whitford to Zacharias, 25 Aug 1954, LADP, box 184, folder 4.

61. A point made in Charles H. Holbrow, "Charles C. Lauritsen: A reasonable man in an unreasonable world," *Physics in perspective*, *5* (2003), 419-472, on 464-465.

did not prevent the wholesale participation of scientists in national defense efforts, but they certainly gave some scientists pause. While Project Vista was largely a failure for its sponsors and the scientists who undertook it and an eventual liability for Oppenheimer, perhaps something positive resulted from that summer in Pasadena: after Vista ended, Oppenheimer's friend DuBridge persevered to establish a more activist agenda for the President's science advisory committee.[62] In this manner, DuBridge and other veterans of studies like Vista helped forge a more direct and durable mechanism to advise the government on matters of science, technology, and national security.

62. Zuoyue Wang, "American science and the Cold War: The rise of the U.S. President's Science Advisory Committee" (Ph.D. dissertation, University of California, Santa Barbara, 1994).

CHAPTER TWELVE

The Fortunate Fox

Richard Polenberg

I want to pose two questions about J. Robert Oppenheimer. First, how would his public reputation have been affected had the Atomic Energy Commission decided to confirm his security clearance, rather than deprive him of it? I hope to show that this counterfactual question is worth asking, and that it enables us to better understand what our culture, including our historians, have done with this singular figure in the history of science and in the history of American politics. A second question: what does the evidence we have of Oppenheimer's intellectual style tell us about the probability of his accepting the discipline of the Communist Party? This question, too, I hope to show is worth asking, for it enables us to better understand the issue of Oppenheimer's involvement, such as it was, with the Popular Front in the pre-war years.

Let us suppose, for a moment, that the Atomic Energy Commission had not found against Oppenheimer in 1954 but had cleared him. Such an outcome is not wholly implausible. After all, Ward Evans, the lone Republican on the Personnel Security Board, and the one who seemed to be the most hostile to Oppenheimer, ended up voting to restore his clearance. Suppose the vote had not gone 2 to 1 against Oppenheimer, but 2 to 1 in his favor? Suppose Kenneth Nichols had then felt obligated to support the Board's decision, and the AEC had decided to go along, reasoning that since Oppenheimer's consultant's contract was going to expire in forty-eight hours, it was more prudent not to condemn him. And, finally, let us suppose that the AEC had decided to publish the transcript of the proceedings anyway.

How then would the scientific community have regarded Oppenheimer? Those who read the hearings would have seen Oppenheimer wilting under Roger Robb's adroit questioning, would have witnessed him falling to pieces when asked about his relationship with Jean Tatlock, about his brother Frank's Communist Party membership, about his conversation with Haakon Chevalier concerning George Eltenton. Even worse, it would have been evident that Oppenheimer had volunteered information about the political affiliations of many of his former students and friends. Here is one depressing passage from early in the hearing, when Oppenheimer was being asked about some of his acquaintances. The questioning began:

> "You say he or she was, as it turned out, a friend of many fellow travelers and communists. Who were they?"

> "Well, Addis was a friend of hers, Lambert was a friend of hers."
>
> "Doctor, would you break them down? Would you tell us who the communists were, and who the fellow travelers were?"
>
> "Lambert was a communist. Addis is reported to be a communist in the Commission's letter. I did not know whether he was a member of the Party or not."
>
> "You knew he was very close, didn't you?"
>
> "Yes, among fellow travelers, Chevalier. Among communists or probable communists, a man and his wife who wrote for *The People's World.*"
>
> "Who were they?"
>
> "John Pittman and his wife. A lawyer called Aubrey Grossman, his wife, she had known."
>
> "Was she a communist?"
>
> "I don't know in the sense of Party membership."
>
> "But very close."
>
> "Close."

The questions were not really meant to elicit information about particular individuals; rather, they were designed to test Oppenheimer's willingness to prove his "loyalty" even if that meant fingering his friends. To preserve his clearance, Oppenheimer was all too willing to do just that. After the above-cited exchange, Oppenheimer asked, perhaps despairingly, "Is the list long enough?"[1]

Not only did Oppenheimer cave in before Robb's questioning and name the names of former students and friends who had been communists or fellow travelers, but the transcript also demonstrated that he fervently supported the use of the atomic bomb during the war. In 1945, as the hearing made clear, he had rejected the advice of the Franck Committee, which had urged a test or demonstration of the bomb so that Japan would have a chance to surrender before the weapon was actually used. Whatever regrets Oppenheimer may have expressed afterward, he never doubted the need to drop the bomb while he was in charge of the Los Alamos laboratory.

So, to conclude this "thought experiment," had Oppenheimer been cleared and the transcript published, his reputation in the scientific community would almost surely have been severely damaged, tarnished beyond repair. Many if not most of his friends would have been dismayed by the evasions, the tattling, the single-minded determina-

1. Richard Polenberg, ed., *In the matter of J. Robert Oppenheimer: The security clearance hearing* (Ithaca, NY, 2002), 75.

tion to see the bomb he had fathered actually used on civilian targets. Had the AEC not made him a "victim," by stripping him of his security clearance, he would not have become the tragic hero idolized by scientists, intellectuals, and liberals appalled by McCarthyism and its excesses. Paradoxically, Lewis Strauss and the AEC would have more effectively destroyed Oppenheimer's image, and his influence, by finding in his behalf, and letting the record do their dirty work for them.

My second question is not counterfactual, but is stimulated by an abundance of warranted information we have about how Oppenheimer's mind worked. As Karl Hufbauer and others have demonstrated, Oppenheimer was not temperamentally suited to pursue a single research topic. For all his brilliance and insight into developments that were transforming the field of quantum mechanics, he never made the intellectual breakthrough that had been expected of a physicist of his caliber, which would have resulted in a Nobel Prize. His achievements as a physicist, important though they were, remained "small and fragmented," especially compared to those of his peers: Fermi, Rabi, Lawrence, Compton, Bethe. Unlike them, Oppenheimer was always inclined to investigate a number of different and not necessarily related phenomena.

This approach to scientific research reflected a more general intellectual style which might best be described as eclectic. Oppenheimer tried his hand at an astonishing array of subjects. Fluent in many languages, he eventually decided to study Sanskrit, the better to read and understand the *Bhagavad-Gita*. He had an avid interest in poetry, particularly the work of John Donne, and allegedly told the Dean at the University of California at Berkeley that he had accepted a faculty position there, rather than at Harvard or Columbia or any of the other universities that had made him offers, because of the library's collection of sixteenth-century poetry. Oppenheimer was interested in philosophy, and in psychiatry, and of course by the mid-1930s was devoting an increasing amount of his time to political causes.

In sum, Oppenheimer epitomizes one of the two contrasting intellectual styles described so brilliantly by Isaiah Berlin in *The hedgehog and the fox,* a book published in 1953. In his discussion of Tolstoy's view of history, Berlin remarks that a great chasm exists between those "who relate everything to a single central vision...in terms of which they understand, think and feel" and those "who pursue many ends, often unrelated and even contradictory." The first group, in which he places Plato and Hegel, Berlin labels "hedgehogs"; the second, which includes Aristotle and Goethe, Berlin terms "foxes." In this typology, Oppenheimer must surely be classified as a fox, someone, in Berlin's words, whose "thought is scattered or diffused, moving on many levels, seizing upon the essence of a vast variety of experience and objects for what they are in themselves, without, consciously or consciously, seeking to fit them into, or exclude them from, any one unchanging, all-embracing...unitary inner vision."[2]

Now, if any theorist could be labeled a hedgehog, it was Karl Marx, for whom such concepts as the class struggle and dialectical materialism were keys to understand-

2. Isaiah Berlin, *The hedgehog and the fox: An essay on Tolstoy's view of history* (New York, 1953), 1-2.

ing all social, economic, and historical change. A casual remark about Oppenheimer made by one member of the Communist Party, Joseph Weinberg, to another, Steve Nelson, therefore takes on more than ordinary significance. Weinberg commented that Oppenheimer is "just not a Marxist." His fox-like nature, his fundamental suspicion of any unitary, all-embracing explanatory model, may help us reformulate one of the questions that have perplexed scholars: was Robert Oppenheimer a member of the Communist Party?

Speakers at this conference have given sharply conflicting answers to this question. Kai Bird and Martin Sherwin explained why their research led them to conclude that Oppenheimer was never a Party member; Gregg Herken and Barton Bernstein explained why their research had led them to believe he was. But the question of Party membership is virtually the only area of disagreement when it comes to Oppenheimer's involvement in radical politics. Everyone agrees that he supported many of the positions the Communist Party took, some of which, as in the case of civil rights before the war, were quite defensible, but others of which, as in the case of foreign policy, were plainly contradictory. Everyone agrees that Oppenheimer's circle of friends in Berkeley included many Party members. Everyone recognizes that he joined numerous front organizations. And everyone agrees that Oppenheimer contributed significant sums of money for the relief of Spanish Civil War refugees, channeling the contributions directly through Communist Party officials.

Historians have focused on the issue of membership for two reasons. First, if Oppenheimer had joined the Party, then he committed perjury at the 1954 hearing when he denied it. Second, and more important, as a member he would have been subject to communist "discipline." To remain in good standing, Party members were sometimes subjected to pressure to do things they did not want to do. Of course, they always had the option to refuse—and then either leave, or be expelled from, the Party. Usually, the pressure simply involved following the Party line through its labyrinthine twists and turns. Yet occasionally, as we will see, it involved a good deal more. Consequently, one way to go beyond the question of Party membership is to ask, hypothetically, whether Oppenheimer would have been willing to accept the discipline that membership required. Here are two examples, taken from 1935, of sacrifices that others were asked to make, and did make.

In September 1935, while a member of the Communist Party, Richard Wright wrote an essay for *The new masses*—"Joe Louis uncovers dynamite"—after the black heavyweight boxer defeated Max Baer. In an evocative, beautifully written essay, Wright, asserting that "Joe was the concentrated essence of black triumph over white," described the spontaneous celebration that erupted among blacks on Chicago's South Side following the knockout victory. But to ensure that the essay had the proper ideological slant—an emphasis, that is, on the need for an alliance of black and white workers—Party officials tacked on a final paragraph, vapid, stilted, exemplifying "agitprop" at its worst. Acquiescing in its inclusion was a price Wright was willing to pay to remain in the Party's good graces, perhaps even to have the article published.[3]

What the Party demanded of Richard Wright was trivial compared to what it asked of Peggy Dennis and her husband, Eugene, a high-level communist functionary. The Dennises had a son, Tim, born late in 1929, who spent most of his childhood in Russia, where his parents were engaged in various Party activities. In January 1935, when the Dennises were preparing to return to the United States, Russian officials told them that they should leave Tim in Russia because the presence of a Russian-speaking youngster would be an embarrassment to the American Communist Party. The Dennises could have remained in Russia; they could even have taken Tim back to the United States and faced expulsion from the Party. Instead they left without him. They saw their son briefly on visits to Russia in 1937 and 1941. They did not see him again until he accompanied Nikita Khrushchev, as a translator, on his visit to the United States in 1959, by which point Eugene Dennis was close to death.[4]

There is nothing in the documentary record to suggest that Oppenheimer, like Richard Wright, would have permitted higher-ups in the Communist Party to tamper with anything he wrote, much less that he would have considered, even momentarily, making the kind of sacrifice the Dennises made. Consequently, the key question is not whether Oppenheimer at some point "joined" the Communist Party, a question that the evidence is unlikely to prove one way or the other. Rather, the question is whether he would have been willing to accept Communist discipline. More specifically, would his commitment to the cause of communism or his loyalty to the Party have led him either to commit, or to protect those who had committed, atomic espionage? In my view, necessarily speculative, a man of Oppenheimer's cast of mind, one so distrustful of absolutes, would not have succumbed to such pressure.

By the late 1940s, Oppenheimer had moved full circle politically. He came to accept the orthodox Cold War view of the threat posed by Soviet totalitarianism. He had distanced himself so completely from his former Popular Front outlook that in 1947 he refused to sign a petition commemorating the twentieth anniversary of the execution of Nicola Sacco and Bartolomeo Vanzetti. Drafted by Gardner Jackson, a devout liberal, the petition criticized the anti-communist hysteria that was developing in the United States and warned that it could turn into the kind of Red Scare that followed World War I. Eleanor Roosevelt signed the petition, as did Reinhold Niebuhr, as did Walter Reuther and a hundred and fifty other notables. But not Robert Oppenheimer, who replied to Gardner Jackson: "Please do not urge that I sponsor the manifesto. It deals with things very far from my field of competence, and where my word neither should nor would have weight. But if my encouragement is of any help, that you have."[5]

3. Richard Wright, "Joe Louis uncovers dynamite," Joseph North, ed., *New masses: An anthology of the rebel thirties* (New York, 1969), 160-164.

4. Peggy Dennis, *The autobiography of an American Communist* (Westport, CT, 1977), 85-87, 244-247, 253-254.

5. Oppenheimer to Gardner Jackson, Aug 1947, Gardner Jackson MSS, Franklin D. Roosevelt

Oppenheimer's cautious outlook helps explain why such rock-ribbed conservatives as John McCloy spoke out in his behalf at the 1954 hearing. Another pillar of the establishment who defended Oppenheimer was Vannevar Bush. Testifying on April 23, 1954, he told the Personnel Security Board that Oppenheimer was being pilloried because he had the temerity to express his independent ideas about the arms race. And Bush added: "I think in fact the republic is in danger today because we have been slipping backward in our maintenance of the Bill of Rights."[6]

Although the Oppenheimer case did not specifically involve First Amendment freedoms of speech and press, in a larger sense Vannevar Bush was right. The cavalier way in which government officials tapped Oppenheimer's telephone, reported on his comings and goings, listened in on his conversations with his lawyer, and placed hidden agents in his entourage—all revealed a frightening abuse of power. Consequently, the Oppenheimer case did raise crucial civil liberties issues. And Vannevar Bush's warning, voiced fifty years ago, remains pertinent because whenever the nation feels threatened the guarantees enshrined in the Bill of Rights are endangered.

Library, Hyde Park, NY.

6. Polenberg, ed. (ref. 1), 206.

PART FIVE

Cultural Resonances

An opera about Oppenheimer? The fact that the most celebrated composer in the United States wrote an opera based on Oppenheimer's life is an indicator of how prominent a personality in American culture Oppenheimer remains, right down to the present. John Adams's *Doctor Atomic*, premiered in 2005, focused on the Day of Trinity in 1945, but this opera, like so much of what has been written about Oppenheimer, was driven by a larger feeling that this particular life was a field on which some of the most compelling aspects of American life in the twentieth century were played out. And not only those aspects having to do with weaponry, military policy, national security, and the appeal of Communism. Oppenheimer was an explorer of cultural terrain, and someone whose life was in turn shaped by his cultural engagements. The three essays in this section address this dimension of Oppenheimer's role in American life, which, as the commissioning and completion of *Doctor Atomic* reminds us, looms as large today as it ever has.

J.L. Heilbron goes after one of the most celebrated and enigmatic of Oppenheimer's cultural connections, his involvement with classical Hindu literature. Heilbron shows the influence on Oppenheimer of the great Berkeley Sanskritist, Arthur William Ryder. Heilbron concludes that the Hindu verses Oppenheimer liked to quote were far from ornamental flourishes, but basic to his sense of himself and his role in the drama of modern history. Charles Thorpe integrates Oppenheimer's post-1945 public life into the larger intellectual history of that era's liberal intelligentsia. He shows that Oppenheimer's trusting identification with the nation's elites diminished his capacity to defend himself against the attacks made on him during the McCarthy Era. Finally, Robert P. Crease analyzes the widespread tendency to depict Oppenheimer as a "tragic" figure, and concludes that this popular image conceals as much as it reveals.

CHAPTER THIRTEEN

Oppenheimer's Guru

J.L. Heilbron

"I have become death, destroyer of worlds." Thus spoke, or may have spoken, the director of the Los Alamos Laboratory as he witnessed the fireball, "brighter than a thousand suns," of the first man-made nuclear explosion. As everyone knows, these phrases come from the *Bhagavad-Gita*, the classic expression of Hindu teaching about duty, resignation, and fate.[1] Oppenheimer frequently quoted from it, sometimes, no doubt, as in these cases, merely for display. Indeed, the quotations describe a display: manifestations, at the request of the warrior prince Arjuna, of the power and glory of God (Krishna). Oppenheimer and Arjuna had more in common than presence at terrifying fireworks. The future designer of nuclear weapons often assimilated himself in situation and character to the ancient Hindu hero. The process began in 1930 or 1931, when Oppenheimer took up the study of Sanskrit under Arthur William Ryder, a tough, opinionated, cocksure, sour stoic devoted to classic Hindu language and philosophy. The Oppenheimer who came under his tutelage was a soft, uncertain product of physics and ethical culture. The flinty teacher steeled the undisciplined student and so may have helped to bring about the conflagration of Hiroshima.

The connection between the *Bhagavad-Gita* and the atomic bomb has been pursued by James Hijiya, professor of history at the University of Massachusetts. He argues that the precepts of the Ethical Culture School in which, for ten years, Oppenheimer received his primary and secondary education, militated against his directing Los Alamos, and a fortiori against his wish to see the weapon the laboratory produced used against a civilian target.[2] The story that underlies the *Bhagavad-Gita* was just

I thank David Cassidy, David Hollinger, Jean-Jacques Salomon, and Nicolas Wittowski for comments and suggestions, and The Bancroft Library for permission to quote from the papers of A.E. Kroeber and W.E. Ritter, and from the oral histories of Dorothea Kroeber and Jean and Kenneth Pitzer.

The following abbreviations are used: JRO, J. Robert Oppenheimer; *JRO Hearing*, United States Atomic Energy Commission, *In the matter of J. Robert Oppenheimer: Transcript of hearing before Personnel Security Board and texts of principal documents and letters* (Cambridge, 1971); TBL, The Bancroft Library, University of California, Berkeley.

1. *Bhagavad-Gita*, 11:32, more commonly translated as "time which destroys all things," as in the edition by Juan Muscaró (London, 1962), 92; "the light of a thousand suns" (11:12), ibid., 90.

2. James Hijiya, "The *Gita* of J. Robert Oppenheimer," American Philosophical Society,

the thing to deaden ethical culture. As the story opens, Arjuna faces a battle against many distinguished men including dozens of his kinsmen. He does not want to kill them. He so informs his charioteer, who soon sets him right, by reason and intimidation. The charioteer is the great lord Krishna in disguise. Krishna teaches Arjuna that everyone must do his duty in his station; Arjuna is a warrior; therefore he must fight. He can no more deny the force of this syllogism than his relatives can escape death at his hands. Krishna has already decided who will die ("I have become death"); Arjuna is only the instrument of fate. To strengthen Arjuna's faith in fate, Krishna shines with the light of a thousand suns.

Ryder published a verse translation of the *Bhagavad-Gita* in 1929, shortly before Oppenheimer came into his classroom. In Ryder's poetical pedagogy, Arjuna thus describes his confusion before the battle (2:7):

> A sinful pity clouds my soul
> I know not wrong nor right;
> Reveal my duty certainly
> Unto my clearer sight;
> Give sure instruction: for I come
> A pupil seeking light.

Krishna wastes no time telling Arjuna he is wasting time. "Wise men pity not/The living or the dead."

There follows a long exposition of doctrine, enforced by Krishna's transformations, during which Arjuna literally sees the light (15:20). He understands his duty. Krishna probes (18:72):

> And have you heard the doctrine with
> A concentrated mind?
> Is all unwise perplexity,
> Bold soldier left behind?

Arjuna allows that he now sees by his own light the need to affirm the will of God. "Perplexity is done." The narrator of the story, Sanjaya, expresses the rewarding conclusion (18:78):

> Where archer Arjun stands, where stands
> The Lord most wise, most pure
> There glory, grandeur, victory,
> And steadfast law are sure.

We have the teaching: Concentrate your mind, banish doubt, do your duty, embrace the steadfast law, and glory will certainly follow.

Proceedings, 144 (2000), 123-167, on 126-127, 147, 152, 159: "To reinforce his determination to build the bomb, Oppenheimer used the *Gita*" (on 129).

Before the *Guru*

Oppenheimer began to teach at Berkeley in the fall of 1929 following a fellowship year in Europe truncated at its beginning by a fit of ill health. He reported to his patron, the Rockefeller Foundation, that he had tuberculosis, of which he was soon and miraculously cured apart from a persistent cough. David Cassidy suggests that Oppenheimer may have been suffering from a psychosomatic condition that used to afflict him when he faced intimidating circumstances.[3] He had reason to feel intimidated. The fellowship year 1928-29 was his second in Europe; the first, in 1925-26, had been a disaster. Then, as a fresh graduate from Harvard, he had pushed himself into the Cavendish Laboratory in the other Cambridge to pursue experimental physics. He had little talent for it and, in any case, would have been overshadowed by the squad of past and future Nobel laureates he found attacking the research front under the command of Ernest Rutherford. Measuring himself against members of Rutherford's confident elite corps, like the former Navy officer Patrick Blackett, Oppenheimer fell into deep depression. By Christmas he was suicidal. It being the season of giving, however, he turned his aggression outward. He tried to strangle an old friend from the Ethical Culture School with whom he went on vacation; and he fantasized (or was it imagination?) that he had left a poisoned apple on Blackett's desk. In this fantasy there may have been a homoerotic crush of the sort that Oppenheimer later felt for Linus Pauling.[4]

Cambridge and experiment having proved disagreeable, Oppenheimer tried Göttingen and theory. Max Born, the presiding theorist there, had met Oppenheimer on a visit to Cambridge, thought him promising, and invited him to join his seminar. Oppenheimer quickly distinguished himself for his wealth of ideas and his bountiful rudeness. His continual interruptions of other students and even of the professor to expose errors or show a better way inspired them to petition Born, in writing, to shut him up. Born did so, tactfully.[5] He also recognized Oppenheimer's ability by recruiting him as a collaborator and supervising his doctoral thesis. Oppenheimer returned to Harvard in the fall of 1926 with a bruised ego and a German Ph.D.

The fellowship of 1928-29, which followed one at Caltech in 1927-28, was to have taken Oppenheimer to Cambridge, Leyden, and Copenhagen as final preparation for teaching European theoretical physics in the United States. Because of the time lost to pseudo-TB, he skipped Cambridge and doubtless felt the better for it. But his cough continued, badly enough to alarm his host in Leyden, Paul Ehrenfest, who urged him to continue his studies not in foggy Denmark but in salubrious Switzerland. Ehrenfest had Oppenheimer's mental improvement in view as well. He diagnosed an inclination to abstractness, vagueness, and dilettantism that could not be cured by exposure

3. David Cassidy, *J. Robert Oppenheimer and the American century* (New York, 2004), 125-126.
4. Ibid., 99-100, 151-152.
5. Max Born, *My life: Recollections of a Nobel laureate* (New York, 1975), 229.

to Bohr's mystical methods. Instead, Oppenheimer should fix up his mind and body in Zurich, with Wolfgang Pauli, whose high standards, powerful mathematics, and cutting criticism had earned him the nickname "Scourge of God." Ehrenfest's kindly incoherent recommendation to Pauli ran as follows: [6]

> The poor devil [Oppenheimer] has been a month and a half with us in Leyden in the rain, rain under a gray sky, and, what is harder, under the pressure of my schoolmasterly style. He always has very clever ideas and almost always gives an especially interesting turn to the discussions in our little group. But then naturally the great calamity arrives that I, as you know, can understand nothing that cannot be visualized. And although he tries with unperturbable good nature and kindness to satisfy my wishes, at the end (except for a few small criticisms where I chanced to be right) I inconvenience him more than I help him. He does not think to complain. But the situation depresses me greatly. So much so that I am really convinced that for the complete development of his great gifts Oppenheimer needs to be knocked into shape [zurechtgeprügelt] RIGHT AWAY AND WITH A LITTLE TENDERNESS. But in contrast to many other young people he thoroughly DESERVES this kindness, for personally he is a very UNUSUAL fine and good fellow....But in rebus physicis there is only ONE Scourge of God! (Thank God!!!...). Therefore if he leaves Leyden I'd very much appreciate it if he can come to you.

Pauli understood that his visitor's problem was deeper than an inability to explain himself in pictures that Ehrenfest could grasp.

In February 1929 Pauli reported on Oppenheimer's progress:[7]

> I think that Oppenheimer feels well here in Zurich, that he can work here, and that scientifically much good can still be got out of him. His strength is that he has many ideas and good ones, and, chiefly, a lot of imagination. His weaknesses are that he satisfies himself too often too quickly with poorly based claims; does not answer his own questions, which are often very interesting, because of lack of persistence and thoroughness (or answers them only provisionally and often incorrectly); and leaves his problems only half digested as conjecture, belief, or disbelief instead of really working and thinking them through. But I definitely expect that through energetic persuasion this can be improved in time, he has much good will and is not obstinate....
>
> Unfortunately Oppenheimer has a very bad characteristic: he comes to me with a rather unconditional belief in authority and regards everything that I say (at least to a very high approximation) as a last and final truth. I know how he came to need outside authorities. They are to solve his problems and answer his questions so that

6. Ehrenfest to Pauli, 24 Nov 1928, Wolfgang Pauli, *Wissenschaftlicher Briefwechsel* (4 vols., 1979-96), Armin Hermann, Karl von Meyenn, et al., eds., vol. 1, 477; cf. Alice Kimball Smith and Charles Weiner, eds., *Robert Oppenheimer: Letters and recollections* (Cambridge, 1980), 121-122.
7. Pauli to Ehrenfest, 15 Feb 1929, Pauli (ref. 6), vol. 1, 486-487.

> he does not have to do it himself. (This connection naturally is not *consciously* clear to him, but only latent in his subconscious.) But how I am to cure him of this, I do not know. Vainly I point to well-documented cases where I have asserted something untrue and confessed my error!...One thing I hope to achieve soon: that Oppenheimer adopts *my* manners, at least in speaking *with me*! That is absolutely necessary, if I'm not to consider myself a hypocrite. For it is an unshakeable dogma with me that formal politeness, like great heresy, must be mercilessly eradicated from social intercourse.

Pauli, who would soon need the attention of a shrink himself (he spent some years in analysis under Jung), picked out exactly the same feature of Oppenheimer's personality that would win over Groves: his willingness, despite his brilliance and arrogance, to submit to authority.[8]

Pauli's cure worked. By June Oppenheimer had almost stopped coughing and had "blossomed spiritually," that is, had shown his great strength as a physicist. He and Pauli worked together on quantum electrodynamics. Oppenheimer remembered this revitalization to the end of his life. The time in Zurich was "just very very good indeed....I got to be not only extremely respectful [!] but also extremely fond of Pauli."[9]

Pauli's shrewd observations about Oppenheimer's reliance on outside authority to give direction to his thoughts and tidy them up echo in observations Oppenheimer made of himself at about the same time. The occasion was a request from the Rockefeller Foundation for an evaluation of the problems faced by young American physicists studying in Europe. The reply must be read as autobiography: [10]

> [The difficulties] are most acute in men who combine a certain weakness, timidity, hesitancy of character with a quite robust vanity—or, perhaps more accurately—with an urgent desire for excellence. In Europe they are removed from their friends, from the pampering of an American university...; they are introduced to the more critical, more disciplined, more professional science of Europe. All this is no doubt excellent; but too often it induces in the victim [!] a state of surrender, and a false metaphysical melancholy which replaces, and makes impossible, an active participation in the European scheme, and an honest attempt to learn from it. The melancholy is presumably [!] unpleasant; it is usually dissipated by return to America, and the consequent revival of pampering. But it acts as a protective coating for an American against that which he was sent to Europe to learn; almost always it is a sterile melancholy. It is the melancholy of the little boy who will not play because he has been snubbed.

8. Charles P. Enz, *No time to be brief: A scientific biography of Wolfgang Pauli* (Oxford, 2002), 192-193, 205-206, 211, 240-249. Cf. Murray Kempton, "The ambivalence of J. Robert Oppenheimer," *Rebellions, perversities, and main events* (New York, 1994), 118-135, which traces Oppenheimer's puzzling cravenness during his hearing of 1954 to deference to authority.
9. Pauli to R. de L. Kronig, 2 June 1929, Pauli (ref. 6), vol. 1, 504; JRO, interview of 1963, Smith and Weiner (ref. 6), 126.
10. JRO to W.J. Robbins, 14 May 1929, Smith and Weiner (ref. 6), 128.

The Pauli cure did not last. Back in lotus-land, pampered in a university that knew nothing of German theoretical physics but what he would teach it, free from criticism and supervision, Oppenheimer's level of discipline declined. "I don't get enough [work] done," he complained to his brother Frank, "because I am busy and lazy, and have not the strength to let other things go to the devil when I should."[11] In accordance with his split-time agreement, Oppenheimer spent the first half of 1930 and of 1931 at Caltech. We next catch sight of him in Berkeley in August 1931. He wrote Frank: "Berkeley is good this year, the department honest and friendly.…I am learning Sanskrit, enjoying it very much, and enjoying again the sweet luxury [authority?] of being taught.…I feel well and strong, and have bought an alarm clock."[12] You must rise early to teach physics and learn a language with 18 vowels, 36 consonants, and, in a simple case, 270 forms of the same verb.

By the end of the fall semester of 1931 the drillmaster of this new discipline, the Pauli and Groves of Sanskrit, or, to speak plainly, Ryder, had become one of Oppenheimer's two closest associates in Berkeley. The other was Ernest Lawrence. That is the obvious inference from Oppenheimer's taking his father to Berkeley in January 1932 especially to meet Lawrence and Ryder.[13] Oppenheimer's mother had died the previous October, leaving a great void in the family. "I am the loneliest man in the world," Robert wrote one of his former teachers at the Ethical Culture School. This loneliness probably drew him closer to Lawrence, who had been most solicitous about his mother's illness, and to Ryder, who offered the *Bhagavad-Gita* as solace to the "reflective mind, depressed and perplexed by the spectacle of life."[14]

The *Guru*

Ryder, born in 1877, was the son of William Henry Ryder, a Congregational clergyman then teaching Greek at Oberlin College in Ohio. The elder Ryder subsequently united his language and his religion as Professor of New Testament Interpretation at the Andover Theological Seminary in Andover, Massachusetts. Arthur Ryder attended the associated Phillips Andover Academy, a secondary school then distinguished for its Hellenism and Puritanism. On graduation in 1894 he proceeded to Harvard, where he studied Greek and Latin literature and, in his last year, Sanskrit. Ryder regarded his Sanskrit teacher, C.R. Lanman, as the exemplar of academic virtues, a selfless, tireless laborer in the vineyard of "oriental studies," a "Napoleon of editors," a one-man moral lesson. The lesson: "nothing but the power of devoted work on worthy

11. JRO to Frank Oppenheimer, 14 Oct [1929], ibid., 135.
12. JRO to Frank Oppenheimer, 10 Aug 1931, ibid., 143.
13. Julius Oppenheimer to Frank Oppenheimer, 18 Jan 1932, ibid., 153; cf. JRO to Lawrence, [3 Jan 1932], ibid., 147.
14. JRO to Lawrence, 12 and 16 Oct 1931, ibid., 144-145; Ryder, preface to *The Bhagavad-Gita*, trans. A.W. Ryder (Chicago, 1929), xvii.

material [is able] to raise a massive memorial." Lanman labored as and for the "joyful recreation of the educated." He stood as a rebuke to piddling perverse pseudo-scholars who forget the power, and miss the joy, of disciplined work. "Hurrah for Professor Lanman! *Laboremus*."[15]

Ryder went off to work, first at the Phillips Academy, as instructor in Latin, and then, from 1899 to 1901, in Germany, where he kept up Greek and Latin, began Old Persian, and wrote a thesis, in German, on difficult Sanskrit texts. The thesis was lean and strong, a taciturn Yankee dissertation, offered with tongue-in-cheek modesty as the second edition of a French treatise published in Paris half a century earlier. But what the French text did in 495 pages, Ryder accomplished in 40, "without sacrificing anything essential."[16] He would become the sort of writer praised in Sanskrit literature, "wherein an author rejoices more in saving the half of a short vowel than in the birth of a son."[17]

In the fall of 1901, Ryder returned to Harvard as an instructor in Sanskrit and assistant to the Napoleonic editor Lanman, to whose Harvard Oriental Series he contributed a translation of the Hindu drama *The little clay cart* (1905). The book has a preface with the usual scholarly apparatus of citations and footnotes; so too do two short notes and a paper published in 1902 and 1906 in the *Journal* of the American Oriental Society.[18] That was the end of Ryder's scholarship. He had realized that burdening literature with footnotes was pointless and indulgent. In the preface to his translation of Dandin's picaresque *Ten princes* (1927), he blasted "those dismal studies in influence and sources [,which] may be securely left in the hands of those who have no love of literature." In contrast, he praised the self-effacing anonymous writer who had completed Dandin's unfinished masterpiece—a writer "who was zealous for art, not for self-exploitation; who stands as a silent rebuke...of any age greedy for scholarship and other stultifying self-advancement." Like Lanman, Dandin and Anonymous had expended their "lofty art with no design save [we have heard it before] the entertainment of the truly cultivated." Ryder allowed very few works into this category. He mentioned besides the *Ten princes* only the operas of Gilbert and Sullivan.[19] Opinionated but not narrow, he liked King Lear and Edgar Lear equally, and detested pomposity. He once described the professor and staff of an academic department, no doubt of literature, as "a sham giant surrounded by real pygmies."[20]

15. G.R. Noyes, "Arthur William Ryder," in A.W. Ryder, *Original poems together with translations from the Sanskrit* (Berkeley, 1939), ed. G.R. Noyes, xvii-xxxix, on xviii; Ryder, "Laboremus," ibid., 37, 39.
16. A.W. Ryder, *Die Rbhu's im Rgveda* (Gütersloh, 1901), 7; F. Nève, *Essai sur le mythe du Ribhavas* (1847).
17. A.W. Ryder, *Dandin's Dasha-Kumara-Churita: The ten princes* (Chicago, 1927), xii.
18. A.W. Ryder, "List of publications," Ryder, *Original poems* (ref. 15), 185.
19. Ryder, *Dandin's Dasha-Kumara-Churita* (ref. 17), x-xi.
20. *Saturday review of literature, 17:8* (16 Apr 1939), 8.

Harvard was not for Ryder. In 1906 he left for the freer air and fewer footnotes of the University of California. (Curiously, two years later his father arrived at Harvard as Professor of New Testament Interpretation.) He moved slowly through the ranks: instructor in Sanskrit, 1906; assistant professor, 1908; associate professor, 1919; professor, 1925. He moved slowly because he would not play the scholar even at Berkeley's level. His friend and colleague George Rapall Noyes, a scholar-translator from Slavic languages, described Ryder's intellectual bent: "Ryder was not...a scholar, a man of letters, or (except by title) a professor; he was a reader, a man of literature, and a teacher. He was a man of genius, but his genius was not of the academic type."[21] He was forceful, learned in languages, sociable, witty ("nothing short of Voltarian"), and aggressive. He could not be ignored; eventually he was promoted.[22]

When young, Ryder formed friendships easily. One friend was William Emerson Ritter, Director of the Scripps Institute for Biological Research at Berkeley, with whom Ryder shared the insight that much in science was empty formalism. "Science is no Moloch," Ryder wrote Ritter, "demanding self-immolation of each generation, but a part of real living, and interpretable to the great public." No Moloch indeed, but Art; for art, when good, differs little from good science. "I get much the same kind of thrill out of Hamlet, Euclid, Hindu philosophy, and the Origin of Species." A bookish and elitist view. "I don't know how one primarily interested in art can have a hostile attitude toward science, for they are at bottom expressions of the same instinct. Bad science, organized, professional science, science pursued for money, is depressing to be done—just as bad art is."[23]

Art soon saw further than science. Ritter sent Ryder a copy of his *War, science and civilization* (1915), which sympathized with the Germans over their clamor for *Lebensraum*. Ritter could have prevented the conflict, if asked. There is no reason, he said, to fight over nature's gifts since science can multiply them without end.[24] Science offers the further assurance that war will pass. Biological evolution has given way to cultural evolution, with a built-in integrative power. Integration culminates in fellow-feeling and altruism; science leads all other activities in its internationalist outlook; science and the "integrative series of human evolution" presage "*progress in civilization...as the differentiation and intensification of love and intellect, and of the intellectualizing of love and the affectionizing of intellect.*"[25] Ryder, who knew Germany and inclined toward the military, would have none of it. "Your optimism as to science applied to politics is refreshing, though a little hard for a non-scientist fully to share. I think you are too generous to the Germans, whom I think to be a permanently inferior people (a pre-bellum judgment much intensified of late)."[26]

21. Noyes (ref. 15), xviii.
22. Ref. 20.
23. Ryder to Ritter, 31 Aug 1911, 11 Feb 1912, TBL, Ritter Papers, box 18.
24. W.E. Ritter, *War, science and civilization* (Boston, 1915), 1, 20-21, 25, 35, 86-87.
25. Ibid., 36, 39, 41-42, 51, 59-60, 92-93 (quote), 98 (quote).
26. Ryder to Ritter, 12 Dec 1915, TBL, Ritter Papers, box 18.

Ryder took the same line with Berkeley's famous anthropologist Alfred Kroeber, who had deduced the character of the eventual peace from general anthropological principles. Ryder thought the argument soft-headed. "You don't convince me about your views on the outcome of the war....It seems to me the decision is already attained, and the terms of peace are now being inflicted on the Germans. I shall be broken-hearted if proved wrong, for I regard the German attitude as deeply debasing and dangerous."[27]

Ryder tried to enlist in the army, studied French conversation to help out at the front, and stayed home because of high blood pressure. He befriended the director of the Military Bureau at the university, Leon Richardson, perhaps with an eye to entering some government service. As the poet turned militarist so the military man turned poet. Richardson admired Ryder's steadfastness, sense of duty, vision, and ambition, however unattainable the goal. "My friend/Although the world should grudge thee its acclaim/Thou art for me, and ever shall be/Great."[28] Ryder had to settle for analyzing historic battles. Dorothea Kroeber (the wife of the anthropologist) was astonished to see him play with her children by the hour, lining up toy soldiers for the Peloponnesian Wars. "He was fascinated by [it], it must have been the mathematics, the exactitude, the exact poetry of war tactics." Late in life, when Oppenheimer knew him, the study of military tactics was one of Ryder's few recreations.[29]

Not all Ryder's warfare was theoretical. In 1921 he opened fire at the Harvard University Press for raising the price of old stock in its Oriental Series, which had been paid for by bequest and marshaled by Lanman. Fie! "You have fixed upon it a mercenary stain." The Press replied that not it but Lanman determined the price. Ryder added cowardice (for switching targets) to his charges. "In my former letter I recorded the contempt I feel, and others feel, for the cynical action of the Harvard University Press. That contempt is now augmented." The Press brought up John F. Moore, one of its syndics and a friend of the Ryder family, as reinforcement. Our hero replied to Moore in the manner of the converted Arjuna: "If I allowed personal sentiment to dictate my course on a matter of right and wrong, I should be a sneak and a coward." The fortress did not yield. Ryder thereupon published the correspondence from his headquarters on Telegraph Avenue, and billed Harvard for expenses in "printing and distributing matter designed to raise the moral tone of the Press."[30]

Despite his self-righteousness, irritability, arrogance, obstinacy, and infallibility, Ryder managed to be good company. He was redeemed by his wide knowledge and

27. Ryder to Kroeber, 7 Jan 1916, TBL, Kroeber Papers, box 26; Dorothea Kroeber, *Alfred Kroeber* (Berkeley, 1970), 95.
28. Noyes (ref. 15), xviii; Richardson, in Ryder, *Original poems* (ref. 15), [v]. Ryder dedicated his translation of *Relatives, being further verses translated from the Sanskrit* (San Francisco, 1919) to Richardson.
29. Dorothea Kroeber, "Oral history," conducted by Anne Brower, 1982, ROHO, TBL, 116; Noyes (ref. 15), xxxviii.
30. A.W. Ryder, "A correspondence with the Harvard University Press" (1921), TBL.

ready wit, and by his devotion to his subject and his students. In the 1930s he was a regular member of a Stammtisch or informal dining club of professors who met on Saturday evenings in Emeryville to eat Italian food and drink homemade wine. Among other regulars was Dorothea Kroeber. Ryder struck her as a "very bright, very peculiar man...very interesting in a group like this." Oppenheimer was another regular. Kroeber again: "He was captivated by Ryder, a well-defined man who considered that he lived by the Eternal Verities; erudite, arrogant, inexorable in his scholarship, sadistically ironic in his disenchantments. He became Oppenheimer's *guru*, and he taught him Sanskrit—the language, its poetry, its ancient wisdom." Ryder probably helped slant the club's discussions of cultural evolution. "The Yogi view was in any case immanent in much of the talk and speculation around the Stammtisch. Those were the years when we dreamed of a world at peace, when the Eastern contemplative way seemed to some of us a possible twentieth-century way."[31]

Ryder kept his classes small. When his course on Hindu philosophical systems became crowded with incompetent students, he cleared the room by requiring four years of study of Latin, Greek, and Sanskrit to enter it. His Sanskrit courses typically had no more than two students. One of Oppenheimer's fellow students, William A.P. White, made a career writing mystery stories under the psudonym Anthony Boucher. His first book, published in 1937, features a professor-detective named Ashwin, which means "rider" in Sanskrit, "accurately reproduced from life." We first meet Dr. Ashwin pacing his lecture room and reading in a booming voice the opening shloka from the *Mahabharata*, the lengthy story of which the *Bhagavad-Gita* is an episode, to his sole student. "'In all literature [Ashwin intoned], I have found only three verse meters which I can read indefinitely without tiring: English blank verse, the Latin and Greek hexameter, and the Sanskrit shloka.'" Boucher: "Like all of Ashwin's frequent ex-cathedra statements, this came forth with an impression of superpontifical infallibility."[32] Ashwin also liked beer and scotch whiskey, the comic verse of Ogden Nash, and jokes about the "somewhat ithyphallic" nudes on the gates to the university gratefully acknowledged as "erected by Jane K. Sather." In Boucher's story, Ashwin's lone student fed him details about murders in the Berkeley hills until Ashwin had solved the case. "I was then in control of all the pertinent facts," he said, in precisely the words that Ryder had used in his relentless battle with the Harvard Press.[33]

The superpontifical professor of Sanskrit imposed his authority and sense of discipline on the impressionable professor of physics. Oppenheimer reported his progress to Frank, who in return questioned about the Hindu notion of dharma (duty). He received this remarkable reply:[34]

31. D. Kroeber (ref. 29), 116, and (ref. 27), 235-236.

32. Anthony Boucher, *The case of the seven of Calvary* (New York, 1937), [iv], 9-11; cf. Noyes (ref. 15), vii, xxvii.

33. Boucher (ref. 32), 285 ff.; Ryder, *Dandin's Dasha-Kumara-Churita* (ref. 17), x.

34. JRO to Frank Oppenheimer, 12 Mar 1932, Smith and Weiner (ref. 6), 155-156.

> You put a hard question on the virtue of discipline. What you say is true, I do value it...more than for its earthly fruit, proficiency. I think that one can give only a metaphysical ground for this evaluation...[as in] the bhagavad gita, Ecclesiastes, the Stoa....This very disparity [in metaphysical formulations] suggests that the fact that discipline is good for the soul is more fundamental than any of the grounds given for its goodness. I believe that through discipline, though not through discipline alone, we can achieve serenity, and a certain small but precious measure of freedom from the accidents of incarnation, and charity. And the detachment which preserves the world which it renounces. I believe that through discipline we learn to preserve what is essential to our happiness in more and more adverse circumstances....Therefore I think that all things that evoke discipline...ought to be greeted by us with profound gratitude; for only through them can we attain to the least detachment, and only so can we know peace.

So far had Oppenheimer made Eastern philosophy his own within two years of meeting his *guru*.

By the fall of 1932 Oppenheimer had advanced to reading some Sanskrit stories with Ryder's help. A year later, in an overcrowded class of three students, he had advanced to the *Bhagavad-Gita*. "It is very easy and quite marvelous." Six months later still, in June 1934: "The Maghaduta I read with Ryder with delight, some ease, and great enchantment; the Veda lies on my shelf, a reproach to my indolence."[35] The indolence, at least in respect of Sanskrit, continued. Physics became more interesting and his students more demanding. In 1936 Oppenheimer met Jean Tatlock, the communist daughter of a Berkeley professor, and added to the sin of forsaking Sanskrit a flirtation with the far left. Ryder could not have approved his student's political awakening. For him the New Deal, which he detested, was already socialism. Ryder probably dropped Oppenheimer at this time. That was his way with friends who failed to meet his expectations of them. By the end of his life almost all of his acquaintance had disappointed him.[36]

Perhaps we catch a glimpse of the disappointed Ryder in a recollection by Tatlock's friend Edith Jenkins of a party in Berkeley in the later 1930s. Oppenheimer, showing off, had just presented the company with an epigram. "[An elderly listener,] deep in his cups and in the solitary wisdom they allowed him, mumbled, 'Never since the Greek tragedies has there been heard the unrelieved pomposity of a Robert Oppenheimer.'"[37] Ryder completed his own tragedy with the farce of dropping dead, at the age of 61, while lecturing to his only student of Sanskrit.[38]

Ryder wrote some English verse that he showed his friends but never published. In one of them, "Tolstoi—a Critical Symposium," he allows creatures of his dour

35. JRO to Frank Oppenheimer, fall 1932, Oct 1933, 4 June 1934, ibid., 159, 165, 180.
36. Noyes (ref. 15), xxvii, xxxviii.
37. Edith Jenkins, *Against a field sinister: Memoirs and stories* (San Francisco, 1991), 23, quoted in Jeremy Bernstein, *Oppenheimer: Portrait of an enigma* (London, 2004), 54.
38. *Daily Cal*, 21 Mar 1938.

imagination to evaluate the great writer turned rustic mystic. They express Ryder's progressive disenchantment with his fellow men. The representative of established religion, who might stand for his father, could see no moral reason that Tolstoy should live eccentrically and uncomfortably. "Religious life is easy, plain." It is also, in Ryder's jaundiced view, deadly in its thoughtless conformity:[39]

> Conventions close to mine and me,
> From testing which we shrink;
> Ah, where would true religion be,
> If everyone should think?

Ryder had no more time for the slavish scientist:

> The world seems very plain to me;
> Full half its joy is missed
> By him who feels no call to be
> An entomologist...
> And here's a man of mighty brain
> Who spreads his writing thin
> O'er such vague themes as love, and pain
> And duty, God, and sin.
> Strange, in a scientific age
> To spend one's efforts thus!
> As if the scientific page
> Were not enough for us.

The conventional clergyman and the blinkered scientist, though not desirable recruits for the Stammtisch, are salts of the earth compared with Ryder's favorite whipping boy, the professor of English literature. He knows much better than "the common herd," plumbs to the bottom of things, and kills all art.

> They are impressed. The critic, no!
> His task, so subtly sure,
> So poignantly elusive, so
> Strange, intimate, obscure,
> Is this: in cultured phrase to tell
> An inattentive earth
> The *pith* of books; to strip the shell
> And pierce to inner work;
> To classify, to eliminate
> All that may touch the heart,
> Inform the mind, improve our state—
> The residue is Art.

39. Ryder, *Original poems* (ref. 15), 9-16.

A mysterious "Olympian" (a thinly disguised Ryder) ridicules academic and drones and closes the symposium:

> The man by whom new fire is brought,
> Is never understood;
> They praise his most imperfect thought,
> Blame him for the good....
> Yet worse, the littler ones are sure
> To organize his truth—Some school or church that shall endure
> In everlasting youth.
> And thus they always kill at length
> The thing they organize;
> The more the body gathers strength.
> The more the spirit dies.
> They of the Inquisition prayed
> To him of Galilee!
> The Renaissance of learning made
> A university!
> None seeking forms, or praise of man,
> Or scholarship, or art,
> Or any social glory, can
> Be quite sincere at heart.

It sums to bitterness and disillusion and an invitation to stoicism. The barb equating the doings of "a university" with the hypocrisy of the Holy Office no doubt was aimed at Harvard.

After the *Guru*

The director of Los Alamos needed discipline. The job required it and Groves's men were on the watch for lapses. Oppenheimer's behavior during 1943 did not disappoint them. In one astonishing indiscretion, he spent the night with his old communist girlfriend Jean Tatlock. During the same year he cooked up his story about Haakon Chevalier with its attendant lies and obfuscations. When asked why he had lied, he replied, "Because I was an idiot."[40] The answer recalls his reason for having left his student and collaborator Melba Phillips in his car wrapped in a rug at three o'clock one cold Berkeley morning. To the policeman who rescued her and found him asleep in his room at the Faculty Club he explained, "I am eccentric."[41] It was also a little eccentric of him (again in that fertile year 1943) to ask his old mentor Pauli, who had come to the United States to sit out Hitler, to undertake as war work writing papers on basic physics in other people's names for publication

40. *JRO Hearing*, 137 (quote), 154.
41. The incident occurred in 1934; Bernstein (ref. 37), 31-32.

in the *Physical review*. Why? To mislead the Germans into thinking that American physicists were busy with science as usual. No one else combined the necessary wealth of ideas with a strong "talent...for burlesque." Pauli declined to have anything to do with Oppenheimer's "Don-Quichotery."[42]

Apart from such lapses and relapses, and despite its forbidding purpose, Openheimer did his duty for Groves enthusiastically. As Krishna teaches (8:48):

> Let none reject his active work,
> Though little to admire;
> For blemish clouds adventures, as
> The smoke beclouds the fire.

A prince must do his job although, inevitably, doing it will have some "blemishes," like killing your friends and relatives. Oppenheimer, the inspirational prince of Manhattan scientists, did his job without flinching. In November 1945, after the smoke had cleared, he told the scientists under his command how he regarded their joint achievement and what he thought it presaged. He observed that, like other people, scientists had difficulty identifying their motives: the scientists at Los Alamos had worked to anticipate the Germans, to win the war, to satisfy their curiosity, to indulge their sense of adventure, to discover whether atomic weapons could be made, and, if so, to develop ones so powerful as to foreclose war among the United Nations. "I believe all these things that people said are true, and I think I said them all myself at one time or another." But none of these motives is the deep one, the root of all the others, the physicists' dharma. "When you come right down to it, the reason that we did this job is because it was an organic necessity. If you are a scientist you cannot stop such a thing. If you are a scientist, you believe that it is good to find out how the world works; that it is good to find out what the realities are; that it is good to turn over to mankind at large the greatest possible power to control the world...."[43]

Oppenheimer began this speech by declaring his ignorance of practical politics. Towards its end he expressed admiration for the Secretary of War, the truly admirable Henry Stinson, and other statesmen. Again he displayed that surprising deference to authority agreeable to Hindu teachings about caste and fate. The use of the bomb was not the business of scientists but of politicians. They are so many Krishnas, the true slayers of men; the bombers of Los Alamos, like Arjuna, are their instruments; the warrior who deploys his weapons in accordance with fate is a hero. We read (11:38):

> Arise, on fame, on victory
> On kingly joys intent!
> They are already slain by me
> Be you the instrument.

42. Oppenheimer to Pauli, 20 May, and reply, 19 June 1943, Pauli (ref. 6), vol. 3, 181-183; cf. Enz (ref. 8), 355-356.
43. Smith and Weiner (ref. 6), 316-317; Hijiya (ref. 2), 135, 137.

Oppenheimer insisted on the distinction between actor and instrument. To mix them up could have dire effects; the instrument might presume to try to influence the course of events. "In the worst instance, [the instrument's claim to being an actor] is used as a sort of screen to justify the most casual, unscholarly and, in the last analysis, corrupt intrusions of scientists into other realms of which they have neither experience, nor knowledge, nor the patience to obtain it."[44] But is it not the duty of the citizen-scientist to express his views through the political process?

In October 1945, Arjuna-Oppenheimer met Krishna-Truman. Secretary of State Dean Acheson chronicled their interview. Arjuna, wringing his hands, to Truman: "I have blood on my hands." Krishna, out of patience, to Acheson: "He didn't set that bomb off. I did. This kind of sniveling makes me sick." Many of Oppenheimer's later statements suggest that the hand-wringing was theater. He saw, or anyway came to see, the matter as Truman did. He said at his hearing in 1954: "I did my job which is the job I was supposed to do. I was not in a policy making position at Los Alamos. I would have done anything I was asked to do...if I thought it was technically feasible."[45] Eventually he fortified this position with an argument from Bohr's corrosive complementarity, which can be misapplied to justify much that cannot be justified. Bohr had resolved the old conundrum of squaring the freedom of the will with predestination (or God's foreknowledge). It is only necessary to observe that action and reflection cannot be accomplished at the same time; hence the feeling of free will and the conviction of necessity both have their truth, though not simultaneously. Oppenheimer: "I may, as we all have to, make a decision and act or I may think about my motives and my peculiarities and my virtues and my faults and try to decide why I am doing what I am." And at Los Alamos? "I was more concerned then with doing what I should than in chattering about it."[46]

Not long after the war Oppenheimer said that he owed his "feeling for the place of ethics" to Ryder, "[who] felt and talked as a stoic." The stoics had a "tragic sense of life, in that they attribute to human actions the completely decisive role in the difference between salvation and damnation."[47] The *Bhagavad-Gita* teaches the same. Ryder: "[It gives] a true answer, satisfying the reflective mind, to the insistent question why formal duties must be performed, even when they appear stupid or cruel." Asceticism or withdrawal does not work; only work works; man must work, doing his duty.

44. Hijiya (ref. 2), 139-140, 143, 145; quote from Oppenheimer, "Physics in the contemporary world," *Bulletin of the atomic scientists, 4:3* (Mar 1949), 67.

45. Oppenheimer, in *New York Times*, 11 Oct 1969, as quoted in Jean-Jacques Salomon, *Science and politics* (Cambridge, 1973), 194n, and in *JRO Hearing*, 236; also quoted in Richard Polenberg, "The ethical responsibilities of the scientist: The case of J. Robert Oppenheimer," William H. Chafe, ed., *The achievement of American liberalism* (New York, 2003), 129-159, on 145, 154.

46. Oppenheimer, *The flying trapeze: Three crises for physicists* (Oxford, 1962), 54, and Thomas B. Morgan, "With Oppenheimer on an Autumn day," *Look* (27 Dec 1966), 67, as quoted in Hijiya (ref. 2), 160.

47. *Time* (8 Nov 1948), 75, in Hijiya (ref. 2), 151.

"Here is the solution to the often-formulated problem of the relation between faith and works. The more this principle is pondered, the more fruitful it appears, the more true." Ryder concluded his explanation of the power of the *Bhagavad-Gita* in words that might have been written for his tormented student. "[The reflective] mind, after bitter experience of the meanness of power, the imperfection of institutions, the sadness of knowledge, inevitably raises the questions: why do one's duty, in such a world as the present?...The full answer, satisfying both intellect and spirit, is given in the Song of the Blessèd One."[48]

Oppenheimer kept the *Bhagavad-Gita* by his desk, gave copies to friends whether they needed it or not, and quoted from it in his speeches.[49] And he had more Eastern philosophy than the Blessed One's Song at his fingertips. Two nights before the test at Alamogordo in June 1945, he talked with Vannevar Bush, who, among his many other wartime responsibilities, served on the committee that oversaw the Manhattan Project. Oppenheimer recited a verse, "which he had translated from the Sanskrit." It struck Bush as so apt to the situation and so characteristic of the man that he wrote it down:[50]

> In battle, in the forest, at the precipice in the mountains
> On the dark great sea, in the midst of javelins and arrows,
> In sleep, in confusion, in the depths of shame,
> The good deeds a man has done before defend him.

What good deeds Oppenheimer had in mind to load the scale against the confusion and shame he felt, or felt he might soon feel, as he stood on the precipice while his laboratory accomplished the release of atomic energy, is not easy to say. Perhaps, in keeping with his remarks to the Los Alamos staff later in the year, he had in mind (besides conventional contributions to good causes) his success in persuading others to pursue that particular scientific goal and good, the "turn[ing] over to mankind at large the greatest possible power to control the world."

To the pile of evidence of Oppenheimer's ongoing fascination with Eastern philosophy I can add a mite. In 1962 Oppenheimer went to Copenhagen to compose an obituary notice of Niels Bohr. I happened to be there, working on the project Sources for History of Quantum Physics with Paul Forman and T.S. Kuhn. Every day for a week some of us had lunch with Oppenheimer in the converted stables in the Carlsberg Brewery that housed Bohr's archives. And every day Oppenheimer reported

48. Ryder, *Bhagavad-Gita* (ref. 14), x-xi, xvii.
49. Hijiya (ref. 2), 130; Bernstein (ref. 37), 193, 201.
50. Vannevar Bush, *Pieces of the action* (New York, 1970), 148. Gregg Herken, *Brotherhood of the bomb* (New York, 2002), 289n, quotes another version, which he says Bush carried about with him for the rest of the war. "In the forest, in battle, in the midst of arrows, javelins, fire/Out on the great sea, at the precipice's edge in the mountains/In sleep, in delirium, in deep trouble/The good deeds a man has done before defend him." Apart from rearrangement of some phrases, Herken's version differs from the one Bush published by the significant replacement of "in confusion, in the depths of shame" with the lesser afflictions, "in delirium, in deep trouble."

the verses of Hindu scripture that had come to his mind while traveling to the stables through the Copenhagen subway system. There he sat before his Elephant beer, ascetic and emaciated, burnishing his image of Eastern sage and perhaps excusing himself from having done his duty, before the lowest members of the academic proletariat.

In the judgment of Isidor Rabi, a sympathetic friend and colleague for forty years, Oppenheimer's addiction to Hindu thought distorted his values and damaged his career. It turned him away from the "hard, crude methods of theoretical physics into a mystical realm of broad intuition." It plunged him into a perpetual "charade." "Overeducated in those fields which lie outside the scientific tradition, such as his interest in religion, in the Hindu religion in particular,... [he ended] in a feeling for the mystery of the universe that surrounded him like a fog."[51] In this accounting, fulfillment of one's potential as a scientist, if at all possible, trumps all else; fuzzy thinking, ascetic and aesthetic impulses, exotic religions, which "lie outside the scientific tradition," plunge the scientist who indulges too freely in them into theatricality and confusion.

There is another way to keep score. As director of Los Alamos, Oppenheimer achieved what not one in a hundred Nobel laureates, or perhaps none at all, could have done.[52] If memory of the horrors of Hiroshima and Nagasaki prevented all out war between the Soviet Union and the United States, it would have been an untellable catastrophe had Oppenheimer not obtained whatever philosophy he needed to complete the work at Los Alamos before the scheduled invasion of Japan. The Eastern lore he accumulated with Ryder supported his directorship at different times and in different ways. At a minimum it identified him with great literature and deep philosophy. Also, no doubt, it helped reinforce his will, stiffen his discipline, provide him justification, and give him consolation when, "in confusion, in the depths of shame," contradictory elements warred in his divided personality and he desperately needed resolve and resolution.[53] On this reckoning, Oppenheimer's engagement with Hindu philosophy and religion has world-historical importance, and his disenchanted *guru* becomes an actor in the creation of nuclear weapons.

51. I.I. Rabi, "Introduction," Rabi et al., *Oppenheimer* (New York, 1969), 3-8, on 7, and as quoted by Cassidy (ref. 3), 179.
52. Cf. Leslie R. Groves, *Now it can be told* (New York, 1970), 63.
53. Some contradictory elements: Cassidy (ref. 3), 99-100, 125-126, 142, 151-152; Abraham Pais, *A tale of two continents. A physicist's life in a turbulent world* (Oxford, 1999), 241; Bernstein (ref. 37), 94, 193; Jean Pitzer, in Kenneth C. Pitzer, "Oral history," conducted by Sally Hughes and Germaine LaBerge, 1996-98, ROHO, TBL, 326: "Oppenheimer could be extremely charming and extremely arrogant, sometimes simultaneously."

CHAPTER FOURTEEN

The Scientist in Mass Society
J. Robert Oppenheimer and the Postwar Liberal Imagination

Charles Thorpe

Introduction

In the wake of the security hearing of 1954, it was widely assumed that Oppenheimer would cease to play any sort of role in public life. It was reported that, reacting to the Personnel Security Board finding, he had told an Australian journalist in a telephone interview, "maybe this is the end of the road for me." Although Oppenheimer denied making the comment, it resonated with the widely held belief that his career was effectively at an end and that the hearing had left him personally a broken man.[1]

However, the hearing did not prove to be the end of the road. It was 1957 before Oppenheimer would agree to be interviewed about the case. When he did so, he told journalist Victor Cohn, "I have tried to prove that a security risk can survive." "I had to establish," Oppenheimer said, "that what was put out as a final judgment about me wasn't the final judgment. And the only way to do this was by surviving."[2]

Oppenheimer had survived in that he was able to keep his position as director of the Institute for Advanced Study at Princeton. However, that role was by no means the sum of his activity after the hearing. His survival equally owed to the way in which he constructed new alliances and, on the basis of these, mapped out a new intellectual role for himself. His survival was to a large extent due to the way in which he embraced his defeat and worked to shape its meaning. In reconfiguring his public role, he drew centrally on the language and the cultural and political orientation of Cold War liberalism. Oppenheimer survived as a symbol and articulator of the embattled position of the intellectual in Cold War America.

The following abbreviations are used: JRO, J. Robert Oppenheimer; JROP, J. Robert Oppenheimer Papers, Manuscript Division, Library of Congress, Washington, D.C.; *NYHT-WP, New York herald tribune-Washington post*, Paris; *NYT, New York Times*; *WP, Washington post.*

1. "Paper, Oppenheimer continue to differ," *NYT*, 5 June 1954, 6.

2. Victor Cohn, "Can a security risk survive?," *Minneapolis Sunday tribune*, 16 June 1957; Cohn, "Oppenheimer survives 'security risk' branding," *Detroit news*, 24 June 1957. Clippings and related correspondence in JROP, box 208.

Social scientists and literary intellectuals pointed to the hearing as indicative of the collective fate of intellectuals in modern America. Sociologist Philip Rieff in an article for *The twentieth century* analyzed the hearing as being indicative of the gulf between intellectuals and the public and between scientific and humanistic intellectuals in an age of specialization, and as indicating the general insecurity of the intellectual in modern America.[3] In the pages of the *Partisan review*, Diana Trilling wrote of Oppenheimer as "a culture hero for American intellectuals, especially for literary intellectuals" and identified his political trajectory as a shared one.[4] For such thinkers, the hearing confirmed Oppenheimer as a symbol of the American liberal intellectual.

It was on the basis of this prepared identity as archetypal intellectual that Oppenheimer reconstituted his public role after the hearing. As he was cut off from his prior role as technocratic governmental advisor, Oppenheimer increasingly took on the persona of the general intellectual and fashioned his identity in these terms—as an interpreter of problems of culture and morality and the public sphere.

Cold War Intellectuals

Oppenheimer found a network of support for this new role in the intellectual milieu provided by Congress for Cultural Freedom (CCF). The Congress, which arose from a conference of intellectuals held in Berlin in June 1950, aimed to solidify the cultural relationship between American and European intellectuals and to advance liberal pluralist ideas as a way of countering the cultural and intellectual influence of Communism. The CCF also promoted the view that intellectuals had a key role to play in the maintenance of liberal democratic values, a view reflecting intense anxiety about mass democracy and mass movements. Popular support for Nazism in Germany confirmed the worst fears of liberal cultural pessimists about the irrationality of the masses and the fragility of liberal values in an age of mass organization. In this context, the defense of liberalism became tied to the preservation of elite cultural values. The idea of mass culture—inherited from conservative thinkers, in particular Ortega y Gasset—became central to Cold War liberalism and to the intellectual outlook of the CCF.[5] This notion was closely tied to the concept of totalitarianism, developed most importantly by Hannah Arendt.[6]

3. Philip Rieff, "The case of Dr. Oppenheimer," Rieff, ed., *On intellectuals: Theoretical studies/ case studies* (New York, 1970), 341-369, esp. 368. The essay was first published in *The twentieth century, 156:930-931* (1954).
4. Diana Trilling, "The Oppenheimer case: A reading of the testimony," *Partisan review, 21:6* (Nov-Dec 1954), 604-635, on 619.
5. José Ortega y Gasset, *The revolt of the masses* (New York, 1932).
6. Hannah Arendt, *The origins of totalitarianism* (London, [1951] 1986); see also Carl Friedrich and Zbigniew Brzezinski, *Totalitarian dictatorships and autocracy* (New York, [1956] 1968). A sociological classic on the relationship between mass culture and totalitarianism is William

The problem of culture was at the heart of the outlook of the famous group of New York intellectuals who dominated the American section of the Congress, the American Committee for Cultural Freedom (ACCF). Clement Greenberg's 1939 essay in the *Partisan review*, "Avant garde and kitsch," identified derivative, popular "kitsch" culture as the hallmark of the fascism and communism. Democratic values depended on the high-cultural innovators of the avant garde. The argument was echoed in Dwight MacDonald's 1960 essay on "Masscult and midcult."[7] Although originally leftist, the New York intellectual milieu came to tie their criticism of mass culture to the defense of liberal pluralism and America's cultural and intellectual leadership. Their formulations became increasingly hostile to any form of upheaval from below and increasingly defensive of the American status quo. At the same time, they saw themselves as formulating a rational and responsible anti-Communist alternative to the politics of McCarthyism. The goal was the protection of what Arthur Schlesinger Jr. termed "the vital center" against "radicalism" on both the left and the right.[8]

CCF-affiliated sociologists and social critics such as Daniel Bell, Edward Shils, Nathan Glazer, and David Riesman identified McCarthyism as populist radicalism of the right and a symptom of "the revolt of masses."[9] The bulwark against McCarthyism in America was seen to be the country's traditions of pluralism and voluntary association first celebrated by Tocqueville.[10] For these pluralist thinkers, science also occupied a key place in the structure of liberal democratic societies. This centrality was expressed by sociologists such as Talcott Parsons and his student Robert K. Merton.[11]

Kornhauser, *The politics of mass society* (London, [1959] 1968). See also Volker R. Berghahn, *America and the intellectual cold wars in Europe: Shepard Stone between philanthropy, academy, and diplomacy* (Princeton, 2001), 115-116; Ian Wellens, *Music on the frontline: Nicolas Nabokov's struggle against communism and middlebrow culture* (Aldershot, 2002), 100-101.

7. Clement Greenberg, "Avant garde and kitsch," *Partisan review, 6* (1939), 34-49; Dwight MacDonald, "Masscult and midcult," *Against the American grain* (New York, 1962), 3-75. See also George Cotkin, "The tragic predicament: post-war American intellectuals, acceptance, and mass culture," Jeremy Jennings and Anthony Kemp-Welch, eds., *Intellectuals in politics: From the Dreyfus affair to Salman Rushdie* (London, 1997), 248-270.

8. Arthur Schlesinger, *The vital center: Our purposes and perils on the tightrope of American liberalism* (Cambridge, 1949).

9. David Riesman and Nathan Glazer, "The intellectuals and the discontented classes," *Partisan review, 22:1* (Winter 1955), 47-72; Edward Shils, *The torment of secrecy: The background and consequences of American security policies* (Chicago, [1956] 1996), esp. 98-104; Daniel Bell, ed., *The radical right* (New York, 1964); see also Richard Hofstadter, *Anti-intellectualism in American life* (London, 1964).

10. Shils (ref. 9); Morton Grodzins, *The loyal and the disloyal: Social boundaries of patriotism and treason* (Chicago, 1956).

11. Talcott Parsons, "National science legislation. Part 2. The case for the social sciences," *Bulletin of the atomic scientists, 3:1* (Jan 1947), 3-5; David A. Hollinger, "The defense of democracy and Robert K. Merton's formulation of the scientific ethos," *Science, Jews, and secular culture: Studies in mid-twentieth-century American intellectual history* (Princeton, 1996), 80-96.

Chemist and philosopher Michael Polanyi, based at Manchester University, was particularly important in aligning science and the defense of scientific freedom with the political and cultural agenda of the CCF.[12] In 1953, Oppenheimer was listed as a sponsor of the CCF conference on "Science and Freedom." Held in Hamburg and organized by Polanyi, the conference brought together prominent natural scientists, social scientists, and humanistic intellectuals. The meeting was important in situating science in relation to Cold War liberal pluralism. Rather than a cultural, artistic avant garde, Polanyi suggested that it was most importantly the scientific community which should be looked to for the preservation of liberal values.[13]

Despite sponsoring the conference, Oppenheimer at this stage still held back from deeper involvement with the CCF. In 1952, he turned down an invitation from Sidney Hook to attend an ACCF luncheon and in September, 1953 was unable to attend an ACCF conference on European anti-Americanism. However, in early March 1954, while preparing for the security hearings, Oppenheimer took the time to try to persuade Einstein to dissociate himself from the group the Emergency Civil Liberties Committee, which was holding a gathering in honor of the physicist's seventy-fifth birthday. Oppenheimer had been contacted by ACCF Executive Director, Sol Stein, who told him that the civil liberties group was a Communist front-organization.[14] Following his intercession with Einstein, the ACCF sent a message to the AEC's Personnel Security Board vouching for Oppenheimer's anti-Communist credentials.[15]

Oppenheimer agreed with the ACCF's stance of trying to foster a responsible, liberal anti-Communism as an inoculation against the excesses of McCarthyism. The Oppenheimer case, however, cast doubt on this strategy. As historian Richard Pells has argued, the ACCF intellectuals had placed faith in the federal executive as a bulwark against the radical anti-Communism of HUAC and McCarthy and of the state-level committees. The fact that Oppenheimer was being persecuted by an execu-

12. Michael Polanyi, *Science, faith and society* (Chicago, 1946). Polanyi was appointed to the CCF's Executive Committee in 1953 and edited the CCF-funded bulletin *Science and freedom.* Frances Stonor Saunders, *The cultural cold war: The CIA and the world of arts and letters* (New York, 1999), 214; Peter Coleman, *The liberal conspiracy: The Congress for Cultural Freedom and the struggle for the mind of postwar Europe* (New York, 1989), 98.
13. Congress for Cultural Freedom, *Science and freedom: The proceedings of a conference convened by the Congress for Cultural Freedom and held in Hamburg on July 23rd – 26th, 1953* (Boston, 1955); American Committee for Cultural Freedom, press release, 11 Nov 1954, JROP, box 110.
14. Sol Stein, Executive Director, American Committee for Cultural Freedom, to JRO, 5 Mar 1954; memo on telephone conversation of JRO with Sol Stein, 6 Mar 1954; transcript of telephone conversation between JRO and Mrs. van Gelder of Mr. Leidersdorf's office, 5 Mar 1954; memo on telephone conversation with Miss Helen Dukas, 9 Mar [1954], JROP, box 32. See also "2 scientists refuse bid to Einstein fete," *New York post,* 11 Mar 1954; Frederick Woltman, "Einstein in middle of new commie hassle," *New York world telegram and sun,* 11 Mar 1954, 1-2, JROP, box 32.
15. American Committee for Cultural Freedom, press release (ref. 13).

tive agency (and further, one associated with scientific-technological rationality and progress) came into conflict with the pluralist view of the forces of anti-liberalism being an upsurge of radicalism from below. The persecution of Oppenheimer led to a loss of faith among liberal anti-Communists and to the drift of liberal intellectuals away from the ACCF.[16]

Ironically, however, the hearings immediately propelled Oppenheimer into the ACCF. It was Daniel Bell who put forward and sponsored Oppenheimer's name for membership at a meeting of the ACCF's Executive Committee, shortly after the AEC verdict.[17] On November 9, Oppenheimer accepted the invitation to join the organization. The ACCF press release, announcing his membership, proudly stated that "[i]n March of 1954, he was...instrumental in helping the cultural freedom group discourage eminent scientists from cooperating with a Communist-line group in this country."[18] Oppenheimer's name immediately began to be mentioned prominently in newspaper reports on the activities and views of the ACCF. It was through the ACCF, and even more importantly through the broader international group, by attending conferences and signing petitions and letters, that Oppenheimer engaged culturally and politically as a public intellectual in the wake of the hearings. The group provided him with a milieu in which he shaped his intellectual identity of these later years.

Oppenheimer's relationship with the group was in many ways a natural fit. His analyses of science and culture, although ostensibly apolitical, meshed closely with the preoccupations of the CCF's Cold War liberalism. Like Polanyi, Oppenheimer presented science as the accomplishment of a cultural elite, necessarily aloof and insulated from the pressures of popular culture. Oppenheimer's liberalism was shot through with a conservative sense of sin and of the futility of attempts to make the world over. The best that could be hoped for, at the present time, was the preservation of high-cultural values within close-knit elite circles.

Science, Elitism, and Cultural Pessimism

Oppenheimer's philosophy of science grew out of the disillusioning of the utopian hopes of the scientists' movement for arms control which emerged at the end of World War II. Whereas in the first two years following the end of the war, Oppenheimer articulated the young scientists' optimism that the peril of the atomic bomb could also be the basis for hope for a new world without war, by 1947 he was presenting a bleaker picture: "The physicists have known sin, and this is a knowledge

16. Richard H. Pells, *The liberal mind in a conservative age: American intellectuals in the 1940s and 1950s* (New York, 1985), 344.
17. Robert Gorham Davis to JRO, 20 Jul 1954, JROP, box 110.
18. Annotation on Davis to Oppenheimer, 20 Jul 1954 (ref. 17); American Committee for Cultural Freedom, press release (ref. 13). See also Michael A. Day, "Oppenheimer on the nature of science," *Centaurus, 43* (2001), 73-112, esp. 74, 109 n. 3.

which they cannot lose." He was now advocating a much more restricted Weberian ethic of responsibility: "The true responsibility of a scientist...is to the integrity and vigor of his science." The failure of early arms control efforts left him with the sense that the values he found in science—such as openness and freedom from coercion—could not be generalized, but rather were in need of protection within the specialized communities of science.[19]

Oppenheimer formulated this defensive conception of scientific community in dialogue with the more optimistic view expressed in J.B. Conant's educational program at Harvard and in his popular histories of science. Conant believed that science would be safeguarded most effectively if it could be demystified for the public. So Conant had introduced *Science and common sense* as a "citizen's guide to the methods of experimental science."[20] Oppenheimer was skeptical of these populist sentiments and, in contrast, dwelt on what he saw as the inescapable gulf of misunderstanding between professional and layman.[21]

These problems of communication between lay and expert communities were compounded by official secrecy. In an article for the journal *Foreign affairs* in July 1953, Oppenheimer alleged that secrecy had created a situation in relation to nuclear issues such that "[t]oday public opinion cannot exist in this field."[22] Oppenheimer's portrait of the effects of official secrecy was precisely the pluralists' nightmare of a binary polity made up of disengaged mass society and a powerfully centralized state. His solution was the pluralist ideal of civil society in which voluntary associations acted as mediating structures, taming the excesses of both the mass and the state.[23] For Oppenheimer, the crucial mediating associations were overlapping specialized communities of experts.

This conception of the place of science in civil society and of science as a model of pluralism was the core message of Oppenheimer's Reith Lectures delivered for the

19. JRO, "Physics in the contemporary world" [1947], *The open mind* (New York, 1955), 81-102, on 88, 91; contrast with JRO, "The new weapon: The turn of the screw," Dexter Masters and Katherine Way, eds., *One world or none* (London, [1946] 1947), 53-60.
20. James B. Conant, *Science and common sense* (New Haven, CT, [1951] 1963), 1. See also Conant, *On understanding science: An historical approach* (New Haven, CT, 1947); and see James G. Hershberg, *James B. Conant: Harvard to Hiroshima and the making of the nuclear age* (Stanford, 1993), 556-559.
21. JRO to James B. Conant, 13 May 1947, JROP, box 27; JRO, "Physics in the contemporary world" (ref. 19), 97-99; Yaron Ezrahi, "Einstein and the light of reason," Gerald Holton and Yehuda Elkana, eds., *Albert Einstein: Historical and cultural perspectives, the centennial symposium in Jerusalem* (Mineola, NY, 1982), 253-278, esp. 259-273.
22. JRO, "Atomic weapons and American policy," *Foreign affairs, 31:4* (Jul 1953), 525-535, on 531.
23. For a later formulation of pluralism in terms of mediating structures, see Peter L. Berger and John Neuhaus, *To empower people: The role of mediating structures in public policy* (Washington, D.C., 1975). For an analysis of pluralism in American thought, see Michael Paul Rogin, *The intellectuals and McCarthy: The radical specter* (Cambridge, 1967), 9-31.

British Broadcasting Corporation in November 1953. Here, he dwelt on the problem of the fragmentation of culture, arising from the specialization of the sciences, and made some rather general and vague proposals for how a common cultural discourse might be renewed.

Oppenheimer suggested not only that science was set apart from lay culture but also that its different branches were equally bounded. Specialization was endemic and irreversible, ruling out the traditional ideal of the cultivated general intellect: "even the best of us knows how to do only a very few things well; and of what is available in knowledge of fact, whether science or history, only the smallest part is in any one man's knowing."[24] The goal of cultivation, the formation of an individual self which would be equal to the scope of human culture, was no longer feasible. In its place, Oppenheimer substituted an ideal of dialogue between diverse communities of specialists.

Just like pluralist political thinkers such as Bell, Shils, and William Kornhauser, Oppenheimer portrayed the ideal citizen and cultural leader as one who, in contrast to the "mass man," was embedded in a complex web of group affiliations. Rather than an idea of common culture as homogeneity, Oppenheimer saw commonality in pluralist terms as the overlapping of a multitude of partial associations. "This," he said, citing Tocqueville, "may be more true of the United States than any other country."[25]

This American pluralism stood against totalitarianism. Oppenheimer commented on the "humourless sort of pun that so powerful a present form of modern tyranny should call itself by the very name of a belief in community, by a word 'communism' which in other times evoked memories of villages and village inns and of artisans concerting their skills and of men of learning content with anonymity." He suggested that the fate of both American democracy and modern science depended on the maintenance of a pluralist pattern of civil association.[26]

As he articulated a conception of science as an instantiation of liberal democratic values, Oppenheimer presented himself as a spokesman for these values and constituted his *own* authority as interpreter of the cultural meaning of science. Giving the Reith Lectures, Oppenheimer was defining his personal role as an intellectual. The lectures were a performance and it mattered that it was Oppenheimer who was delivering them. Though he announced the death of the cultivated man, these lectures were nothing if not a display of cultivation; and while he pointed to the splintering of culture into specialized segments, he demonstrated his own ability to transcend such divisions. Oppenheimer constituted himself as an embodied solution to the very problems of cultural fragmentation which he was pointing out.

Yet Oppenheimer's lectures were in some ways a practical example of his point about the gap between science and common understanding. Many among the British audience who listened to the lecture on the radio complained that they could not follow

24. JRO, "The sciences and man's community," *Science and the common understanding* (London, 1954), 92-110, on 98.
25. Ibid., 102.
26. Ibid., 105, 103.

the speaker. Amidst the controversy of the security hearings the *Manchester guardian* recalled that the Reith Lectures had been "controversial only in one sense: they could not be understood, so many complained, by the common educated man."[27]

The Reith Lectures prepared the part that Oppenheimer would assume in the wake of the security hearing and supplied him with the vocabulary which he would use to re-make himself. While the hearing closed off to Oppenheimer the possibility of a technocratic role in the state, it left available to him the cultural role which he had begun to map out in the Reith Lectures. A *Herald tribune* reviewer of the published version of the lectures—*Science and the common understanding*—told readers that "a few paragraphs" of this book "may tell more of the essential faith and nature of this man than columns of testimony."[28] It was in terms of the cultural problems addressed in the Reith Lectures that Oppenheimer made sense of his new position. He presented his new status as victim as manifesting the more general defeat of the intellectual in a fragmented and degraded modern culture.

On December 26, 1954, some six months after the AEC's finding against him, Oppenheimer gave a lecture over the Broadcasting System of Columbia University. He recapitulated many of the ideas which he had first articulated in the Reith Lectures. But this new version was without the optimistic notions of pluralism and community which he had outlined just a year earlier. Where the Reith lectures had resonated with the optimism of American liberal pluralists like Shils and Bell, his Columbia address was infused with the cultural pessimism of Max Weber and Ortega y Gasset. Here, Oppenheimer emphasized personal feelings of weakness and of being overwhelmed by a world beyond one's control. His communitarianism now took the form of a defensive retreat:[29]

> This is a world in which each of us, knowing his limitations, knowing the evils of superficiality...will have to cling to what is close to him, to what he knows, to what he can do, to his friends and his tradition and his love, lest he be dissolved in a universal confusion and know nothing and love nothing.

The lecture was shot through with fear of the dissolution of self, of being unable to hold one's own against others:[30]

> If a man tells us that he sees differently than we, or that he finds beautiful what we find ugly, we may have to leave the room, from fatigue or trouble; but that is our weakness and our default. If we must live with a perpetual sense that the world and the men in it are greater than we and too much for us, let it be the measure of our virtue that we know this and seek no comfort.

27. "London letter," *Manchester guardian weekly* (22 Apr 1954), 6.
28. Lewis Gannet, "Book review, *Science and the common understanding*," *New York herald tribune*, 22 June 1954.
29. JRO, "Prospects in the arts and sciences," *The open mind* (New York, 1955), 133-146, on 144.
30. Ibid., 145.

Oppenheimer painted a picture of the individual confronting others as alien and hostile and experiencing the social world itself as a vast and opposing power. He presented not only the individual, but culture itself as on the defensive against this great anonymity. "Never before today," he said, "has the integrity of the intimate, the detailed, the true art, the integrity of craftsmanship and the preservation of the familiar, of the humorous and the beautiful stood in more massive contrast to the vastness of life, the greatness of the globe, the otherness of people, the otherness of ways, and the all-encompassing dark."[31]

Pervading the lecture was a profound sense of homelessness, of the "artist's loneliness" and of the scientist working at the boundaries of knowledge who finds himself "a very long way from home." Oppenheimer described this sense of homelessness, above all, in relation to mass culture. This meant the ersatz products channeled through what he called the "superhighways" of the "mass media." Echoing Clement Greenberg, Oppenheimer linked popular culture with totalitarianism. The "superhighways" ranged "from the loud-speakers in the deserts of Asia Minor and the cities of Communist China to the organized professional theater of Broadway. They are purveyors of art and science and culture for the millions upon millions." While making us aware of events across the globe, these "superhighways" were ultimately destructive of genuine solidarity. "[T]hey are," he said, "the means by which the passivity of the disengaged spectator presents to the man of art and science the bleak face of unhumanity."[32] Degraded mass culture left no room for an authentic intellectual role.

Oppenheimer's response was therefore to urge that the intellectual retreat from modern society to the pre-modern "villages" which he thought remained the true sites of cultural creation. The task of the intellectual was to preserve "these gardens in our villages," far from the superhighways, "keeping open the manifold, intricate, casual paths, to keep these flourishing in a great, open windy world."[33]

Alistair Cooke, who had followed the security hearing for the *Manchester guardian*, wrote that in his Columbia address Oppenheimer both spoke about and himself instantiated and embodied "the isolation of the specialist" in modern society. From "the loneliness of his own exile" Oppenheimer articulated the condition of "the lonely man in society." "So in the end Oppenheimer came to identify the world's plight with his own: that of a natural recluse too much bruised by the public world which repulsed the great gifts he felt he owed it."[34]

Oppenheimer's view of science continued to be characterized by a kind of defensive elitism. In a 1958 lecture, "Knowledge and the structure of culture," at Vassar College, Oppenheimer presented a defense of scientific specialization as a bulwark against the dilution of knowledge by mass culture: "We who live in universities, artists, people who spend their life trying to find out a little more about the world...have a kind of high

31. Ibid., 144.
32. Ibid., 140, 138, 142-143.
33. Ibid., 146.
34. Alistair Cooke, "Isolation of the specialist in America," *Manchester guardian*, 30 Dec 1954, 2.

duty to insist on being difficult, and insist on doing things that maybe not everybody can readily do, and insist on being recondite and honest and intimate." But that still was small remedy for the sense of commonality Oppenheimer thought was essentially lacking in modern life.[35] Oppenheimer told an audience at Chapel Hill the following year that the very intimacy of small specialized communities, as much as it brought these few people together, by the same token rendered them "isolated" from the rest of culture and humankind. Whatever the benefits of such intimacy among specialists it did not touch the "public sector of our common life."[36]

Science was in some important senses anathema to the public sector. The growth of science and its cultural prestige have been "rather adverse to our common discourse, to the excellence of our common discourse and the public sector." In particular, the image of objectivity attaching to science, and the fetishism of objectivity arising from the prestige of science, are corrosive of the kind of intersubjective discourse which constitutes the public sector. Oppenheimer argued that "We have to some extent lost the confidence in the value of talking with one another in a common discussion where verifiable truth in the sense of the sciences, and objectivity in the very special sense of the sciences, is not attainable." The sciences themselves have been "encysted and broken off from the great stream...of our common life. The situation today, in fact, seems to involve an alienation between the world of science and the world of public discourse." The fetishism of objectivity "has emasculated, impoverished and intimidated the world of public discourse...has...denied to public discourse its element of legitimacy, has given it a kind of arbitrary, unrooted, and unfounded quality." The fascination with objectivity has, paradoxically, led to a degradation of public discourse into subjectivism, a situation in which no basis can be found for the establishment of principles which transcend the individual and bind individuals into a community: "Thus any man may say what he thinks, or almost...there is no way of arriving at a clarification or a concensus, that is as to the meanings, as to the purposes, as to the beauties of life." What was lacking, Oppenheimer argued, was a recognition "of a difference between the kind of use and value which public discourse has as its high ideal and the kind of criteria by which the sciences themselves must judge themselves."[37]

Oppenheimer's writings and speeches expressed widely recognized dilemmas in the life of the mind in the mid-twentieth century. In his emphasis on cultural fragmentation and discontinuity, Oppenheimer was expressing key themes of the era.[38] In 1959, C.P. Snow delivered his massively influential Rede Lecture at Cambridge University, attacking the division between the "two cultures" of science and the hu-

35. JRO, *Knowledge and the structure of culture: The Helen Kenyon lecture at Vassar College* [29 Oct 1958] (Poughkeepsie, NY, 1958), 14.

36. JRO, *Some reflections on science and culture* (Chapel Hill, NC, [1959] 1960), 5, 8.

37. Ibid., 8-10.

38. Silvan S. Schweber points to this in describing Oppenheimer as "almost 'postmodern' during the last decade of his life": Silvan S. Schweber, *In the shadow of the bomb: Oppenheimer, Bethe, and the moral responsibility of the scientist* (Princeton, 2000), 34.

manities.[39] Historian of ideas Jacques Barzun likewise lamented the divided condition of the "House of Intellect," criticizing academic specialization which rendered scholars "mutually incommunicado."[40]

In the same year, Oppenheimer contributed an essay to a volume on *Issues in university education*, edited by philosopher Charles Frankel. This was referred to several years later by University of California President Clark Kerr as he argued for a new academic "multiversity." Kerr wrote: [41]

> The faculty world seems to sense a loss of unity—intellectual and communal unity. In a large measure this can be attributed to the "overwhelming predominance of things that are new over things that are old" and to what Robert Oppenheimer calls "a thinning of common knowledge." Knowledge is now in so many bits and pieces and administration so distant that faculty members are increasingly figures in a "lonely crowd," intellectually and institutionally.

Kerr presented the multiversity as a pluralistic solution, allowing loose cooperation between different knowledge-communities. It was an image strikingly close to Oppenheimer's portrayal in the Reith Lectures of the many rooms in the ramshackle "House of Science." But there were also worries that the multiversity would be, as historian David Kaiser aptly puts it, just a sprawling academic "suburb," yet another lonely crowd.[42]

If the public sector was increasingly fragile and unable to stem the fragmentation of intellectual life, Oppenheimer instead thought that integration could take place on a smaller scale. The significance of the CCF for him was in the potential which its gatherings and conferences offered for this integration. On its tenth anniversary, Oppenheimer sent a message of congratulations to the Congress: "By its example, its pronouncements, its publications, and its many, varied gatherings and discussions, it has upheld the dedication of those who enjoy a wide measure of liberty, has given hope to many who have been denied it, and has given courage to all who treasure culture and prize freedom. May its future decades be increasingly devoted to the culture of freedom."[43] Shils, in a similar message, explicitly linked the function of the Congress to overcoming the problems of cultural fragmentation, "specialisation," and intellectual

39. C.P. Snow, *The two cultures* (Cambridge, [1959] 1993).

40. Jacques Barzun, *The house of intellect* (New York, 1959), 11.

41. Clark Kerr, *The uses of the university* (Cambridge, [1963] 1972), 101. The first two quotations within Kerr's text are from JRO, "Science and the human community," Charles Frankel, ed., *Issues in university education: Essays by ten American scholars* (New York, 1959), 48-62, on 56 and 58. The last is clearly a reference to David Riesman's *The lonely crowd: A study of the changing American character* (New Haven, CT, 1950).

42. David Kaiser, "The postwar suburbanization of American physics," *American quarterly*, *56* (2004), 851-888.

43. Untitled document, 11 Apr 1960, published in "A word about the Congress" [ca. 1965]; Nicolas Nabokov to JRO, 1 Apr 1960, JROP, box 253.

"isolation" about which Oppenheimer was so concerned. The Congress represented "[a]n intellectual community, bound together by the sense of affinity which arises from a respect for universal standards of truth....*In the whole of intellectual life today, it seems that the Congress for Cultural Freedom is doing more than any other body to foster the growth of this world-wide intellectual community, cutting across the boundaries which separate academic disciplines, professions, levels of development and political and economic systems.*"[44]

Some months after Snow's lecture, Oppenheimer gave a talk in Rheinfelden, Switzerland, for the CCF in which he returned to the theme of the decline of the public sector and of public discourse. The diplomat George Kennan was there, as was Shils, who described Kennan and Oppenheimer as two "icily lofty American saints."[45] The French political philosopher and writer Raymond Aron had initially asked Oppenheimer to give a talk introducing "Western policy in the atomic age." But Oppenheimer said, "I would prefer the somewhat wider and more interesting theme of the effects of contemporary science on Western culture and politics, of which the problem of nuclear weapons is a principal but rather special example."[46] In contrast to Snow's call for more science against Oxbridge humanism, Oppenheimer argued that the public culture had been stunted by "an overemphasis...of the role of certitude," based on the prestige of science. This was particularly the case regarding atomic weapons which were now usually addressed in terms of rational-choice models rather than ethics. "What are we to think of such a civilization," Oppenheimer asked, "which has not been able to talk about the prospect of killing almost everybody, except in prudential and game-theoretic terms?"[47]

It was, however, unclear what position on nuclear weapons Oppenheimer was suggesting as an alternative. The philosopher Karl Jaspers articulated a common complaint about the difficulty of pinning down Oppenheimer's ethical and political views. Referring to Oppenheimer's quasi-religious appeal at the end of "Prospects in the arts and sciences" to "love one another," Jaspers wrote: "In such sentences I can see only an escape into sophisticated aestheticism, into phrases that are existentially confusing, seductive, and soporific in relation to reality."[48]

Novelist Mary McCarthy took Oppenheimer's frequent but vague talk about love as a sign that he had finally lost his marbles. Writing to Hannah Arendt about the CCF conference "Progress in Freedom" held in Berlin in June 1960, she said, "Another feature of the Congress was Oppenheimer, who took me out to dinner and is, I discovered, completely and perhaps even dangerously mad. Paranoid megalomania and sense of divine mission." At one point, Oppenheimer turned to CCF Secretary-General Nicolas Nabokov "and said the Congress was being run 'without love'. After

44. Shils, in "A word about the Congress" (ref. 43). Italics in original.
45. Shils, quoted in Coleman (ref. 12), 121.
46. JRO to Nicolas Nabokov, 3 June 1959, JROP, box 253.
47. JRO, "In the keeping of unreason," *Bulletin of the atomic scientists, 16:1* (Jan 1960), 18-22, on 22.
48. Karl Jaspers, *The atom bomb and the future of man* (Chicago, 1961), 202.

he had repeated this several times, I remarked that I thought the word 'love' should be reserved for the relation between the sexes."[49]

Aron recalls his impressions of Oppenheimer from the 1960 conference: "Devoured by an internal flame or by the battles he was fighting with himself, he tended to take any episode of his existence not seriously but tragically. I can see him in his room, with his wife, discussing the latest conference presentations with me, as though he were disturbed by their possible banality."[50] To others, however, it appeared that Oppenheimer and his compatriots had avoided banality by succumbing to melodrama.

The core theme of the 1960 conference was the problem of mass culture, addressed by figures such as Shils, Manès Sperber, Bertrand de Jouvenal, the British cultural and literary critic Richard Hoggart, and Irving Kristol. Not all were uniformly pessimistic. Nevertheless, if the CCF's goal was to engage in cultural cold-warfare, fighting communism on the cultural front and asserting Western hegemony, it seemed to *Time* magazine that it was doing a poor job, particularly when it came to advertising the benefits of the Western capitalism and democracy to African and Asian leaders. *Time* reported from Berlin that "[t]he Afro-Asians came expecting leadership, and found only hand-wringing." They came "want[ing] only some cars and some irrigation ditches and some good technical ideas from the gloomy Westerners." Instead, "the spokesmen for the sophisticated societies spent most of their time reproaching themselves or apologizing." Friedrich Luft announced that amid West German affluence "culture is dead." Raymond Aron lamented the weakness of parliamentary rule in France. Mary McCarthy "moaned" that "Western literature is the mirror on the ceiling of the whorehouse" and Oppenheimer "apologized for all the wrongs he said science has done."[51]

In his inaugural address in Berlin, Oppenheimer had warned of the dangers inherent in the increase in nuclear stockpiles. If there were to be another great war, he said, "none of us can count on having enough living to bury our dead." In keeping with the tenor of the conference, Oppenheimer framed the problems of nuclear weapons as being primarily cultural rather than political and ended the talk by returning to the theme of specialization and the decline of the public sector.[52]

Politics, Culture, and the Intellectual

Three months after the Berlin conference, Oppenheimer made what was perhaps his most significant foreign tour, to Japan. This trip demonstrated the way in which,

49. Mary McCarthy, quoted in Saunders (ref. 12), 333.
50. Raymond Aron, *Memoirs: Fifty years of political reflection* (New York, 1990), 175.
51. "Intellectuals," *Time, 76:1* (4 Jul 1960), 17. See also Coleman (ref. 12), 174-176; Berghahn (ref. 6), 108-113. The conference program is in JROP, box 253. See also *Congress for Cultural Freedom news* (Sep-Oct 1960), JROP, box 119.
52. "Oppenheimer deplores increase in deadliness of atom stockpiles," *NYT*, 17 June 1960; clipping in JROP, box 253.

through his activities with the CCF, Oppenheimer had become effectively rehabilitated as a public figure. Despite the fact that the U.S. government still officially viewed him as a communist security risk and that in the months immediately following the hearing, the FBI had taken seriously the idea that he might defect to the Soviet Union, Oppenheimer presented to the rest of the world the civilized face of American intellect and culture.[53] Oppenheimer's worldwide intellectual status and CCF affiliation were of considerable value and significance at a time when American elites took seriously the idea that the Cold War was a struggle for intellectual influence.

The two-week trip, sponsored by the Japan Committee for Intellectual Interchange (CII) of the International House of Japan, took place against the recent background of large left-wing protests and riots in Tokyo in May and June against the signing of a new security treaty with the United States. The protests had led President Eisenhower to cancel a planned trip to Japan. Although successful ratification of the treaty led to a restoration of outward calm, tensions in American-Japanese relations would not have been far from the surface during Oppenheimer's visit.[54]

In addition, Oppenheimer's visit was potentially a direct and painful reminder both of wartime defeat and of post-war subordination to the U.S. This was especially the case because of the central place in post-war Japanese national consciousness of their being the only nation to be the victim of atomic weapons. The sense of atomic victimization was reinforced by an incident in 1954 when a Japanese fishing boat, the *Lucky dragon*, was covered with fallout from an American nuclear test at the Bikini Islands.[55]

On arrival in Tokyo on September 5, Oppenheimer gave a crowded and chaotic impromptu press conference. When asked by a journalist whether he felt remorse for his role in building the bomb, Oppenheimer replied: "I do not regret that I had something to do with the technical success of the atomic bomb. It isn't that I don't feel bad. It is that I don't feel worse tonight than I did last night."[56] In reply to questions about whether he would be visiting Hiroshima, he said that, though he would like to, he would not have

53. For concern that Oppenheimer might defect, see Aug 1954 documents in JRO FBI security file (microfilm) (Wilmington, DE, 1978), 100-17828, Sec. 49.

54. Nicholas Evan Sarantakes, "Alliance in doubt: American reaction to the 1960 US-Japanese security treaty crisis," *American diplomacy, 4:3* (Autumn 1999), http://www.unc.edu/depts/diplomat/static/archive.html; George R. Packard III, *Protest in Tokyo: The security treaty crisis of 1960* (Princeton, 1966).

55. John W. Dower, *Embracing defeat: Japan in the wake of World War II* (London, 1999), 487-495, esp. 492; Lise Yoneyama, *Hiroshima traces: Time, space, and the dialectics of memory* (Berkeley, 1999), esp. 14; James J. Orr, *The victim as hero: Ideologies of peace and national identity in postwar Japan* (Honolulu, 2001), esp. 36-71; Lawrence Badash, *Scientists and the development of nuclear weapons: From fission to the Limited Test Ban Treaty, 1939-1963* (Atlantic Highlands, NJ, 1995), 92.

56. *Time* (19 Sep 1960), 47; quoted also in Peter Goodchild, *Oppenheimer: Shatterer of worlds* (Boston, 1981), 274. The Japanese *Asahi evening news* reported his comments with a somewhat different emphasis: "I feel just as sorry tonight as I felt last night": "Oppenheimer to give series of talks here," *Asahi evening news*, 6 Sep 1960, 3. Clippings and correspondence relating to JRO's trip to Japan are in JROP, box 41.

time and that it was not on his itinerary: "I have no quasi-official duties in Hiroshima." The atomic bomb was not what he wanted to talk about. Instead, he asked the newspaper men to "stress that I am glad to be here to talk about the things I know we can work on together: science and common cultural problems which underlie political problems."[57]

During Oppenheimer's trip, twenty-one-year-old American anti-nuclear protestor Tim Reynolds traveled from Hiroshima to Osaka to meet the physicist and express disappointment that he would not be visiting the city on which the first atomic bomb was dropped. Oppenheimer then explained that his sponsoring body feared that he would receive a hostile reception in the city.[58] The young American was the son of Earle Reynolds, who, with his family, had carried out a landmark protest against nuclear testing by sailing their yacht, the *Phoenix*, into the area around the Marshall Islands restricted for American nuclear tests. Tim Reynolds's appeal to Oppenheimer to visit Hiroshima made explicit the connection between Oppenheimer's visit to Japan, the current controversy around nuclear testing, and the atomic bombings at the end of World War II.

But Oppenheimer hoped to distance his visit from the issue of the atomic bomb —preferring, as he said, to emphasize "common cultural problems." He was welcomed to Japan, not as father of the atomic bomb, but as a representative of science, of the pure science ideal, and of liberal intellectual culture. Science was a key motif in the post-war reconstruction of Japan as democratic, western-oriented, and allied with America.[59] Oppenheimer's visit was symbolic of that connection. At a time when Japanese intellectuals had been at the forefront of opposition to the security treaty, Oppenheimer put forward and embodied a model of the intellectual role in which science and western-oriented liberalism were interwoven.[60] His lectures in Japan repeated the lessons of the dangers of cultural fragmentation and mass culture and of the necessity of elite intellectual community which he extrapolated from the American experience. While Oppenheimer's vision of the role of the intellectual was ostensibly apolitical, its meaning was in fact highly political in the context of Cold War culture and relations between the U.S. and Japan.

Oppenheimer gave a two-hour lecture on the physics of elementary particles at the Kyoto Institute for Fundamental Physics, but most of his speeches during the trip were on the broader cultural meanings of science. For example, to a gathering of more than a thousand in Osaka he gave a two-and-a-half hour talk on "Tradition and discovery" and in Tokyo spoke on "The future of civilization in the scientific age."[61]

57. *Asahi evening news*, 6 Sep 1960 (ref. 56).

58. "Oppenheimer met by American youth," *Mainichi daily news*, 18 Sep 1960; Earle Reynolds to JRO, 16 Sep 1960, JROP, box 41.

59. Dower (ref. 55), 494.

60. On the role of Marxist-oriented "progressive intellectuals" in the movement against the security treaty, see Packard (ref. 54), 26-31, 138-141, 270-278.

61. "Dr. Oppenheimer gives lecture in Kyoto Univ.," *Mainichi daily news*, 15 Sep 1960, 3;

On September 9, Oppenheimer spoke in Tokyo's Bunkyo Public Hall for the International House, on the topic of "Science and culture." The talk repeated Oppenheimer's standard themes: the growth and specialization of the sciences, the increasing divorce between science and common understanding and the decline of the "public sector," concluding with a discussion of "intellectual community" versus "mass culture." He suggested that popular culture was a form of gluttony: "One does not eat well unless one is hungry; there is a certain frugality to the best cooking; and something of this sort is wrong with the mass media." He added, though, "one must be critical, but... one must not be a snob; one must be rather tolerant and almost loving." Yet for cultural values not to be lost amid the overwhelming "noise level" of mass culture, they needed to be nurtured within an intellectual community. This meant, firstly, protecting communities of specialists (scientific or artistic)—"we have, all of us, to preserve our competence in our own professions...to preserve what we know intimately, to preserve our mastery. This is, in fact, our only anchor in honesty." Secondly, it meant bridging the divisions between specialists to form an overarching intellectual community, to "reknit the discourse and the understanding between the members" of this community. As he began his talk, Oppenheimer apologized that he would be talking about what he knew, American and western European, rather than Japanese culture. But he suggested that, as science and advanced industrialism became universal, so was this western experience increasingly generalizable. America's cultural problems were common ones, as were the solutions. Oppenheimer appropriated the discourse of the New York intellectuals and American Cold War liberalism and made it a universal discourse. As a gift to his Japanese host, CII chairman Yasaka Takagi, Oppenheimer him sent a copy of the *Partisan review*.[62]

The paradox of Oppenheimer as security risk at home and American cultural ambassador abroad was partially resolved when he was officially rehabilitated by the new Kennedy administration. Oppenheimer was a figure of admiration among the young technocrats who flocked to work in Washington. Following Kennedy's assassination, it was Lyndon Johnson who presented Oppenheimer with the AEC's Fermi Award on December 2, 1963, calling the award "one of President Kennedy's most important acts." Although officially rehabilitated, however, Oppenheimer was still barred from secret government work. As the *New York herald tribune* put it, the award was "an honor, not an indemnity."[63] Nevertheless, the prize signaled that he was no longer a pariah.

"Oppenheimer in Kyoto," *Asahi picture news*, Sep 1960; "1,000 hear Oppenheimer give talk," *Asahi evening news*, 19 Sep 1960, 1; "Oppenheimer to talk in Tokyo on Sept. 21," *Asahi evening news*, 15 Sep 1960, 33; "Oppenheimer stresses meaning of 'science'," *Asahi evening news*, 22 Sep 1960; JRO, "The future of civilization in the scientific age," *France-Asie* (Mar-Apr 1961), 1807-1815.

62. JRO, "Science and culture," International House of Japan, *Bulletin*, *6* (Oct 1960), 2-10, 21, on 10, 21; JRO to Yasaka Takagi, 27 Oct 1960; Yasaka Takagi to JRO, 26 Nov 1960, JROP, box 41.

63. Goodchild (ref. 56), 274-276; *New York herald tribune* (8 Apr 1963), quoted in John

Oppenheimer described his reaction to being chosen for the award by saying: "Most of us look to the good opinion of our colleagues and to the good will and confidence of our Government. I am no exception."[64]

Oppenheimer did not in any way see himself as an outsider critical intellectual. Rather, he continued to align himself with the elite and with the establishment. Portraying public culture as anti-intellectual and stultifying, he was instead attracted by the notion of retreating into a small and elite circle who could protect and carry high culture through a dark time. An opportunity to realize this ideal came when Agnes Meyer, widow of Eugene Meyer, the former owner of the *Washington post*, was seeking to donate her Mount Kisco estate, in upstate New York, as a venue for cultural and intellectual activities. The Princeton Jefferson scholar Julian Boyd persuaded Meyer to entrust the plans to Oppenheimer and the Institute for Advanced Study. This was on the basis, Boyd said, of the Institute's "zeal in cherishing both the humane spirit and the lonely intellect...[and] in drawing forth from many lands what Jefferson called 'the aristocracy of virtue and talent.'" Oppenheimer helped to draft Mrs. Meyer's plan —emphasizing that the gatherings would bring together "scholars, scientists, artists, statesmen and teachers to stay...in an atmosphere conducive to scholarly thought, discussion and common understanding." The CCF secretary-general, composer Nicolas Nabokov, was listed with Boyd, Oppenheimer, and architect Wallace Kay Harrison (who designed the UN building in New York) on the advisory committee to the project.[65] In a letter to Nabokov Oppenheimer enthused about the "open, hilly country" of the farm and the house, which he described to Nabokov as "extremely fine, in perfect taste, embellished with what can only be regarded as a great deal of loot in the way of art from China, and a good deal of modern European art as well."[66]

Organized by Oppenheimer and partly supported by the CCF, the first of these Seven Springs Farm conferences took place in the summer of 1963 with philosophers Stuart Hampshire from Oxford and Jeanne Hersch from Geneva, University of Edinburgh psychiatrist Morris Carstairs, Los Alamos veteran and the first head of the President's Scientific Advisory Committee George Kistiakowsky, the poet Robert Lowell, Kennan, and Harrison, alongside Boyd, Nabokov, and hostess Agnes Meyer. Numbers were kept small in order, as Meyer put it, "to maintain intimacy of discussion."[67]

The theme of the conference was allusively suggested to invitees with an extract of Oppenheimer prose:[68]

Major, *The Oppenheimer hearing* (New York, 1971), 292.

64. *NYT*, 4 Apr 1963, quoted in Philip M. Stern, *The Oppenheimer case: Security on trial* (New York, 1964), 455.

65. Julian Boyd to JRO, 3 Aug 1961; Draft of letter to be written by Mrs. Meyer to the Eugene and Agnes E. Meyer Foundation [ca. June-July 1963], JROP, box 66.

66. JRO to Nicolas Nabokov, 13 Apr 1962, JROP, box 66.

67. Jeanne Hersch, "Colloquy on Mount Kisco," *Congress news*, Winter 1964, 1-2; Mrs. Eugene Meyer, letters of invitation, 27 Feb 1963, JROP, box 66.

68. Statement by JRO attached to draft letter of invitation by Nicolas Nabokov, 12 Oct 1962,

> For myself, there are at least two things that I should look for: an image of the world that responds to the changing needs and hopes of men without resort to the institution of war, indeed without its possibility; and a world in which, despite the vastness, the complexity, the rapid change in the circumstances of our life, and the knowledge which underlies it, the scope and depth of human responsibility and human nobility grow instead of shrinking.

Oppenheimer chaired the discussion, but as Hersch put it, he "was hardly a 'chairman' in any conventional sense. He simply, from time to time, suggested a neglected aspect of a question, softened a particular dogmatism, cast doubt on such and such an alleged bit of evidence or dropped a word indicating a complexity which might lie hidden beneath an over-simplification resulting from facile hope or impatient pessimism."[69] Meyer was in no doubt that the event relied on Oppenheimer's personal and intellectual "leadership."[70] And Oppenheimer's talk, in which he dwelt on Bohr and the meanings of complementarity, was punctuated by her comments, "That's lovely Robert" and "That's wonderful, Robert."[71] She wrote to him that when she later read and re-read the script of his talk, "I re-lived the deep emotions with which I heard you speak. And yet, there was something lost because the warmth of your voice and personality gave the experience of listening to you an ambience which cannot be recaptured."[72] She later wrote to Kitty about the annual Seven Springs Farm event that "I cannot imagine our meeting without Robert."[73]

The intimate converse of a cultural and political elite, which Oppenheimer found in the CCF and at Seven Springs Farm, was the realization of his ideal of a select few preserving cultural values against a hostile mass society. As Hersch put it, "I had the feeling that such perfection of setting and circumstance had come into being in order to give truth a chance to exist for a moment, through an exchange of human words."[74] However, the cultural, political, and indeed military, conflicts of the 1960s threatened to fragment this elite liberal milieu in which Oppenheimer felt comfortable in his later years and which he had hoped to play a role in nurturing.[75]

The Vietnam War gave rise to difficult questions regarding the compact between the liberal intelligentsia and American global power. In April 1966, as part of a series of exposé articles on CIA covert operations, the *New York Times* revealed the extent to which the agency used the distribution of funds to influence academic and intellectual activity. Among the revelations was that the CCF and its magazine *Encounter* had been bankrolled by the CIA. In the ensuing controversy, Oppenheimer came to

---

JROP, box 66.

69. Hersch (ref. 67), 1.

70. Mrs. Eugene Meyer to Lester B. Pearson, 28 Jan 1967, JROP, box 66.

71. Original transcript of JRO's Seven Springs Farm talk, on 1, 2, 3, JROP, box 282.

72. Mrs. Eugene Meyer to JRO, 17 Jul 1963, JROP, box 66.

73. Mrs. Eugene Meyer to Kitty Oppenheimer, ca. Mar 1966, JROP, box 66.

74. Hersch (ref. 67), 2.

75. See also Saunders (ref. 12), 359-362.

the defense of the Congress. In a letter published in the *Times* on May 9, Oppenheimer, Kennan, John Kenneth Galbraith, and Schlesinger stated that, based on their experience with the Congress, they had "no question regarding the independence of its policy, the integrity of its officials or the value of its contribution. In our experience the congress...has been an entirely free body, responsive only to the wishes of its members and collaborators and the decisions of its Executive Committee."[76] The statement gave the impression of denying CIA involvement without literally so doing. Dwight MacDonald called it "an evasion, not a lie, but not meeting the issue either." Stuart Hampshire, in Princeton at the time, said that "Oppenheimer was amazed that I was amazed, and amazed that I was upset at the *New York Times* revelations. But I was upset, yes. There were people who were put in a terrible position. Oppenheimer wasn't amazed because he was half in it himself. He knew full well. He was part of the apparat [Hampshire's name for the CCF's controlling group]. I don't think it bothered him morally. If you're imperially-minded, which the Americans were at the time, you don't think much about whether it's wrong or not. It's like the imperial British in the Nineteenth Century. You just do it."[77]

Oppenheimer's complacency regarding the CIA involvement in intellectual life is revealing, for the CCF was the closest instantiation for Oppenheimer of his ideal of intimate, cohesive, and influential intellectual community. This intimacy and cohesiveness was disrupted by the Vietnam War. Agnes Meyer worried that the 1966 Seven Springs Farm recontre, aimed at understanding the "growing lack of lack of sympathy" of European intellectuals towards the United States, might be marred by a "knockdown fight about Viet Nam." Since a number of the attendees, most prominently Senator William Fulbright, were against the war, Meyer was worried that the gathering might appear "like an anti-administration meeting." She herself was against the war, but felt unable to speak out because of ties to the Johnson administration. Oppenheimer promised her "not to say a word on the subject," although he too was uneasy about U.S. involvement in the war.[78]

In December, in an article for a special edition on "Europe and the United States today" of the *Herald tribune*'s European edition, Oppenheimer gave voice to his concerns. Characteristically, his wording was oblique and the argument was framed in terms of culture and morality rather than as direct political criticism. His premise was that the cultural problems of the age were shared by Europe and America. At a time when the world was divided along the Cold War axis of East and West, Oppen-

76. "C.I.A. spies from 100 miles up; satellites probe secrets of Soviet Union," *NYT*, 27 Apr 1966; "Triple pass: How C.I.A. shifts funds," *NYT*, 19 Feb 1967; "Units linked with C.I.A.," *NYT*, 19 Feb 1967; J.K. Galbraith, George Kennan, Robert Oppenheimer, and Arthur Schlesinger Jr. to the Editor, "Record of Congress for Cultural Freedom," *NYT*, 9 May 1966.
77. MacDonald and Hampshire quoted in Saunders (ref. 12), 378-379.
78. Mrs. Eugene Meyer to JRO, 26 May 1966; JRO to Mrs. Eugene Meyer, 1 June 1966; Mrs. Eugene Meyer to JRO, 2 May 1966, JROP, Box 66. See also Mrs. Eugene Meyer, draft letter of invitation, Oct 1965, JROP, box 66.

heimer traced the historical roots of America and American freedom to the cultural and religious traditions of western Europe. But this tradition was now under strain both in Europe and in America. He said, "I doubt whether, at this writing, there is among us any wide conviction that France, or the United States, Germany or England, now has a government even remotely competent to the problems of the time, or in fact has available for those problems the human resources, the insight, the wisdom and skill, and the underlying stoic confidence for which they call." The article was suffused with a sense of entropy and decline. So it was with "haunting nostalgia" that Oppenheimer quoted Lincoln's belief "that the past actions and influences of the United States were generally regarded as having been beneficial toward mankind." Oppenheimer observed that few were likely to think that about current U.S. foreign policy. Vietnam represented the sacrifice of America's moral innocence, and a historical connection between new American and old European colonialism. No longer an immaculate City on a Hill, America now shared in the sins of the Old World. Oppenheimer said, "the sense of what it is like to live with a government held wrong in a moral matter is a new and desperate bond between European peoples and those of the United States." It was in the cultural and intellectual bond between America and western Europe that Oppenheimer placed his faith. In a reference to the traditional role of the intellectual, speaking truth to power, Oppenheimer concluded, "This is a time in need, not of delusion but of hope, when we, not least in Europe and America, who have taught hope so willingly and widely, must bear true witness."[79]

Oppenheimer accepted a position on the CCF's Board of Directors in autumn 1966 and was due to attend the Board meeting in March of the following year. Indeed, he told CCF Executive Director Michael Josselson, "There is nothing to which I more look forward than playing a part in the Congress' new life with you and other friends." Josselson was eager that Oppenheimer's *Herald tribune* article should provide a basis for discussion at the meeting. In a memorandum, he informed Board members that "recently, Robert Oppenheimer (in a statement criticizing U.S. involvement in Vietnam) castigated most of the Western governments for their incompetence in the face of major problems of our time." Josselson suggested to Oppenheimer that they should distribute in advance of the meeting a copy of Oppenheimer's article and that at the meeting "have you comment as briefly as you wish on how you see the Congress's role in the framework of your statement."[80] However, by February, Oppenheimer was in too much pain, under treatment, and his speech and hearing were impaired. On the 13th, only five days before his death, he wrote, "I think that it

79. JRO, "The forbearance of nations," *NYHT-WP*, 6 Dec, 1966.
80. JRO to Josselson, Feb 13 1967, JROP, box 42; "Exploratory memorandum. Subject: The seminar program of the Congress," enclosed with Josselson to JRO, 17 Jan 1967. On JRO's position on the CCF Board of Directors, see "Ford Fund helps Paris art group," *NYT*, 2 Nov 1966, and JRO to Minoo Masani, 7 Dec 1966, JROP, box 119.

would be appropriate to regard my seat on the Board as empty, though I hate to put these words in writing."[81]

Reports exposing covert CIA funding of cultural activities continued to pour out through the press in 1967. *Ramparts*, a left-wing Californian journal, produced more revelations about CIA front-organizations, particularly the agency's influence in student organizations and academic and intellectual life. These were picked up by the mainstream press, ensuring that the scandal over the CCF remained in the news that year.[82] The Congress was dissolved in 1967.[83]

Conclusion

In the wake of the hearing, Oppenheimer connected his individual fate with the tragic fate of the intellectual in mass society. In common with Cold War liberal pluralism, Oppenheimer interpreted contemporary problems as having above all to do with culture: fragmentation, atomization, specialization, and the new cultural influence of the masses. His outsider status after the hearing became iconic of the isolation and marginalization of the intellectual in an age of mass media and over-specialization. Yet in this outsider status, Oppenheimer claimed a new kind of insider role—as a key participant in the cultural Cold War and member of the Cold War liberal elite of the CCF. His assumed status as victim of a "revolt of the masses" solidified his place in the imagination of Cold War liberalism and certified his status as spokesman for liberal cultural values.

However, the Cold War intellectuals' analyses of contemporary issues as problems of culture seem, on reflection, to hide more than they illuminate. In his classic work *The intellectuals and McCarthy*, Berkeley political scientist Michael Rogin argued that the pluralist view of McCarthyism as a populist revolt of the masses was a distortion. In fact, McCarthyism drew on traditional American conservatism and relied on the acquiescence, and very often the support, of business, political, administrative, and intellectual elites. The Oppenheimer case clearly instantiates this. McCarthy lurked in the background. But it was Eisenhower who ordered Oppenheimer's dismissal, Admiral Strauss's Atomic Energy Commission which carried out the purge, and scientific colleagues, such as Edward Teller and Luis Alvarez, who twisted the knife. Oppenheimer never publicly confronted the events in these terms and so maintained his faith in

81. JRO to Raymond Aron (Chairman of the Board of Directors, CCF), 13 Feb 1967, JROP, box 119.

82. "U.S. officials confirm C.I.A. aid to students," *WP*, 15 Feb 1967; "CIA probe to take in new areas," *WP*, 19 Feb 1967; "Students assail CIA whitewash," *NYT*, 25 Feb 1967; "O what a tangled web the CIA wove," *WP*, 26 Feb 1967; "The CIA: Culprit or fall guy," *NYHT-WP*, 20 Feb 1967; Walter Lippmann, "The CIA affair," *NYHT-WP*, 22 Feb 1967; "Johnson backs CIA's secret subsidies," *NYHT-WP*, 24 Feb 1967, 1; Saunders (ref. 12), 381-390.

83. It continued under the title of the International Association for Cultural Freedom until 1979.

the ultimate wisdom and authority of elites. Rogin has described Cold War pluralist thought as underpinned by "the hope that if only responsible elites could be left alone ...[they] would make wise decisions."[84] This applies equally as a description of Oppenheimer's core outlook—his outlook both before and, paradoxically even more so, after the hearing. In his identification with elites and his consequent failure to develop real insight into his own political fate, Oppenheimer can be seen to represent many of the contradictions of Cold War intellectual life.

84. Rogin (ref. 23), 248-260, on 282.

CHAPTER FIFTEEN

Oppenheimer and the Sense of the Tragic

Robert P. Crease

Bruce Perry, the biographer of Malcolm X—another famously complex individual—once observed that everybody has their own Malcolm. The same could be said of J. Robert Oppenheimer: everybody has their own Oppie.

The result is what physicists might call a "Many Oppie Problem." Oppenheimer inspired diametrically opposed reactions even among people who knew him well: for some he was a secular saint, a 6-foot Gandhi, an Isaiah; for others he was a brutal and often petty colleague, pretentious, even a fool. I heard one historian of science remark that Oppenheimer was the second most loathsome and mendacious American physicist, losing out in that category only to Edward Teller. Abraham Pais, who wrote that "[i]n all my life I have never known a personality more complex than Robert Oppenheimer,"[1] compared his attitude toward Oppenheimer to people's attitudes toward New York City. Some people love it, some hate it, neither group really understands it. The proper attitude toward New York City is a love-hate relation. Everyone has their own New York.

I want to talk about one particular aspect of the Many Oppie Problem: the sense of tragedy associated with him. Oppenheimer once remarked to a *Washington post* journalist that the hearing was more farce than tragedy; this, of course, was dissembling. The Oppenheimer story is routinely described as tragic by friends and commentators. The word crops up everywhere, and not just in connection with the 1954 hearing over his security clearance. Years before the 1954 hearing, in 1946, David Lilienthal had called Oppenheimer a "tragic figure."[2] Years after, at Oppenheimer's memorial service, George Kennan noted "that strong element of tragedy which all who knew him sensed, I think, in his situation."[3] And now we have the excellent biography by Bird

The following abbreviations are used: APP, Abraham Pais Papers, currently in New York; JRO, J. Robert Oppenheimer, *JRO Hearing*, United States Atomic Energy Commission, *In the matter of J. Robert Oppenheimer: Transcript of hearing before Personnel Security Board and texts of principal documents and letters* (Cambridge, 1971); JROP, J. Robert Oppenheimer Papers, Manuscript Division, Library of Congress, Washington, D.C.

1. Abraham Pais, *J. Robert Oppenheimer: A life*, with supplemental material by Robert P. Crease (New York, 2006), 139.

2. David E. Lilienthal, *The journals of David E. Lilienthal* (New York, 1964), *2*, 69.

3. Hans A. Bethe, Henry DeWolf Smyth, and George Kennan, "Three tributes to J. Robert Oppenheimer," pamphlet issued by the Institute for Advanced Study.

and Sherwin with that word in the title: *American Prometheus: The triumph and tragedy of J. Robert Oppenheimer.*[4] Clearly, something about the Oppenheimer story invites comparison with the tragic. But what is meant by the word here? Surely something more than "unfortunate." Do we, for instance, mean a story that is able to disclose a deeper meaning about life if we place it alongside tragic and dramatic literature in history and fiction, and analyze it as we would them? Or something else?

Moreover, if the Oppenheimer story invites comparison with the tragic, this does not necessarily mean that the invitation is fulfilled. Alongside many other historiographical questions having to do with Oppenheimer—What about that J.? Was he really a Member? What words were exchanged with Chevalier in that pantry?—I think we have a duty to analyze the origin and legitimacy of this widespread urge to seek comparisons between his life and career and tragic literature.

Historical figures to whom Oppenheimer has been compared include Xerxes, the Persian king who audaciously built a bridge of ships across the Hellespont; Galileo, the scientist who was tried and convicted by the church; Alfred Dreyfus, the French officer whose court martial was incited by anti-Semitism; Warren Hastings, who was impeached while serving as the first Governor-General of India; and T.E. Lawrence, another man with a scholarly background who turned revolutionary leader before meeting a political downfall. Literary figures to whom Oppenheimer has been compared include Thomas Becket, in *Murder in the cathedral*, whose well-intentioned actions were undermined by arrogance and pride; Faust, who made a pact with the devil; Lear, who became more human in the end; the Sorcerer's Apprentice, who unleashed forces he could not control.[5] Occasionally Oppenheimer is called a mixture; Holton called Oppenheimer "part Ulysses, part Faust, part Prince Myshkin."[6]

Some writers, such as the Alsops, merely invoked a historical analogy—in their case, with Dreyfus—while others sought to elaborate these analogies in depth. These include Giorgio de Santillana, who compared Oppenheimer's trial with that of the subversive and articulate Galileo. One similarity is that each scientist went into his hearing convinced that he would prevail over his less intelligent, more politically minded accusers, and was utterly shocked when he did not. Another similarity was

4. Kai Bird and Martin J. Sherwin, *American Prometheus: The triumph and tragedy of J. Robert Oppenheimer* (New York, 2005).

5. The Xerxes comparison was made by Richard Polenberg (who likens the hearing to "a latter-day Greek tragedy" on xxviii) in his conclusion to Richard Polenberg, ed., *In the matter of J. Robert Oppenheimer: The security clearance hearing* (Ithaca, NY, 2002), 397. Of the most famous comparisons, that to Galileo was made by Giorgio de Santillana in *The reporter*, 26 Dec 1957, while the Dreyfus comparison was made by Joseph and Stewart Alsop in *We accuse! The story of the miscarriage of American justice in the case of J. Robert Oppenheimer* (New York, 1954). Victor McElheny suggested the Hastings comparison to me in conversation. The Thomas Becket, T.E. Lawrence, Faust, Lear, and Sorcerer's Apprentice comparisons have been mentioned in various articles and discussions.

6. Gerald Holton, "The Oppenheimer case," *New York Times book review*, 4 Jan 1970, 3, 16, on 3.

that an important trigger in each case was the fear on the part of the accusers of the charisma of the accused. Santillana quotes the Fiorentine ambassador:[7]

> He reminded me that Galileo wrote exquisitely, and had a marvelous capacity for persuading people of whatever he wanted to, and there was a danger that through his influence some fantastic opinion might take hold among those Fiorentine wits, which are too subtle and curious.

In the Washington, D.C., of 1954, the "fantastic opinion" of which Oppenheimer's enemies were afraid was "that one does not immediately have to rush ahead to make any weapon that can conceivably be made."[8]

Oppenheimer reminded Freeman Dyson of M.F., the protagonist of Auden and Isherwood's *F6*, "a Hamlet-like figure compounded of arrogance, ambiguity and human tenderness...an intellectual polymath, expert in European literature and Eastern philosophy"[9] who, because of his gifts, is offered the chance to lead a politically imperative expedition up a difficult mountain peak, initially balks, eventually accepts, and dies at the summit.

But the sheer variety of tragic heroes to whom Oppenheimer is compared is one clue that something is awry—that he does not fully resemble any. Also, the tragic conceit dovetails so comfortably and conveniently with certain myths as to make us suspicious, myths about scientists and the state, about leftists in Cold War America, or about what Barton Bernstein has called the vendetta interpretation, in which Oppenheimer was the victim of wicked enemies, Lewis Strauss, the Air Force, McCarthyism, or all the above.[10] Finally, with a little reflection we can come up with other dimensions of the Oppenheimer story that also seem to disqualify it from being truly tragic.

In what follows, I shall try to make a beginning towards understanding the tragic dimension of the Oppenheimer story—and it is no more than preparing the ground—by outlining some reasons why I think his life and career do invoke the tragic, and others why it does not.

What is Tragic

There are several reasons why the Oppenheimer story invokes the tragic. A first is the traditional notion that the plot—the "soul" of tragedy, according to Aristotle—involves a reversal of fortune. Much about the Oppenheimer hearing indeed suggests the classic Greek formulation of tragedy in which a well-intentioned leader "of great

7. De Santillana (ref. 5), cited in Holton (ref. 6), 3.
8. Holton (ref. 6), 16.
9. Freeman Dyson, *Disturbing the universe* (New York, 1979), 77-78.
10. Barton Bernstein, "The Oppenheimer loyalty-security case reconsidered," *Stanford law review, 42* (1990), 1383-1484, on 1385.

reputation and good fortune," as Aristotle says, is brought down by mistakes due to character—usually pride or aggressiveness—in a story involving reversal and recognition. While enemies assist in the downfall, the fault ultimately lies in the protagonists themselves. They trip themselves up by their own virtues, their personalities so powerful that only they are strong enough to do themselves in. While the tragic denouement of the Oppenheimer hearing does not involve the usual fate of tragic heroes—death—the reversal (conversion into the opposite) was surely thorough, complete, and extreme enough to qualify. The wildly successful and indispensable leader of the most advanced military project ever, excluded from military service. The passionate architect of a scheme for world community, banished from government circles. The man who loved his country, made its scapegoat.

A second reason why is our sense that the action bringing about the reversal is not a mere slip-up, an isolated decision, but has something to do with Oppenheimer's whole character and circumstance. That is, it involves what Lloyd K. Garrison, who directed Oppenheimer's defense in 1954, referred to as "a problem of human nature, which can't be cut up into little pieces," by which he meant the nature of a loyal and trustworthy person—and as well with what security board members Gordon Gray and Thomas Morgan, in answering Garrison, referred to as "the whole question," by which they meant how to operate in a world in which security is such a pressing issue.[11] Slip-ups and isolated decisions may have played a role in the reversal, but a much bigger context was at work in bringing it about.

Yet a third reason why the Oppenheimer story appears tragic is that it seems to contain a basic message about the human condition. Richard Polenberg noted of the hearing that "like a latter-day Greek tragedy, the transcript…offer[s] insight into such timeless traits of human character as honor, fortitude, and humility, and, sadly enough, their less admirable counterparts: treachery, timidity, and pride."[12] For Gerald Holton, it reveals another recurrent lesson: the danger that the government, in the face of an external threat, will purge its own through "the 'security system,' under which Government agencies investigate the actions, opinions, associations, private lives and political beliefs of individuals," a system which ironically serves in the long run "to endanger the security of the nation, and of the world."[13] That point seemed current when Holton wrote those words in 1970, and remains current today. But the Oppenheimer story even seems to reveal the deepest truth that we can learn from the tragic, what philosopher Dennis Schmidt refers to as the mutability of human praxis: "We learn that things may well, without warning, convert into their opposite. We learn that what was once divine can, in a flash, become monstrous….What becomes visible here is the expanse of the human condition—the possibilities belonging to praxis—an expanse so great that it is capable, at any given moment, of converting its situation into

11. *JRO Hearing*, 990.
12. Polenberg (ref. 5), xxviii.
13. Holton (ref. 6), 3.

its other."[14] The tragic figure, in the words of Jean Vernant and Pierre Vidal-Naquet, "discovers himself to be enigmatic, without consistency, without any domain of his own or any fixed point of attachment, with no defined essence, oscillating between being the equal of the gods and the equal of nothing at all. His real greatness consists in the very thing that expresses his enigmatic nature: his questioning."[15]

What is Untragic

But the Oppenheimer story also has untragic sides, which I will now try to enumerate. In what follows, please remember that I am speaking of what we know about Oppenheimer—which, after all, is what seems to evoke the tragic—not what may have happened in private.

One is the seeming lack of an aspect that, as I mentioned, Vernant and Vidal-Naquet consider essential to tragic protagonists: self-questioning. When Oppenheimer tells Gray apropos of the Chevalier incident, "I wish I could explain to you better why I falsified and fabricated," this rings true. As Philip Stern notes, Oppenheimer reveals that "he had never come to a full understanding of his own behavior."[16] To be sure, that is typical of tragic heroes—they cannot see themselves clearly, either, and this failure of self-understanding helps do them in. But what seems different is that Oppenheimer—again, in the purportedly tragic face that we see—appeared to have lacked even curiosity about himself. Despite his seemingly candid remark to the effect that his life was open book, he did not act, either to himself or others, as though this were the case; and he did little to help others interpret what could be read in it. He left no personal diary, no autobiography. He repeatedly—and clearly deliberately—frustrated efforts by interviewers to elicit personal feelings and reflections about what he had been through, refusing to lapse into the confessional mode. The man who preached the virtue, even necessity of openness sought successfully to prevent the performance of a play based on his life. We are used to more self-reflection—attempts at self-reflection, anyway—from our tragic heroes.

A second untragic side to the Oppenheimer story is the sense that he was always playing at a role. Rabi has pointed to this dimension of Oppenheimer's personality in various interviews, and Dyson has also discussed it. "He had a talent for self-dramatization," Dyson wrote, "an ability to project to his audience an image larger than life, to bestride the world as if it were a stage."[17] Oppenheimer seemed to court a sense of

14. Dennis Schmidt, "Stereoscopic thinking and the law of resemblances: Aristotle on tragedy and metaphor," Walter Brogan and James Risser, eds., *American continental philosophy* (Bloomington, IN, 2000), 66-91, on 77.

15. Jean-Pierre Vernant and Pierre Vidal-Naquet, *Myth and tragedy in ancient Greece* (New York, 1990), 139, cited in Schmidt (ref. 14), 77.

16. Philip M. Stern with Harold P. Green, *The Oppenheimer case: Security on trial* (New York, 1969), 360.

17. Dyson (ref. 9), 77.

his being bigger than life, of being remote, up on the mountain, through what Pais called his "priestly, solemn" manner. Oppenheimer tried to evoke tragedy, whether he said so or not, and regardless of the quality of his material. Pais wrote: "O's talent: to withhold a piece of information which then makes a statement elevated & mysterious while in fact there is not that much to it."[18] Post-hearing, the role that Oppenheimer discovered himself playing changed. He was more aware than ever that he bore what James Baldwin, in another context, famously called the burden of representation—that the words he uttered were not those of just another scientist, but political or social statements that were more than about him and addressed to a historical audience. Oppenheimer's refusal to talk in the confessional mode was clearly deliberate. It was as if he was aware of his forthcoming historical role as a martyr, that he was as an actor on the world stage, that all he had to do was to step back and let it this story unfold, and that the confessional mode was unnecessary to this role and could only sully it. He would only speak, as he put it in the farewell speech at Los Alamos, of "what has happened to us" rather than "what has happened to me."[19] In some respects this was salutary. It made him less prone to the petty animosities, defensiveness, and score-settling that permeate, say, Teller's autobiography. But it also made Oppenheimer's words often sound like pronouncements issued by a public agency. And public agencies are not the stuff of tragedy.

Yet a third non-tragic dimension is Oppenheimer's detachment. His description of the hearing as a "train wreck" implies he saw himself merely as a passenger who had nothing to do with building the rails or running the train. His remark that the tragic element of the hearing came not from his being a protagonist but from the "chorus," those around him, likewise repudiated his agency in the affair. His comment to a reporter that a story existed behind the story was true—but he himself was a key player in that story behind the story. He acted as though what happened to him at the hearing was not part of who he was. He seemed to lack, in short, that dimension of tragedy that Aristotle called recognition. The tragic protagonist is not mere victim, buffeted by events beyond his control, but fully embraces the world and takes personal responsibility. Even while struggling against the world he acknowledges that he is a participant in it, and that he has no choice but to embrace it. But Oppenheimer often appeared unworldly, detached from events around him. Writer John Leonard noted this exasperating quality—"We pursue him, and he eludes us"—in a review of a collection of Oppenheimer's letters and recollections:[20]

> Even the jacket photograph of this handsome book from Harvard raises more questions than it answers. A dreamy Oppenheimer, as if doped or deliquescent, stares

18. Abraham Pais, untitled handwritten note, miscellaneous notes file, APP.
19. JRO, "Speech to the Association of Los Alamos Scientists," Alice Kimball Smith and Charles Weiner, eds., *Robert Oppenheimer: Letters and recollections* (Cambridge, 1980), 315-325, on 315.
20. John Leonard, review of Smith and Weiner (ref. 19), *New York Times*, 12 May 1980.

> back at us, sad that we don't know what he is tired of thinking about. His expression combines condescension and misadventure; we would never understand. We ask of him a few notations on sin. He replies with a machine that is noiseless, when it isn't speaking Sanskrit. I am sure, after reading so many books on Oppenheimer, that we were stupid to deny him a security clearance, but I wonder whether he ever belonged to our world at all.

We are used to more worldly commitment from our tragic heroes than what Oppenheimer reveals.

Multiple Tragic Dimensions

So far I have focused on that part of Oppenheimer's life connected with the hearing. But Kennan's reference to "that strong element of tragedy...in his situation" refers to more than just that. Oppenheimer's life had multiple tragic dimensions, and I want to mention two others.

One was his failure to achieve in physics what his capacity seemed to promise. Once in the 1950s, during the oral part of the physics qualifying exam at the University of Wisconsin, a student was asked what J. Robert Oppenheimer had contributed to physics. "I don't know," the student answered—and was informed that was the correct answer.[21] The remark was caustic and flip. Any physics student should have been able to cite at least the Born-Oppenheimer approximation, the Oppenheimer-Phillips effect, and black holes. Nevertheless, the remark was true in spirit to Oppenheimer's ambiguous scientific legacy. These three contributions, which testify to his outstanding mind, and which would have ennobled any career, nevertheless seemed to fall short of the expectations left by Oppenheimer's formidable reputation, ambition, and ability. This is why neither the student nor the examiner thought they sounded correct.

Oppenheimer's strength was the ability to integrate all aspects of a complex situation and articulate them in a way that preserves the intricacy. The correlative weakness was that he did this so well that it seemed to leave him unable to propose a way forward, which would involve a crass simplification doing an injustice to the intricacy of what he so well mastered. But it left listeners satisfied he had outlined the problem brilliantly.

My favorite example involves the discussion at the sixth Rochester conference about the tau-theta puzzle, and his remark, "The τ meson will have either domestic or foreign complications. It will not be simple on both fronts."[22] This delighted the conference participants, who repeated it without knowing quite what to make of it. It was classic Oppenheimer, who when addressing difficult problems often crafted enigmatic utterances that sounded insightful not because they added anything new,

21. Robert K. Adair, personal communication.
22. Sixth Annual Rochester Conference, High Energy Nuclear Physics, *Proceedings* (New York, 1956), VIII-22.

but because they allowed each listener to interpret them in the way he or she wanted to. I once asked four physicists what this remark meant, and got the answer—four of them, in fact. Oppenheimer's contrast between the domestic and foreign front, they confidently declared to me, respectively, referred to the contrast between the properties of the J and those of other particles, between strange and nonstrange particles, between field-theoretic and non-field-theoretic approaches, between normal and revolutionary science. No doubt Oppenheimer did not intend any one of these meanings, but rather the ambiguity itself.

Still, we expected more from him, and he expected more from himself. As Pais wrote, Oppenheimer "could never find enough satisfaction with his signal achievements" and "forever was compelled to reach for more."[23] This, rather than the hearing, Pais wrote, was his real tragedy. In another note, Pais wrote, "His tragedy is that he was almost a genius."[24] At the hearing, Rabi, citing Oppenheimer's contributions, remarked sarcastically, "What more do you want, mermaids?"[25] Something about J. Robert Oppenheimer indeed made us expect mermaids.

A final tragic dimension has to do with the sense I think is widely shared that his character was inconsistent. S.S. Schweber described it as "fractured." Holton has written of the strong "magnetic field" needed to line up the various fragments of his complex personality, one present at Berkeley and Los Alamos but not during other periods of his life.[26] We find this fragmentation, I think, to have affected his public essays, where it prevents him from being more daring. Here, too, he displays an uncanny ability to take stock of a situation, to grasp it in all its complexity, and to express it in a way to preserve the intricacy. But again as in his physics he did this so that it left him unable to propose an oversimplifying way forward. His voice is simultaneously oracular—fascinated by the meaning of a situation, the potential for novelty, and the possibility of revolution—and cautious and humble—impressed by what is known, the obstacles in the way to progress, and the weight of tradition. One FBI agent described it as his "typical 'egotistically modest' manner."[27] The essays end up being both stirring and insubstantial, dangling before us the prospect of a coming breakthrough and future harmony but ultimately leaving us empty-handed. They often suggest an implausible complementarity between suffering and joy, loneliness and belonging, past and present; they exhibit manufactured grandeur and simulated sincerity; and they give rise to expectations of personal revelation but fail to come through. The notes of depth sound just a little hollow, the notes of optimism ring just a little false, so that they seem less

23. Pais, *J. Robert Oppenheimer* (ref. 1), xxii.

24. Pais, manuscript of talk on JRO, Coral Gables, FL, 17 Jan 1979, APP.

25. *JRO Hearing*, 468.

26. S.S. Schweber, *In the shadow of the bomb: Bethe, Oppenheimer, and the moral responsibility of the scientist* (Princeton, 2000), xv; Gerald Holton, "Success sanctifies the means: Heisenberg, Oppenheimer, and the transition to modern physics," *The advancement of science, and its burdens* (Cambridge, 1986), 141-162, on 155.

27. W.A. Branigan to A.H. Belmont, 5 Jan 1955, JRO, *FBI security file* (Wilmington, DE, 1978), microfilm, Vol. 54.

a personal expression than a statement issued by a public agency. Yet Oppenheimer was acute enough to have heard this himself.

Oppenheimer was a modern-day Comtian with soul and sophistication; he never believed that one science could be reduced to another, nor sought to underestimate the complexity of the human world. While in the later years he seemed discouraged about the possibility of what, in the Los Alamos farewell address, he had called the "community of interest" needed to meet the upcoming cultural and political challenges,[28] he did not explore or inquire into the discontents and tensions. What would have been truly profound would have been to seek to articulate this discontent, rather than merely point to it and then ultimately deny it via the metaphor of an underlying complementarity. Oppenheimer's own experiences leading up to the hearing would have been an excellent starting point, because the hearing was a paradigmatic experience of the kind of disruption in the relation between science and society he was alluding to. A curious, distressing, even ominous aspect of the later essays is his absence of reflection about an event that was not only one of the deepest things to shake up the relations between science and society, but also the deepest thing to shake his soul. Again, however, he remained impersonal, speaking of what happened to us rather than what happened to him. He almost acts as though he were not a part of the world in which *that* event took place, in *that* part of the "open house" of which he sometimes spoke.[29] This is what gives his later essays such an unsettling serenity.

This aspect, no doubt, is one reason why, when thinking of a literary analogue for Oppenheimer's life, Pais thought of Henry James's somewhat sentimental novel *The beast in the jungle.* Its protagonist, John Marcher, lives his life keeping a distance from those around him, in the belief that he has been selected for some prodigious event—in his dreams it takes the form of an encounter with a beast lying in wait—only to find, in the end, that the event in question has already taken place, and thanks to his aloofness has passed him by. This aspect may also be why, when Pais interviewed Melba Phillips, she suggested that he read J.P. Marquand's book *The late George Apley*, whose eponymous hero, born into wealth and privilege, takes himself far too seriously and never quite lives up to his own expectations. Both of these books resonate with an important aspect of Oppenheimer's life, but neither are tragedies.

We wanted more from him—much more. But we only expected so much because he came so close to delivering. For he gave the appearance as Weisskopf once put it, of representing "the spirit and the philosophy of all that for which we are living."[30] In several different ways, therefore, Oppenheimer's life attracts yet repels the characterization "tragic." It is a tragedy without a tragic hero; or, to put it another way, the tragedy of J. Robert Oppenheimer was that he was *almost* a tragic figure.

28. Smith and Weiner (ref. 19), 315-325, on 320.
29. JRO, "The open house," N. Metropolis, Gian-Carlo Rota, and David Sharp, eds., *Uncommon sense* (Boston, 1984), 67-77.
30. V.F. Weisskopf to JRO, 5 Apr 1954, JROP, box 202.

PART SIX

The Public Oppenheimer

After 1945, Oppenheimer was on display. Uneasily, still, his public presence never settled into a comfortable mold. Either Oppenheimer was the nuclear physicist writ large, bearing the burdens of science and history; or he was an original, peerless, a figure unto himself. His image matters, of course, because the atomic bomb has become a chief metaphor for science's impact on our world. In public understandings of Oppenheimer we can discern much about our own age's fascinations. Depending on setting and audience, these stretch from intellectual profundity to personal charisma to straight-ahead nuclear devastation.

Daniel J. Kevles places Oppenheimer within an expanded twentieth-century story of scientists and the military. Kevles displays how the seeming uniqueness of the nuclear physicist's dilemma was foreshadowed in the First World War's experience with chemical warfare. He shows, too, that the same story can capture important features of the post-Cold War era. Taking a different tack, S.S. Schweber places Oppenheimer into contrast with the century's iconic public scientist: Einstein. Analyzing the conceptions of science, self, and community implicit in the lives of the two men, Schweber reveals much about Oppenheimer's troubled self-presentation. The section closes with the transcript of a conference discussion on documentary film, responsible for some of the most highly public versions of Oppenheimer and his era. Jon Else and Peter Galison, both well-versed in nuclear-age documentaries, reflect on their experiences, choices, and constraints in bringing figures like Oppenheimer to the broadest of audiences.

CHAPTER SIXTEEN

Scientists, Arms, and the State
J. Robert Oppenheimer and the Twentieth Century

Daniel J. Kevles

No thoughtful person today can fail to be aware that the association of scientists, arms, and the state is fraught with troublesome questions. Many of the questions center on the scientist's obligation to the state, particularly whether to deploy his or her expertise to hazardous, potentially destructive purposes and to defending against them. These troublesome issues pervade the history of science in the twentieth century, from World War I, when it was first systematically enlisted for military purposes, then more insistently in World War II and the Cold War, and now unsettlingly in the war against terrorism.

It is in part because of the vexing vitality of these issues that J. Robert Oppenheimer continues to compel our attention. The perils and possibilities of the scientist's involvement with the military state are powerfully symbolized in the career of Oppenheimer—who was transformed by World War II from, as legend has it, an otherworldly theoretical physicist into the internationally renowned creator and sage of American nuclear strength, and who was then politically humiliated and destroyed by the vicious and bare-knuckled politics that marked his security hearing in 1954 and the judgment that he was a security risk.

At the outbreak of the Second World War, Oppenheimer was a politically knowledgeable citizen, a dedicated antifascist, and, like most American physicists at the time, eager to deploy his physicist's expertise in the war against the Nazis. His commitment led him to the Manhattan Project and, ultimately, to the mesa in New Mexico and the directorship of the Los Alamos laboratory. Apprehensive along with other physicists that the Germans might develop an atomic bomb, Oppenheimer threw himself into the task of producing a nuclear weapon as rapidly as possible, succeeding brilliantly after the German surrender in May 1945. In the several years following Hiroshima and Nagasaki, the sociologist Philip Rieff wrote, "Oppenheimer became a symbol

The following abbreviations are used: CBW, chemical and biological warfare; CWS, U.S. Army, Chemical Warfare Service; JRO, J. Robert Oppenheimer.

of the new status of science in American society. His thin handsome face and figure replaced Einstein's as the public image of genius."[1]

Oppenheimer was a complicated symbol, a rueful and reluctant participant in the postwar weapons effort. At the end of August 1945, having been informed by the War Department of the need to push the nuclear weapons program fully ahead, he told a fellow physicist that he had allowed "that we felt reluctant to promise that much real good could come of continuing this atomic bomb work just like poison gases after the last war." He added that he was feeling a "profound grief...and a profound perplexity about the course we should be following."[2] In November 1947 he publicly reflected that "physicists felt a peculiarly intimate responsibility for suggesting, for supporting, and in the end, in large measure, for achieving the realization of atomic weapons," and he continued, "Nor can we forget that these weapons, as they were in fact used, dramatized so mercilessly the inhumanity and evil of modern war. In some sort of crude sense which no vulgarity, no humor, no overstatement can quite extinguish, the physicists have known sin; and this is a knowledge which they cannot lose."[3]

But not all physicists who had worked on the atomic bomb felt the same. Ernest O. Lawrence, who won a Nobel prize for his invention and development of the cyclotron at Berkeley and who was a good friend of Oppenheimer's at least for a number of years, privately protested Oppenheimer's rueful observation of 1947, declaring: "I am a physicist and I have no knowledge to lose in which physics has caused me to know sin."[4] And Oppenheimer's own knowledge of sin did not dull his willingness to continue developing atomic weapons for the state. From 1945 until the end of 1953, when the loss of his security clearance cut him off from intimate connection with the affairs of national security, Oppenheimer was a dutiful soldier of science, emphasizing to a War Department official in 1945, for example, that "of course...all of us would earnestly do whatever was really in the national interest, no matter how desperate or disagreeable."[5] He accepted appointment to a diverse array of government scientific advisory committees concerned with national defense. His most important service was as chair of the General Advisory Committee of the new Atomic Energy Commission, where "the principal job," as he later recalled, "was to provide atomic weapons and good atomic weapons and many atomic weapons."[6]

Oppenheimer thus presents something of a paradox, embodying at one and the same time a sense of sin associated with the forging of nuclear weapons and a commit-

1. Philip Rieff, "The case of Dr. Oppenheimer," Philip Rieff, ed., *On intellectuals: Theoretical studies, case studies* (Garden City, NY, 1969), 314-340, on 319.
2. Quoted in Barton Bernstein, "Four physicists and the bomb: The early years, 1945-1950," *Historical studies in the physical and biological sciences, 18:2* (1988), 231-263, on 242.
3. JRO, "Physics in the contemporary world," *Bulletin of the atomic scientists, 4* (Mar 1948), 66.
4. Quoted in Herbert Childs, *An American genius: The life of Ernest Orlando Lawrence* (New York, 1968), 405.
5. Quoted in Bernstein (ref. 2), 242.
6. Ibid., 251.

ment to improving and multiplying those weapons for the sake of national security, a task that could lead to further sin. It is a paradox that troubled many members of the Los Alamos generation, Ernest Lawrence notwithstanding, and it has pervaded the life of weapons-related science since World War II. However, the paradox is only of recent vintage: the imperatives of national security have not always called forth the gall of sin in the hearts of the weapons innovators, including those who advanced chemical warfare in World War I, the first major enlistment of science in arms for the state.

Chemical Warfare in World War I

Chemical warfare developed in a moral context of international agreement that dated from the Hague Conventions of 1899 and 1907. At the time the conventions were concluded, there were no chemical weapons, but the delegates were evidently aware that some nation might forge and use them. In 1899, the drafting committee proposed that the contracting parties would "agree to abstain from the use of all projectiles the sole object of which is the diffusion of asphyxiating or deleterious gases."[7]

Objections were raised by both the British and the Americans, who held that their nations should not deprive themselves of a class of weapons that might be advantageous. The conference nevertheless adopted the ban. It apparently did not do so because of any belief that gas was somehow an unacceptably cruel weapon in and of itself. Rather, the ban rested on the fear that chemical weapons could be used against cities and civilians, including women and children.[8] The Hague Convention's prohibition against gas shells thus was an assertion, widely supported even in military circles, that at the opening of the twentieth century, civilian populations should not be fair game in warfare among the advanced civilized nations.

At the outbreak of World War I, in August 1914, none of the belligerents had made any preparations for chemical warfare. Outside Germany, most lacked the necessary industrial-scale technology and adequate chemical industries. But German industrial chemistry was formidable. Its organic dyestuff industry dominated world markets, and its new Kaiser Wilhelm Institutes, established in 1910 on fifty acres in the leafy Dahlem suburb of Berlin, were distinguished world centers for research.[9]

The Institute for Chemistry was headed by Fritz Haber, the father of chemical warfare. In 1909, with Carl Bosch, he had devised the means of fixing nitrogen from the air, an achievement that would win the Nobel Prize. Haber was the son of a Jewish businessman who had become wealthy by trading in chemical products. Evidently eager to assimilate to the circles of German academia and industry, the younger Haber

7. Quoted in L.F. Haber, *The poisonous cloud: Chemical warfare in the First World War* (Oxford, 1986), 18.
8. Richard M. Price, *The chemical weapons taboo* (Ithaca, NY, 1997), 33-34.
9. Unless otherwise noted, the account of chemical warfare in World War I is based on Haber (ref. 7) and Price (ref. 8).

converted to Lutheranism and got rid of his given middle name, Jacob. He was an ardent German nationalist and in 1914 joined in signing the declaration of the 93 German intellectuals—the list also included Max Planck, Wilhelm Ostwald, and Walther Nernst— defending their country's prosecution of the war.[10]

Haber and his institute were already involved with the military in programs to produce nitric acid for munitions. By early 1915, he saw in gas warfare another way to assist his country. The war had bogged down in the seesaw battles of the trenches. Haber argued that the stalemate could be broken by the release of chlorine gas from cylinders, thus generating a cloud that would drive the enemy out of the trenches into the open. Chlorine was plentiful in dyestuff plants. Since the gas would be released from cylinders, it would not violate the letter of the Hague Convention's prohibition against delivering it by projectiles, i.e., shells. That nicety, together with the fact that Germany was short on artillery shells and was desperate for a decisive breakthrough, led to approval of the proposal by the German General Staff. The gas attack would take place at Ypres, in Belgium, because the commander in the area was the only one who was willing to tolerate it.

Haber himself led the attack along with a scientific team, part of the gas troops called *Pionierkommandos*, that included James Franck, Gustav Hertz, and Otto Hahn. The gas commandos were dug in on a four-mile front at Ypres facing French territorials and an Algerian division of the French army. One late April afternoon, the Germans, wearing protective masks, opened the valves of 5,730 cylinders, each containing two hundred pounds of liquid chlorine, and released the contents, emptying the cylinders within about ten minutes. A thick green-yellow wave of gas about five feet high wafted westward on a breeze towards the Allied trenches. Soon hundreds of men were choking and vomiting and dying while the rest of the troops fled in panic to the rear.[11]

The German command considered the attack successful enough to warrant an extensive chemical warfare program. Haber's institute became the center for research and development in offensive and defensive chemical warfare, an enterprise facilitated by Haber's excellent industrial and academic contacts as well as solid organizational talents.

The rest of the belligerents, all of them unprepared for chemical warfare, hastened to establish similar efforts. In 1915-16, the balance of the R&D effort went to defense—notably, clothing treated to protect the skin and gas masks with effective absorbent filters. In the first weeks of the gas war, British troops were advised to breathe through socks steeped in urine. Eventually, reliable respirators were developed that used charcoal as the absorbent.

The United States, isolationist in its policies, had been in denial about the military demands it might face in the war. When it became a belligerent in April 1917, it had neither a gas shell in its arsenal nor a gas mask in its standard equipage. The

10. The text of the manifesto and list of signers is in Georg F. Nicolai, *The biology of war* (New York, 1918), xi-xiii.
11. John Cornwell, *Hitler's scientists: Science, war, and the devil's pact* (New York, 2003), 61.

government, with the assistance of the National Academy of Sciences and its new National Research Council, quickly mounted an effort in chemical warfare research that involved some seven hundred chemists at the Bureau of Mines and more than twenty academic institutions. In 1918, both the stateside research effort and the gas field services in France were consolidated in a new Chemical Warfare Service.[12] By the end of the war, counting all the belligerents, at least 5,500 people were engaged in chemical warfare R&D. Between a quarter and a third had university degrees. The rest were technical assistants.[13]

In the spring of 1916, in the Verdun sector of the battlefront, the French initiated the use of gas shells, with the result that gas dispersed by explosion came to replace the gas cloud driven by the wind. The Germans quickly followed suit, as did the rest of the belligerents. The move violated not only the spirit of the Hague Convention, as Haber had done, but its letter as well. But no matter to the belligerents after almost two years of slaughter: shells allowed offensive forces control of targeting instead of being victimized by wind and weather.

In 1917-18, the balance of R&D shifted to the development of weapons, partly no doubt because the shift to shells invited tailoring the type of gases they carried to the terrain, target, and tactical purpose. In 1917, the belligerents began to use mustard gas (dichlorodiethyl sulfide), which Germany pioneered, also at Ypres, in July. It was soon joined by a variety of arsenical compounds. Their comparative battlefield effectiveness was measured by a figure called "Haber's constant," the product of concentration of gas and time required to cause death. Gas shells were also now hurled towards the enemy from so-called Livens projectors, a British invention, mortar-like launchers that could fire a thirty-pound gas shell 1,500 yards. By November 1918 both sides between them had deployed some twenty-two different chemical agents. They ranged from lung and skin irritants to blood poisoning chemicals, all delivered by shells, mortars, grenades, and aerial bombs. By the end of the war, as much as one third of the artillery munitions of the belligerents—in the case of the Germans one half—was devoted to chemical warfare.

Perhaps the most reliable estimate of gas casualties puts the number at about 560,000 personnel. The number may have been relatively small compared with the millions who died from bullets and high explosives, but falling victim to gas was not a pleasant experience. Soldiers exposed to it rarely died immediately. Rather, they suffered inflammation of the bronchia and air sacs that resulted in coughing, retching, vomiting, and violent spasms. After three or four hours, they either began to recover or to die from heart failure brought on by pulmonary edema. Mustard gas, an astringent irritant on the skin, lingered on surfaces and the ground, posing its own hazards. One soldier had his buttocks and crotch badly burned when relieving himself at night. Faced with the unseen threat, soldiers tended to suffer anxiety from imagining that mustard gas or its remnants were in their area.

12. Daniel J. Kevles, *The physicists: The history of a scientific community in modern America* (Cambridge, 1995), 132.
13. Haber (ref. 7), 106-107.

Condemnations

When the Germans introduced poison gas warfare in 1915, Allied commentators condemned them, not because gas was regarded as a particularly horrible weapon but largely because they had violated the spirit and intent of the Hague Convention. The condemnations virtually disappeared once the Allies deployed their own chemical weapons, not least because Britain and France imposed a news blackout on the subject in mid-1917. Then in 1918, as an Allied victory loomed, gas was hailed as a triumph of Allied industry. Behind the blackout and the celebrations, after gas warfare escalated into its new, more diversely offensive phase, chemical weapons were increasingly recognized in military and policymaking circles as a serious threat to both troops and their morale. In the twenty-four months following the first attack at Ypres, British troops suffered 20,000 gas casualties. In the sixteen months following the second attack, with mustard gas, they suffered 160,000 casualties.

Even before the news blackout was lifted, artists and writers, some of them veterans of the European front, called searing attention to how soldiers experienced gas attacks. Perhaps the most famous painting to emerge from the chemical war was John Singer Sargent's *Gassed*, of 1918, a rendition of a file of troops blinded by mustard gas. In August 1917, following a gas attack, the British poet Wilfred Owen wrote *Dulce et decorum est*, one of the most powerful indictments of chemical weapons to come out of the war: "And watch the white eyes writhing in his face, / His hanging face, like a devil's sick of sin; / If you could hear, at every jolt, the blood / Come gargling from the froth-corrupted lungs, / Obscene as cancer, bitter as the cud..."[14] Later, other writers, including the German Erich Maria Remarque, dramatized how the protections of uniform, coat, and gas mask stimulated fears of being cut off from fresh air and one's friends.

A growing theme in the mounting criticism of chemical warfare was its threat to civilians. Gas, capriciously following the wind, sometimes drifted as much as ten miles behind the front lines. French and Belgian townspeople tried to protect themselves against it. At least 5,500 civilians fell victim to gas during the war. In a letter to the *Times* shortly after the Armistice, the leaders of the British medical profession condemned gas because it was uncontrollable, could injure non-combatants, and with new gases might "blot out towns and even nations."[15]

Even more worrisome was the prospect that chemical warfare might be joined to rapidly developing aerial weapons and inflict devastating damage on civilian populations. Indeed, with an eye to these two technologies, the most dramatic developed during the war, the United States and Britain had forged the means and plans to assault cities from the air with chemical bombs, had the conflict continued into 1919. But even some advocates of chemical warfare held that while they were acceptable for use against combatants, employing them against civilians would be

14. Quoted in ibid., 79.
15. Ibid., 294.

"barbarous and inexcusable," to quote a report from the League of Nations.[16] The graphic depictions of gassed soldiers, melded with the idea that civilians, including women and childen, might be gassed to death, led to international attempts to ban chemical weapons after the war.

Attempts at control

Outside the Chemical Warfare Service, high-ranking American military officers still considered attacks against civilians and cities inadmissible. Gas was declared a reprehensible weapon by the General Board of the U.S. Navy, Army Chief of Staff Peyton C. March, who had seen civilian gas casualties in a field hospital in France, and General John J. Pershing, who spoke for all of them in saying that "chemical warfare should be abolished among nations, as abhorrent to civilization."[17] At the urging of Secretary of State Charles Evans Hughes, the treaty resulting from the Washington conference on limitation of arms in 1922 included a categorical ban against chemical weapons. The ban was supported almost unanimously in a poll of the American public, prompting an advisory committee to the treatymakers to note that "the conscience of the American people has been profoundly shocked by the savage use of scientific discoveries for destruction rather than for construction."[18] The Senate ratified the treaty by a vote of seventy-two to zero. It never went into effect, however, because the French refused to ratify it.

But the outcries of artists, the calls to conscience, and the demands for a ban generated a reaction—the formation of what has been called a "gas lobby." When the Wilson administration moved to abolish the Chemical Warfare Service in 1919, the lobby—an active coalition of CWS officers, American chemists, and the American chemical industry—campaigned to have it continued. Critics might charge that poison gas was a barbaric weapon, but General William Sibert, the wartime chief of the CWS, liked to point out that while some twenty-four percent of casualties from rifles and artillery had died, fewer than two percent of those from gas had not come back. "Gas," Sibert concluded, "is twelve times as humane as bullets and high explosives."[19]

At hearings on Capitol Hill, Senator Duncan U. Fletcher of Florida had been wondering: if we "kept on with these scientific means of destruction," would we not run the danger of "wiping out the whole human race?" General Amos A. Fries, a West Pointer who had been Chief of Gas Service in the American Expeditionary Forces in France, declared that as a weapon that placed noncombatants in jeopardy, gas was not

16. Price (ref. 8), 71.

17. Pershing quoted in ibid., 82; March, testimony, U.S. Congress, Senate, Subcommitee of the Committee on Military Affairs, *Hearings, reorganization of the Army,* 66th Cong., 1st and 2nd sess., 1919, I, 93.

18. Quoted in ibid., 85.

19. "Report of the director, Chemical Warfare Service," U.S. War Department, *Annual report, 1919,* I, 4391.

unique. For Fries, the issue was not some new form of warfare but the historic one of war itself. The more "deadly" the weapons, he argued, "the sooner...we will quit all fighting." If war could be made "so terrible" that that it would not last "more than five or ten minutes," then war would never begin." The nation's developing chemical arsenal, Fries declared the following year, would go a long way toward "deterring" other countries from starting wars.[20]

To the end of strengthening chemical deterrence and preparations against future "German monstrosities" should deterrence fail, the gas lobby called for the development of a vigorous organic chemical industry and—in this era before extensive federal patronage of academic and industrial R&D—the establishment of high protective tariffs to achieve that end, which they obtained in 1922. Under a clause in the Versailles Treaty, the Allies also gained access to the patents and trade secrets of the German chemical industry.[21]

The gas lobby won several other victories. Congress restored the Chemical Warfare Service, making it a permanent part of the army. At Geneva in 1925, an international conference on the arms trade adopted a "Protocol for the prohibition of the use in war of asphyxiating, poisonous or other gases, and of bacteriological methods of warfare." World War I veterans joined the gas lobby in opposing the Protocol, contending that soldiers counted gas an essential and humane weapon that had brought the war to an earlier end. The Protocol died in the Senate, a victim of an absence of leadership on its behalf by the administration of Calvin Coolidge and of the gas lobby's advocates and arguments. The specter of gas bombing civilians had been reversed from its role in the Washington Treaty—from "it must be banned" to "it cannot be banned so we had better prepare for it."[22]

In the event, the preparations were weak. The gas chemists were honored in various ways, and in the United States they served on various advisory committees for chemical warfare; but they did not become an admired new power group like the atomic physicists after World War II. They displayed no moral anguish about their wartime role. This was in part perhaps because of their claim that gas was a more humane weapon than bullets or high explosives—a "higher form of killing" as Haber once put it.[23] It was no doubt also because chemical weapons had only been used against combatants, not civilians. When Haber received the Nobel Prize in 1919 for nitrogen fixation, the French protested the award, and he does not seem to have been revered in Germany for his role in chemical warfare.[24] Germany had, after all, lost the war. Moreover, as the military leadership in all the belligerent powers knew, gas, although an increasingly significant weapon after the introduction of mustard gas, had

20. Ibid., I, 178, 364-365.
21. Price (ref. 8), 72; Haber (ref. 7), 287.
22. Price (ref. 8), 79, 91-97.
23. Cornwell (ref. 11), 62.
24. Dietrich Stoltzenberg, *Fritz Haber: Chemist, Nobel laureate, German Jew* (Philadelphia, 2004), 213.

been a troublesome and never a decisive weapon in the war. The professional military continued to dislike chemical warfare and increasingly worried about the adverse public reaction to its potential use against civilians. After the Washington Treaty, the U.S. War Department ordered an end to the manufacture of poison gas and limited research to only defensive purposes. The CWS, short on funds and staff, busied itself with devising peaceful uses for its chemicals, including insecticides.[25]

World War II

During the interwar period, however, a great deal changed in the technology of war, and, with the rise of fascism the horrific prognostications of 1918-19 were increasingly realized. The world glimpsed the realities of air attack when Italy, which had subscribed to the Geneva Protocol, dropped five-hundred-pound mustard gas bombs on Ethiopian towns, killing numerous women and children as well as an estimated 15,000 soldiers. The British prime minister remarked apprehensively that "if a great European nation, in spite of having given its signature to the Geneva Protocol against the use of such gases, employs them in Africa, what guarantee have we that they may not be used in Europe?" During the Spanish Civil War, Italian and German planes bombed Spanish cities with high explosives, including Guernica, its suffering memorialized in Picasso's gripping painting, and the Japanese waged war from the air against Chinese cities. The air assaults in both countries were condemned in the United States, where a *New York Times* reporter declared that bombers "could destroy centuries of civilization in a few minutes." In an article in the widely read magazine *Life* in 1939, the humorist James Thurber deployed his signature irony in writing of a future war: "This time destruction was so complete, that nothing at all was left in the world, except one man, and one woman, and one flower."[26]

In the 1930s, American scientists, by reason of the internationalism of their profession, tended to be less isolationist than their countrymen. A growing fraction of them being Jews, refugees, or both, many had special reasons for wanting to halt Hitler. On the whole, the American scientific community formed part of the vanguard in favor of military preparedness once World War II broke out and favored strong aid to the Allies, including the kind of aid that they were uniquely qualified to provide. It was clear from the beginning that World War II would be an unprecedentedly technological conflict. The Germans seemed formidable, marshaling tanks and bombers to defeat Poland in 1939, wage successful *Blitzkrieg* in the spring of 1940, and then in the summer challenge Britain's capacity to fight with bombing assaults on its cities and submarine attacks on its Atlantic supply lines. The circumstances established a

25. Edmund Russell, *War and nature: Fighting humans and insects with chemicals from World War I to* Silent Spring (Cambridge, 2001), 61-64.

26. Pauline Maier et al., *Inventing America: A history of the United States* (New York, 2002), 795.

new imperative for the social responsibility of science: do whatever possible to meet the technological threat from fascist aggressors by forging an all-out technological response in the democracies.

In the way that World War I had been a war of chemists, World War II proved to be one of physicists, the principal progenitors of microwave radar, rockets, and the proximity fuze as well as the atomic bomb. To be sure, during World War II, nations on both sides invested significantly in the development of gas weapons and means of defending against them, notably filters for gas masks. Nevertheless, neither side used their chemical arsenals. Their professional military leaders still found poison gas a cumbersome weapon, and both sides feared retaliation, especially by assault from the air on civilians in cities.[27]

The atomic bomb

Why then the willingness to develop an atomic bomb, a dramatically unconventional innovation that promised to wipe out entire cities? It was not, as in the case of chemical warfare, to score a decisive breakthrough in a military stalemate. It was to forestall the possibility that Werner Heisenberg, who had chosen to remain in Germany, might build an atomic bomb for Hitler. It was to prevent being beaten to the punch by the Nazis.

But then, once the bomb was developed, Germany was already defeated. Why use it against Japan, which was almost defeated, too? Entire books have addressed that question.[28] Suffice it to say here, from the perspective of the Hague Convention, that, starting with the German bombings of London and Coventry, and then the Allied bombings of Hamburg, Dresden, and Tokyo, to name just a few cities devastated from the air, moral sensibilities about bombing civilians had been progressively weakened, among scientists as well as policy and opinion makers. Microwave radar, developed initially to assist conventional warfare between combatants, was soon exploited for these bombing campaigns. By the time the atomic bombs were ready, the moral sensibilities that had prompted the generation of the Hague Conference, World War I, and the Washington Treaty to recoil at the idea of dropping poison gas on cities and civilians had been almost completely shattered.

Some Manhattan Project scientists lobbied to prevent the atomic bombing of Japan, calling it inhumane and, amid the growing tensions with the Soviet Union, likely to provoke a postwar nuclear arms race. Such scientists were in the minority. The majority were represented by, among others, J. Robert Oppenheimer, who in June 1945 wrote for an advisory group asked to respond to the dissenters: "We recognize our obligation to our nation to use the weapons to help save American lives...[and] we see

27. Russell (ref. 25), 109.
28. For a concise entree to the debates and literature on the atomic bombings of Japan, see J. Samuel Walker, *Prompt and utter destruction: President Truman and the use of atomic bombs against Japan* (Chapel Hill, NC, 1997).

no acceptable alternative to military use."[29] In a sense, as I once heard Oppenheimer remark in a private gathering, Hiroshima and Nagasaki comprised another step in a long, seemingly ineluctable march through the first half of the twentieth century to truly total war.

Among most of the Allied populations, the nuclear end of the war provoked a jumble of ambivalent feelings, not only joy and relief but doubt and fear that perhaps science had finally gone too far, especially in giving policymakers the tools to destroy cities and civilians. But whatever ambivalence the bomb provoked, the postwar circumstances differed from those of the 1920s. If the chemical bombing of cities had only raised fictional apprehensions, the nuclear bombing of Hiroshima was graphically real, searingly reported in John Hersey's 1946 book *Hiroshima*.[30] If the United States could indulge in isolationism in the 1920s, it could not do so now, with Europe and Japan in ruins and the Soviet bear menacing. Amid the emerging Cold War, the argument that had helped defeat the Geneva Protocol—that war had to be deterred by a commitment to R&D in the relevant weapons—now prevailed, sealing a consensus, rapidly emerging even before Hiroshima, that national security required ongoing, vigorous programs of military research and development.

The Cold War

Oppenheimer and nuclear arms

As the Cold War developed during the postwar years, physical scientists, especially physicists, renewed their enlistment in the machinery of arms and the state. The threat to American security was palpable, given the existence of long-range bombers and, in the near-term offing, intercontinental ballistic missiles. Like the United States at large, many physicists held that the unpreparedness of the 1930s had led to World War II, and they were committed to maintaining national security now so as to avoid a World War III. Thus historical and technological circumstances—the power of nuclear weapons, the reach of new delivery systems, the utter vulnerability of cities, the potential combustibility of the Cold War—forced Oppenheimer and his fellow scientists to embrace their paradox, to accept both the anguish of their sin and the continuing responsibilities of national security.

With the responsibilities came rewards, of course, including power, esteem, and handsome funds for research and training by defense and defense-related agencies. Like

29. Science Panel of the Interim Committee, "Recommendations on the immediate use of nuclear weapons," 16 June 1945, Robert C. Williams and Philip Cantelon, eds., *The American atom: A documentary history of nuclear policies from the discovery of fission to the present, 1939-1984* (Philadelphia, 1984).

30. The popular response to the bomb is recounted in Paul Boyer, *By the bomb's early light: American thought and culture at the dawn of the atomic age* (New York, 1985).

the chemists after World War I, physicists and their allies linked a vigorous system of scientific research and industrial development to the requirements of national security. And, now, in contrast to the 1920s, the federal government invested heavily in the academic and industrial enterprises of science.[31]

Oppenheimer noted that physics was booming in America, but he permitted himself ironic meditations upon the distortions that the devotion to security visited upon the nation's universities. In 1949, he remarked that "it is a great liberal university"—he meant the University of California—"that is the only place in the world, as far as I know, that manufactures, under contract with the United States government, atomic bombs." And he added, "I have sometimes asked myself whether we can find any analogy to this situation in the practice of the monastic orders that devote a part of their attention and derive part of their sustenance from the making of their private liqueurs."[32] Physical scientists and engineers became part of a system—in academic, industrial, and governmental labs—that was driven by powerful incentives to produce new weapons. The thoughtful physicist and policy adviser Herbert York has observed that the scientific advisory system makes it possible for "the most exuberant and persuasive of our technologists to promote ideas and sell hardware that often take us far beyond the point that mere prudence requires."[33]

The enthusiasm of many scientists and engineers for new weapons technologies rapidly came to be driven by far more than social responsibility. It no doubt expressed some combination in different people of professional ambition, opportunism, or simply desire for a job, of ideological conviction or concern for national security—and a taste for technical sweetness, a powerful technological drive, a consuming desire to bend the resources of the state to satisfy a Promethean will. Salient examples come to mind from the rolls of the postwar defense research and advisory establishment, including, for instance, Edward Teller, with his obsessive determination to build a hydrogen bomb, and Wernher von Braun, who, consumed by the aim of building giant rockets and accomplishing space flight, happily built V-1s and V-2s for Hitler, then missiles for the United States.[34]

Unlike the technological enthusiasts, many Cold War scientists held that greater security was not to be gained through an endless quest for superiority in arms but through a halt in the arms race, or at least by its restraint. They campaigned for international control of nuclear weapons, a nuclear test ban, arms limitation, and other measures designed to achieve the same ends. Oppenheimer himself, together with almost the entire General Advisory Committee of the Atomic Energy Commission,

31. Kevles (ref. 12), chapts. 21-23.

32. JRO, *Uncommon sense*, N. Metropolis, Gian-Carlo Rota, and David Sharp, eds. (Cambridge, 1984), 30.

33. Herbert F. York, *Making weapons, talking peace: A physicist's odyssey from Hiroshima to Geneva* (New York, 1987), 87.

34. Michael J. Neufeld, *The rocket and the Reich: Peenemünde and the coming of the ballistic missile era* (New York, 1995); Richard Rhodes, *Dark sun: The making of the hydrogen bomb* (New York, 1995).

argued eloquently against proceeding with a crash program for the hydrogen bomb. In 1945, and again in 1949, he judged the H-bomb an immoral weapon because, with its unlimited power, it could cause unlimited human disaster.

After President Truman ordered work on the H-bomb to proceed with all possible speed, Oppenheimer thought about resigning from the General Advisory Committee but was persuaded not to do so—possibly, as James B. Conant, the chemist, Harvard president, and another key opponent who did not resign either, recalled, "because I did not want to do anything that seemed to indicate we were not good soldiers and did not do what we could to carry out orders of the President!"[35] Some may say, perhaps, that, as a matter of conscience, he should have resigned, that he should have quit being a good soldier—or, indeed, that he should never have been a good soldier in the first place.

The German playwright Heinar Kipphardt advanced precisely that view in his play of the 1960s titled after the security hearings, *In the matter of J. Robert Oppenheimer*. His Oppenheimer says, "I will never work on war projects again. We have been doing the work of the Devil, and now we must return to our real tasks."[36] But in the real—as distinct from the melodramatic—world, Oppenheimer commented, "What I have never done is to express regret for doing what I did and could at Los Alamos: in fact, on varied and recurrent occasions, I have reaffirmed my sense that, with all the black and white, that was something I did not regret."[37] Oppenheimer displayed no regret either about the considerable postwar service he had done the state. Nowhere in his public writings did he raise questions about whether scientists ought to work for the military. It seemed obvious to him that they should, when and where their services were needed.

At the time of the decision on the hydrogen bomb, no one knew how to build a thermonuclear bomb. But when in early 1951 Teller and Stanislaw Ulam devised an ingeniously practical way actually to do so, Oppenheimer signed on. His decision and that of others who had opposed the H-bomb was contingent on circumstances. The Korean War was on, the Soviets seemed malevolently bent on conquest, and surely, like the Americans, the Soviets could solve the technical puzzle, too. Still, Oppenheimer and many other physicists of the era remained a voice for restraint and the avoidance of nuclear terror.

Chemical and biological weapons

All through the nuclear arms race, the United States maintained a highly secret program of research in chemical and biological warfare. The biological part, begun during World War II, was defensive in intent, but as in chemical warfare a necessary

35. Quoted in James G. Hershberg, *James B. Conant: Harvard to Hiroshima and the making of the nuclear age* (New York, 1993), 482.
36. Heinar Kipphardt, *In the matter of J. Robert Oppenheimer* (New York, 1968), 126-127.
37. Quoted in David Cassidy, *J. Robert Oppenheimer and the American century* (New York, 2004), 433.

complement to that were inquiries into offensive biological weapons; one had to know what one might need to defend against. Funding for the programs remained relatively low through the 1950s. But then defense analysts contended that the nation required the means to fight local brushfire wars without resorting to nuclear weapons. John Kennedy embraced that position, and in the early 1960s funding for CBW skyrocketed, reaching more than $295 million annually in 1964. The products of this program helped enable the use of herbicides and defoliants in Vietnam.[38]

Precisely such uses, however, coupled with the dissent against the war as such and mounting suspicion of the military's connections with science, helped stimulate growing protests against the country's CBW programs. A movement emerged in the public at large and in the Congress for the United States to ratify the Geneva Protocol and establish an additional, sweeping ban on biological weapons. Strong support for the movement also came from influential members of the biology community, fearful that the rapidly developing field of molecular biology might be exploited to produce virulent new organisms and substances for military purposes. The military itself endorsed a ban, echoing the views toward chemical warfare of the generals in World War I that such weapons were not tactically or strategically advantageous. The ban won support among policymakers, too. They reasoned that biological warfare constituted a poor man's weapon: it seemed advantageous to deny such weapons to nations that could not afford the weapons systems of the United States and the Soviet Union.[39]

In 1972, the leading nations, including the United States, signed the Convention on Biological and Toxin Weapons, and in 1975 the United States ratified the Convention as well as the Geneva Protocol. The Convention went beyond the Geneva Protocol in that it not only banned the use of biological weapons, but also prohibited possession of them. President Nixon ordered the destruction of all the country's stocks of biological weapons and ordered, in accord with what the treaty permitted, research in CBW defense only. Appropriations for research in chemical weapons plummeted (in 1972 dollars) from $70 million in 1970 to about $23 million by 1976.[40]

However, by the late 1970s, fears revived that the Soviet Union was secretly violating the Convention and pursuing research on lethal biological agents. Worries also surfaced that the poor men of the Middle East were developing CBW (worries that, as proved to be the case in Iraq, were not unfounded). The Reagan administration reversed the decline in CBW research, pushing the budget for it back up to the levels of the mid-1960s. With an eye to maintaining the American commitment to the Convention on Biological and Toxin Weapons, it fostered only defensive research. But preparing for defense of course entailed research into likely offensive means.[41]

38. Charles Piller and Keith Yamamoto, *Gene wars: Military control over the new genetic technologies* (New York, 1988), 48. For an overview of its subject, see Robert Harris and Jeremy Paxman, *A higher form of killing: The secret history of chemical and biological warfare* (New York, 2002).
39. Ibid., 58-59.
40. Ibid., 58-61, 162-166.
41. Ibid., 60-61, 155.

Security in the Era of Terrorism

The horrendous events of September 11, 2001, placed bioterrorism high on the national security agenda. The USA PATRIOT Act and subsequent regulations imposed severe restrictions on access to pathogenic organisms in university laboratories, prohibited selected foreign scientists and students from working on such organisms, and established restrictive visa rules that often prevented foreign scientists, including students, from even entering the country. The federal government has caused acute anxiety by proposing to restrict the publication of sensitive basic research without prior review. The reason is that a good deal of biomedical research is double-purpose: it may serve the needs of health, but it can also assist bioterrorism.[42]

In the end, the security restrictions may be largely beside the point. Unlike early Cold War nuclear physics, which was confined to a handful of countries, contemporary biology is a global enterprise. It is unlikely that any government will really ever be able to control the flow of scientific information, not with five thousand or more journals in the world, and the Internet available to anyone who wants to use it. And locking up lethal biological agents is perhaps as difficult as locking up information about them. In the early Cold War, controlling uranium and plutonium, the key materials of nuclear security, was much easier. Raw uranium ore was hard to obtain, and turning it into fissionable fuel involved large and costly processing plants and reactors. But the agents of bioterrorism require comparatively small-scale production plants. Moreover, they are ubiquitous in their natural forms, and new forms—not to mention old ones like poliovirus—can be manufactured from DNA sequences readily available on the Internet with automated tabletop technologies.[43]

The new wave of security restrictions, questionable though they may be, is reminiscent of the tidal wave that swept over physics in the late 1940s and early 1950s. Yet as in the Cold War and physics, along with the restrictions has come rich funding for biology—$1.7 billion to the National Institutes of Health in fiscal year 2003—for research into bioterrorism.[44] Like physicists before them, many biomedical scientists are flocking after the money and for similarly diverse reasons. But doing research, while necessary, is probably not enough. The issue is not the deterrence of nations but of shadowy, underground groups, cells, and networks. The challenge posed by terrorism is unprecedented in the history of science, arms, and the state. What one hopes for is not only scientific knowledge and technology but wise guidance. One hopes, in short, for the courage, tenacity, and vision with which J. Robert Oppenheimer pursued the end—to quote the majority report concerning the hydrogen bomb—"of providing by example some limitation on the totality of war and thus limiting the fear and arousing the hopes of mankind."[45]

42. Daniel J. Kevles, "Biotech's big chill," *Technology review, 106* (Jul-Aug 2003), 40-49.
43. Ibid.
44. Ibid.
45. General Advisory Committee to the Atomic Energy Commission, "Report on the 'Super,'" 30 Oct 1949, Williams and Cantelon (ref. 29), 125-126.

CHAPTER SEVENTEEN

Intersections
Oppenheimer and Einstein

S.S. Schweber

Introduction

Both Einstein and Oppenheimer became iconic figures at a certain stage of their life—in ways whose similarities and divergences speak volumes about the changing significance of physics. After 1919, Einstein was hailed in the public sphere as a genius, "the greatest scientist of the world" and "the most revolutionary innovator in physics" since Newton—a "universe maker."[1] In a public address on the occasion of Einstein's sixtieth birthday in 1939, Oppenheimer himself could state that "[Einstein's] name is perhaps more widely known than that of any other living scientist; to many millions of people it has come to stand for science itself, and for all we admire in the way of life and thought of the scientist."[2] For his part, Oppenheimer was the "boy wonder" of the American physics community during the 1930s. After 1945, having overseen with General Groves the making of the weaponry that terminated the Second World War and changed the planet's balance of power and its conception of terror, he became identified in the public mind not only with the awesome instrumental power of science, but also as someone struggling to define the responsibilities of scientists in the new world they had helped create. As I.I. Rabi noted in his eulogy of 1967, "[his] rise to public eminence and recognition was quite sudden and was not presaged by events preceding the war. At that time Oppenheimer's reputation and influence were centered around the small and close circle of physicists....[Yet later] Oppenheimer, after Einstein, emerged as the great charismatic figure of the scientific world."[3] After the revocation of his clearance and his "trial" in 1954, Oppenheimer was seen as a

I am indebted to Cathy Carson, Paul Forman, Snait Gissis, Geri Holton, and Skuli Sigurdsson for their help and criticisms.

The following abbreviations are used: AEA, Albert Einstein Archives, National Library, Jerusalem, Israel; JRO, J. Robert Oppenheimer; JROP, J. Robert Oppenheimer Papers, Manuscript Division, Library of Congress, Washington, D.C.

1. Isaiah Berlin, *Personal impressions* (Oxford, 1982), 281.
2. JROP, box 256, folder Einstein's 60th Birthday—3/16/39. See also "Celebration of the sixtieth birthday of Albert Einstein," *Science, 89* (1939), 335-336.
3. Isidor I. Rabi, Introduction, in Rabi et al., *Oppenheimer* (New York, 1969), 3-8, on 4.

tragic figure—a symbol of the fallen hero to many of his fellow physicists and to liberal intellectuals, at times compared to Galileo and to Dreyfus.[4] For Einstein, too, there is a tragic element in the last twenty-five years of his life, for he stopped questioning the validity of his approach to the unification of gravitation and electromagnetism, and lost his connection to the physics community.

The intersecting trajectories of these two physicists through the twentieth century hold many lessons for us. Einstein had a high regard for Oppenheimer and had only respectful things to say about him. If he knew of some of Oppenheimer's less admirable qualities and of some of his deplorable deeds, he nonetheless would have said in a Spinozian spirit, "I cannot hate him because he must do what he does." Oppenheimer, though deeply admiring of Einstein, was at times publicly highly critical of him. I shall give an account of Oppenheimer's recorded statements regarding Einstein. I shall then address the question: "Why the antagonism on Oppenheimer's part?" and attempt to give some answers by looking at Oppenheimer's and Einstein's relation to their roots and their visions of the unity and the future of science. As we shall see, there were striking differences in the character of these two off-scale individuals, in their approach to physics, in their relation to the communities they were part of, and in their presentation of self.

The Einstein-Oppenheimer Interaction

Oppenheimer had first met Einstein in January 1932, when Einstein visited Caltech on his round-the-world trip of 1931-32. Then in January 1935, Oppenheimer went east to attend a meeting of the American Physical Society. While in New York he took the opportunity to go to Princeton to visit the Institute for Advanced Study, where he had been invited to spend a year as a visiting member. In a letter to his brother Frank, he conveyed his reaction to Fine Hall, the building where both the mathematics department of Princeton University and, until 1938, the Institute were located. He wrote: "Princeton is a madhouse: its solipsistic luminaries shining in separate & helpless desolation. Einstein is completely cuckoo....I could be of absolutely no use at such a place, but it took a lot of conversation & arm waving to get Weyl to take a *no*."[5]

Undoubtedly Oppenheimer's strong reaction, already evident in the mid-1930s, was due to the contrast between his and Einstein's working style, their sharply differing attitudes to quantum mechanics, and their conflicting views of what constituted the important contemporary problems in physics. Einstein's working style is well known. In

4. See the essays in this volume by Charles Thorpe and Robert P. Crease. It has been suggested that one aspect of tragic figures is that they discover themselves to be enigmas, without any fixed point of attachment. Crease suggests that Oppenheimer never questioned himself and never came to an understanding of himself, and thus casts doubt on the validity of considering him a tragic figure.
5. JRO to Frank Oppenheimer, 11 Jan 1935, Alice Kimball Smith and Charles Weiner, *Robert Oppenheimer: Letters and recollections* (Cambridge, 1980), 190.

his personal life as well as in his scientific activities, he was very much a loner. This was already true in Berlin when he was director of the Kaiser Wilhelm Institute for Physics in the early 1920s.[6] After he came to Princeton in 1934, his researches, which were primarily concerned with his unified field theories, were carried out with his assistant. By contrast, Oppenheimer in the 1930s created at Berkeley the most important school of theoretical physics in the United States. As Bethe noted in his eulogy, Oppenheimer "always knew what were the important problems, as shown by his choice of subjects. He truly lived with these problems, struggling for a solution, and he communicated his problems to his group."[7] And the members of the group, consisting of a half-dozen or so graduate students and a half-dozen postdoctoral fellows, helped one another to find solutions by meeting together daily, with Oppenheimer presiding. This form of cooperative investigation that made use of the collective knowledge of the group became the characteristic mode of distributed inquiry in research groups—but only much later. Yet already in the late 1920s Oppenheimer had recognized the transformation taking place in theoretical physics, and in physics more generally: physics had become a cooperative enterprise.

The difference in Einstein's working style and his own was made explicit in Oppenheimer's radio address on the occasion of Einstein's sixtieth birthday in 1939. He began with laudatory remarks—"Few men have contributed so much to our understanding of the Physical World, to our ability to predict and follow and control its behavior," he noted—and went on to declare: [8]

> But if few scientific workers would quarrel with the fact that Einstein is in many ways a perfect symbol of their work, there are many who would feel that there is something a little false and fabulous in the way that he is thought of....[T]here is...a general impression, supported in part by his eminence, that his work has been qualitatively different from that of his fellow workers; that it is abstruse, and remote, and useless. This seems to me a very strange ground for admiration. And of course it is not true; and the truth is much better than the fable.
>
> All discoveries in science grow from the work, patient and brilliant, of many workers. They would not be possible without this collaboration; nor would they be possible without the constant technological developments that are necessary to new experiments and new scientific experience. One may even doubt whether in the end they can be possible except in a world which encourages scientific work, and treasures the knowledge and power which are its fruits.

6. See Giuseppe Castagnetti and Hubert Goenner, "Directing a Kaiser-Wilhelm-Institute: Albert Einstein, organizer of science?", Giuseppe Castagnetti, Hubert Goenner, Jürgen Renn, Tilman Sauer, and Britta Scheideler, *Foundations in disarray: Essays on Einstein's science and politics in the Berlin years*, Max-Planck-Institut für Wissenschaftsgeschichte, Preprint 63 (1997), 55-80.
7. H.A. Bethe, "J. Robert Oppenheimer," National Academy of Sciences, *Biographical memoirs, 71* (1997), 174-219, on 184.
8. "Celebration of the sixtieth birthday of Albert Einstein" (ref. 2), 335-336.

Oppenheimer related the possibility of scientific knowledge to material practices. Being closely connected with Lawrence's Radiation Laboratory and the Kellogg Lab at Caltech, it was clear to him that material practices were institutionally embodied. I believe that his remarks also reflected the political views he held at the time.

By the same time, in an interesting coincidence, Einstein and Oppenheimer had come into their closest proximity in questions of physics. From the start, Oppenheimer's focus had been quantum mechanics. The discipline that he loved, and that he propagated so effectively to his students, was foreign to Einstein's sensibility. However, Oppenheimer's explorations in the 1930s brought him closer to Einstein's concerns. Through general relativity, each became enwrapped in his own way in what Richard Feynman would presciently call "the historical question of physics."[9] In the late 1930s Oppenheimer concerned himself—if not with the cosmological problem, though deeply interested in it—with a historical question: namely, what happens to heavy stars when they exhaust their nuclear fuel? And with his students, George Volkoff and Hartland Snyder, he had discovered that general relativity predicted the existence of strange dense objects—which John Wheeler later called "black holes." But the importance of this work was not recognized at the time, neither by the authors nor by the astrophysical community at large, because it could not be observationally corroborated, and because with the outbreak of World War II other problems became physicists' primary concern.[10] Einstein, for his part, was hostile to the idea of black holes, as he did not believe that any solution of the general relativistic field equations that contained a singularity was physically acceptable. Nor did he ever contemplate the physical possibility of black holes. According to Freeman Dyson, Oppenheimer too in later life was uninterested in black holes—although Dyson thought that, in retrospect, they were his most important contribution to science.[11]

In the fall of 1947 Oppenheimer became the director of the Institute for Advanced Study and a professor of physics there. He and Einstein thus became colleagues. Their interactions became colored by two separate considerations, one intellectual, the other political. After his own brief excursion into the field, Oppenheimer did not believe that either general relativity itself—because of lack of observational data—or quantized versions thereof—because of conceptual and technical difficulties—were areas of physics that would reward theorists' efforts. He went so far as to discourage, if not forbid people from taking up these problems while at the Institute. At the same time, Oppenheimer's interactions with Einstein were also polarized by the difficulties he was having with the House Un-American Activities Committee and potentially even greater ones with Joseph McCarthy's Senate committee. Oppenheimer felt under

9. R.P. Feynman, *The Feynman lectures* (Reading, MA, 1963), vol. 1, chapt. 3, p. 9.

10. The one text on general relativity written during the 1940s does nor refer to Oppenheimer's work. That book was written by Peter Bergmann, a close associate of Einstein. See also Karl Hufbauer's essay in this volume.

11. Freeman Dyson, "The scientist as rebel." The essay is part of Dyson's introduction to John Cornwell, ed., *Nature's imagination: The frontier of scientific vision* (Oxford, 1995).

great pressure not only to dissociate himself from persons who might be considered fellow travelers, but also to insulate the Institute from charges that it was harboring such individuals. Oppenheimer thus distanced himself from Einstein, who had made his political stand and his commitment to the protection of civil liberties clear.[12]

If in his 1939 radio address and then his public eulogies Oppenheimer's criticism of Einstein was subdued and respectful, it became explicit in the talk he gave in 1965 to commemorate the tenth anniversary of Einstein's death. Already ill with the throat cancer that was to kill him a year and a half later, he accepted an invitation from UNESCO to speak in Paris. He went there just for a day to deliver the lecture. He justified giving it by noting that he had known Einstein for over thirty years and that after he became the director of the Institute they "were close colleagues and something of friends."[13]

In his opening remarks he indicated that he thought "it might be useful, because I am sure it is not too soon, and for our generation it is a little too late, to start to dispel the clouds of myth [that surround Einstein's genius] and to see the great mountain peak that these clouds hide. As always, the myth has its charm; but the truth is far more beautiful." What was true of the myth, Oppenheimer was quick to call attention to, was Einstein's extraordinary originality: "Einstein was a physicist, a natural philosopher, the greatest truly of our time." Although others would surely have come to formulate the meaning of the Lorentz invariance of Maxwell's equations in the way Einstein had done and would have understood the necessity of this viewpoint also for mechanics, still "this simple, brilliant understanding of the physics could well have been slow in coming, and blurred had he not done it for us." Oppenheimer continued:[14]

> The general theory of relativity, which, even today, may very well prove wrong, no one but he would have done for a long time. It is in fact only in the last decade, the last ten years, that one has seen how a pedestrian and hard-working physicist, or many of them, might reach that theory and understand this singular union of geometry and gravitation, and we can do that today only because some of the "a priori" open possibilities are limited by the confirmation of Einstein's discovery that light would be deflected by gravity.

12. Einstein had made headlines in the *New York Times* for his defense of individuals who had been subpoenaed to appear before Senator McCarthy's and Senator Jenner's committee, and his advocacy of civil disobedience in these matters had been deemed questionable by a *New York Times* editorial.

13. Unless otherwise noted, quotations are from the original unedited transcription in JROP, box 235, folder UNESCO (Paris) Einstein, 12/13/65 draft. An edited version, toned down in places, was published as "On Albert Einstein," *New York review of books*, 17 Mar 1966, 4-5.

14. On receiving the transcript, Oppenheimer edited the first sentence to read, "The general theory of relativity which, even today, is not well proved experimentally, no one but he would have done for a long, long time," as appears in the published version.

Oppenheimer then pointed to another facet of Einstein's works: the deep elements of tradition embedded in them. He enumerated three such elements, not necessarily the only ones. One was the statistical tradition going back to Maxwell and Boltzmann, which had led Einstein to the laws governing the emission and absorption of light by atomic systems. The second was his "total love of the idea of a field and the following of physical phenomena in minute and infinitely subdividable detail in space and time….It was this which made him see that there had to be a field theory of gravitation, long before the clues to that theory were securely in his hand."[15] The third element was a philosophic one, which Oppenheimer characterized as a "form of the principle of sufficient reason." He pointed to Einstein's demand for care when stating what we mean, for exactitude about what we can measure, and for clarity about what elements in physics are conventions. For it was Einstein, Oppenheimer said, who recognized and insisted that conventional elements could have no part in the real predictions of physics.

But, Oppenheimer went on, in the last twenty-five years of Einstein's life, the years he spent at Princeton, in a certain sense his tradition failed him. "And this, though a source of sorrow, should not be concealed. He had a right to that failure." Einstein spent those years first in trying to prove that the quantum theory had inconsistencies in it; Oppenheimer remarked that "no one could have been more ingenious in thinking up devilish examples."[16] But the inconsistencies were not there—and eventually Einstein could only say that he didn't like the theory, that he didn't like the abandonment of continuity and of causality. "[T]o see them lost, even though he had put the dagger in the hand of the assassin by his own work, was very hard on him." He also struggled with an ambitious program to amalgamate the understanding of electromagnetism and gravitation in such a way as to explain "what he regarded as the semblance—the illusion—of discreteness of particles in nature."[17] But, Oppenheimer declared,

> I think that it was clear then, and believe it to be obviously clear today, that the things that this theory worked with were too meager, left out too much that was known to physicists but had not been known in Einstein's student days. Thus it looks like a hopelessly limited and historically rather accidentally conditioned approach.

"Oppenheimer view of Einstein warm but not uncritical": The tenor of what followed was accurately captured by the headline with which the *New York Times* led off its coverage. Oppenheimer's most critical statements, as the *Times* reported them, were a litany of Einstein's limitations: that "Einstein's early work was paralyzingly

15. Later edited to "It is this tradition which made him know that there had to be a field theory of gravitation..."
16. Later edited to "No one could have been more ingenious in thinking up ~~ingenious~~ unexpected and clever examples."
17. The interpolation "the illusion" was added upon editing.

beautiful but full of errors," whose correction had delayed the publication of Einstein's collected work by ten years (and "A man whose errors take ten years to correct is quite a man"); that Einstein founded no school and that he did not have many students; that he did not play a vital role in the development of the atomic bomb; and that his letter to President Roosevelt in which he sought to make him aware of the possibility of a fission bomb (to quote the *Times*) "was not important."[18]

And yet, in the rest of his lecture Oppenheimer again extolled Einstein for his goodwill and humanity, for his ever-present concern for not doing harm. He described Einstein's attitude toward human problems by the Sanskrit word *Ahinsa*, not to hurt, harmlessness. He commended him for his consequential stand against violence and cruelty wherever he saw it, noting, "after the war he spoke with deep emotion and I believe with great weight about the supreme violence of...atomic weapons" and the need to "now make a world government."

Oppenheimer's ultimate remark was revealing of both his subject and himself. "In his last years, as I knew him," Oppenheimer concluded, "Einstein was a twentieth-century Ecclesiastes, saying with unrelenting and indomitable cheerfulness 'Vanity of Vanities, all is Vanity.'" The statement demonstrated, on the one hand, Oppenheimer's empathy. Einstein early on had come to appreciate "the nothingness of the hopes and strivings which chases most men restlessly through life."[19] And on the other hand, it reflected the fact that both of them had accomplished tasks at the limit of human capabilities and had received a commensurate degree of fame. And both had to confront their inability to match these heights thereafter, yet kept on seeking what turned out to be futile ways to do so. As Einstein had once told Oppenheimer, and as the latter now repeated, "You know, when it's once been given to a man to do something sensible, afterwards life is a little strange."

Incidentally, Oppenheimer's ambiguous UNESCO talk had immediate repercussions. Upon reading the article in the *Times*, Abraham Pais wrote him saying euphemistically that it made him "slightly uncomfortable."[20] And in the note Oppenheimer attached to the copy of the full text of his speech he sent to various friends, he indicated that a number of his colleagues "suggested that I had been out of my mind" after reading the *Times* coverage.[21] He must have gotten word that Helen Dukas, Einstein's long-time secretary, was deeply upset by what she had read of his talk as well. He thereafter wrote her saying, "When I saw the story in the *New York Times* last month I shuddered for you". With his note he enclosed his edited version of the talk—which was later published in the *New York review of books*—and added disingenuously, "I

18. Henry Kamm, "Oppenheimer view of Einstein warm but not uncritical," *New York Times*, 14 Dec 1965.
19. Einstein, "Autobiographical notes," Paul Arthur Schilpp, ed., *Albert Einstein: Philosopher-scientist* (New York, 1949), vol. 1, 2-95, on 3.
20. Abraham Pais to JRO, 20 Dec 1965, JROP, box 285, folder UNESCO-Einstein Correspondence.
21. JROP, box 285, folder UNESCO (Paris) Einstein, 12/13/65 draft.

hope that this [the edited version], which is what I did say, will seem closer to the truth, and more welcome."[22]

Still, the views that Oppenheimer had expressed in Paris were widely disseminated in Europe. The *Gazette de Lausanne* gave the story extensive coverage, while the *Nouvel observateur* translated the speech and printed it in full in the December 22 issue. Gerard Bonnot, a reporter for *L'Express*, interviewed Oppenheimer right after his lecture—while he was sipping some whiskey—and *L'Express* devoted a page of its December 20-26 issue to their exchange. The views Oppenheimer expressed in fluent French during the interview were even sharper than in the lecture:

> During all the end of his life, Einstein did no good [*Einstein n'a rien fait de bon*]. He worked all alone with an assistant who was there to correct his calculations. . . . He turned his back on experiments, he even tried to rid himself of the facts that he himself had contributed to establish....He wanted to realize the unity of knowledge. At all cost. In our days, this is impossible.

But interestingly, Oppenheimer stated that "he was convinced that still today, as in Einstein's time, a solitary researcher can effect a startling [*foudroyante*] discovery. He will only need more strength of character [*force d'âme*]." Bonnot concluded his interview by asking Oppenheimer whether he had any regrets? Any nostalgia? A smiling Oppenheimer, disarmed somewhat by the whiskey, answered "Of course, I would have liked to be the young Einstein. This goes without saying."[23]

Roots and Tradition

> The art of renunciation is an act of courage—it requires the sacrifice of things universally desired (not without hesitation and regret) for matters that are great and incomprehensible.
>
> Zbigniew Herbert[24]

Simone Weil perceptively noted,[25]

> To be rooted is perhaps the most important and least recognized need for the human soul. It is one of the hardest to define. A human being has roots by virtue of his real, active and natural participation in the life of a community which preserves in living

22. JRO to Helen Dukas, 20 Dec 1965, JROP, box 285, folder UNESCO-Einstein Correspondence.
23. George Bonnot, "Oppenheimer et Einstein," *L'Express*, 20-26 Dec 1965.
24. Zbigniew Herbert, "Spinoza's bed," *Still life with a bridle* (New York, 1991), 145.
25. Simone Weil, *The need for roots: Prelude to a declaration of duties toward mankind* (London, 1952), 41.

> shape certain particular treasures of the past and certain particular expectations for the future. This participation is a natural one, in the sense that it is automatically brought about by place, condition of birth, profession and social surroundings. Every human being needs to have multiple roots. It is necessary for him to draw wellnigh the whole of his moral, intellectual and spiritual life by way of the environment of which he forms a natural part.

In contrast to Einstein, for whom roots without the communal aspects became sufficient, the rootless Oppenheimer was always in need of a community. Their similarities and contrasts in this respect bring out important features of both men's characters.

Both Einstein and Oppenheimer were born into emancipated, non-observant Jewish families. The nature of Einstein's ties to Judaism and to the Jewish culture is well-known enough that it need not be detailed here.[26] However, I want to point to an interaction of the young Albert which I believe was deeply influential and which signals a commonality with the experiences of Oppenheimer. This relationship was with Gustav Maier, a friend of the Einstein family in Ulm and an avuncular figure to the young Albert. Maier was the manager of the Ulm branch of the Deutsche Reichsbank until 1881, at which time he moved to Frankfurt am Main to become the manager of Reichsbank branch there. Before leaving Ulm he published a little pamphlet entitled "More light! A word on the 'Jewish question' to our Christian neighbors" ["*Mehr Licht! Ein Wort zur 'Judenfrage' an unsere christliche Mitbürger*"] in which he defended both reform Judaism and socialism against the charge of hostility to Christianity. In 1886 Maier was one of the founders of the Frankfurt Peace Union. In 1891 he withdrew from the business world, moved to Switzerland—and devoted himself to writing on social and economic issues. By virtue of common political leanings and cultural interests, he and Jost Winteler met and became good friends, the two of them helping found the Swiss Ethical Culture Society in 1896, whose publication Maier edited until 1919. It was Maier who helped the young Albert, whom he considered a *Wunderkind*, a boy wonder, obtain the permission to take the university entrance examination required of applicants to the Swiss Federal Institute of Technology (Eidgenössische Technische Hochschule, ETH) without a secondary-school certificate.[27] It was also he who arranged Albert's stay with the Winteler family in Aarau for him to complete his high school education. Albert often visited Maier while he was a student at the ETH from 1896 until 1900.

In 1898, moreover, Maier published a very influential book entitled *Social movements and theories up through the modern labor movement* [*Soziale Bewegungen und Theorien bis zur modernen Arbeiterbewegung*] that went through nine editions. In nine chapters and a little less than 150 pages it gave a succinct account of the history of communism, socialism, and anarchism. And in an epilogue Maier stressed the need

26. See, for example, Max Jammer, *Einstein and religion* (Princeton, 1999).
27. See Albert Einstein, *The collected papers of Albert Einstein*, John Stachel et al., eds. (Princeton, 1987), vol. 1, 10-14.

for social justice and for educational opportunities for the working classes in the new industrial and urbanized world. It was probably Maier and Winteler who planted the seeds of socialism in Einstein—and Einstein repeatedly made his own commitment to socialism well known.[28] Nor were his ties to Maier and Winteler, and theirs to the Ethical Culture Society, forgotten, as evidenced by the fact that Einstein agreed to write a message to be read at the celebration of the seventy-fifth anniversary of the founding of the Ethical Culture Society in January 1951.[29]

As is well known, Oppenheimer was educated at the Ethical Culture School in New York City. The Ethical Culture Society was a parallel development to the Social Gospel movement of American Protestantism. It too embodied a response to the dismal conditions of the lower classes brought about by the industrialization and urbanization of the last third of the nineteenth century. However, in the United States the Ethical Culture Society, its support coming from emancipated middle and upper classes Jews, never formed ties with socialism.

The Ethical Culture Society was to provide an outlet for the tensions created by the inability of some well-to-do, emancipated Jews to identify as American Jews. It clearly did not do so for the young Robert.[30] After Oppenheimer left the Ethical Culture School and went to Harvard (1922–25), he began seeking a more satisfying approach to religious thought in the Hindu classics, albeit in English translation. "Later," notes James Hijiya, "he called the *Bhaghavad-Gita* 'the most beautiful philosophical song existing in any known tongue.'...In 1963, *Christian century* magazine asked him to list the ten books that 'did most to shape your vocational attitude and your philosophy of life.' Along with Shakespeare's *Hamlet* and Eliot's *Waste land*, Oppenheimer listed the *Gita*."[31]

And in contrast to Einstein's relationship to Spinoza, in whose *Ethics* he delighted, I do not believe that Oppenheimer found a similar anchor in the *Gita* and the actions of Arjuna. Jonathan Israel in his masterly *Radical enlightenment* stresses "the unity, cohesion, and compelling power" of Spinoza's system, a feature with which Einstein

28. See, for example, Albert Einstein, "Why socialism," *Ideas and opinions* (New York, 1954), 151-158.

29. Two years earlier Oppenheimer had been the principal speaker at the installation of Richard Boyd Ballou as director of the Ethical Culture School in Riverdale.

30. When the young Robert was recuperating from a severe case of dysentery, his parents asked Herbert Smith, his remarkable English teacher at the Ethical Culture School who became a sort of surrogate parent, to accompany him on a trip to the Southwest of the United States. Smith later told Alice Kimball Smith and Charles Weiner that he became aware of "how deeply Robert felt the fact of being Jewish when he asked to travel west as Smith's brother." And his close friend from the Ethical Culture School, Francis Fergusson, believed that for Robert going to New Mexico was partly to escape from "his Jewishness and his wealth, and his eastern connections." Smith and Weiner (ref. 5), 9. Oppenheimer did not escape these tensions at Harvard, nor was the atmosphere at Berkeley and at Caltech free from anti-Jewish prejudice.

31. James A. Hijiya, "The *Gita* of J. Robert Oppenheimer," American Philosophical Society, *Proceedings, 144* (2000), 123-167, on 130. See also J.L. Heilbron's essay in this volume.

surely resonated. Some have suggested that Einstein's radical reductionism has its origin in Spinoza's view that "there can be but one substance and therefore but one set of rules governing the whole of reality which surrounds us and of which we are part,"[32] and similarly that his commitment to strict determinism derives from Spinoza's Proposition XXIX of Part I of the *Ethics,* that "in nature there is nothing contingent, but all things have been determined from the necessity of the divine nature to exist and produce an effect in a certain way," and, moreover, that "things could have been produced by God in no other way, and in no other order, than they have been proposed." The parallel between Einstein's comportment and his way of life, particularly after 1930, and that of Spinoza has often been commented upon.[33] To this should be added that both developed impressive political perspicacity.[34] While giving up his religious affiliation with Judaism, Einstein, like Spinoza, nurtured his Jewish roots and identified as a member of the cultural Jewish tradition.

Oppenheimer too gave up his religious affiliation with Judaism, *qua* religion, but in addition he uprooted himself from the Jewish cultural tradition. I believe he hoped to find with the *Gita* if not the possibility of planting new roots, at least some comfort in committing himself to some aspects of the Hindu tradition. The need to distance himself from being identified as Jewish then intensified Oppenheimer's nationalism. Identifying himself as an American was the one continuous, constant aspect of Oppenheimer's changing identities. The contrast with Einstein's rejection of nationalism is clear-cut.

I believe that Einstein's willingness to identify himself as a Jew, to be active in the support of Zionism, to readily and openly discuss his religious beliefs annoyed, if not exasperated, Oppenheimer. Similarly, their sharply differing views of nationalism and socialism during the early phase of the Cold War, when Oppenheimer believed the United States stood for the defense of cultural freedom and saw socialism as having led to totalitarianism in its Stalinist and Hitlerian manifestation, created a chasm between them.

Philosophy

I have indicated some of the factors that might have motivated Oppenheimer in making his critical public statements regarding Einstein, particularly at the UNESCO conference. But I believe there is an additional reason for the sharpness of some of his

32. Jonathan Israel, *Radical enlightenment: Philosophy and the making of modernity* (Oxford, 2001), 231.
33. See Michel Paty, "Einstein and Spinoza," Marjorie Grene and Debra Nails, eds., *Spinoza and the sciences* (Dordrecht, 1986), 267-302. See also Jammer (ref. 26).
34. Recall, for example, Spinoza's analysis of Cromwell's regime in the *Tractatus theologico-politicus* and Einstein's analysis of militarism written shortly after the end of the Second World War.

remarks. Einstein was not only a great physicist, but also a extremely influential philosopher of science. Oppenheimer had aspired to make a mark as a philosopher—but failed—and his philosophical stance was deeply at variance with Einstein's.

The drive towards "unity" was a constant in all that Einstein did, as though he had adopted the Heraclitean saying "What is wise is one" as a guiding principle throughout his life. A much-cited letter to his friend Marcel Grossmann in 1900 highlighted his desire "to recognize the unity of a complex of appearances which…seem to be separate things."[35] His philosophy of unity extended beyond science. So he put it in a letter he wrote in 1950 to a grieving father who had lost a beloved son to illness: [36]

> A human being is a part of the whole, called by us "Universe", a part limited in time and space. He experiences himself, his thoughts and feelings as something separated from the rest—a kind of optical delusion of his consciousness. The striving to free oneself from this delusion is the one issue of true religion. Not to nourish the delusion but to try to overcome it is the way to reach the attainable measure of peace of mind.

In science, most theoretical physicists of Oppenheimer's generation thought that Einstein was wrongheaded in his approach to unification. Einstein and Oppenheimer belonged to two different generations, with the advent of quantum mechanics delineating the boundary. Nonetheless the vision of unification was nurtured by Einstein's efforts.

Except for a brief period after the advent of general relativity stretching from 1915 until roughly 1926-27, physicists addressing "fundamental" issues before the 1960s were primarily concerned with the elucidation and representation of four separate, seemingly independent realms. The experimental and theoretical researches of the 1930s made it possible to identify four fundamental interactions: 1) gravitation, 2) electromagnetism, 3) the (strong) nuclear forces, and 4) the weak-decay interactions. After the advent of quantum mechanics, Einstein was one of the few physicists who attempted to give a unified description of the forces of nature—in his case a unified description of gravitation and electromagnetism, and this at the classical level.[37]

The modern program of unification traces its origins to Newton and his realization of the universality of gravitation, and thence to Maxwell with his unification first of electricity and magnetism, then of optics, i.e., light, with electromagnetism. With the work of Kirchhoff and others, it became clear that the laws of physics discovered on earth are also valid in distant stars. With Einstein the vision became all-encompassing. Einstein advocated unification coupled to a radical form of theory reductionism. In 1918, on the occasion of Max Planck's sixtieth birthday, he stated, "The supreme test

35. Cited in Gerald Holton, *The advancement of science, and its burdens* (Cambridge, 1998), xxxi.

36. Einstein to Robert S. Marcus, World Jewish Congress, 12 Feb 1950, AEA 60-424. Einstein's German draft is AEA 60-425.

37. Oskar Klein is one of the few during the 1930s; Schwinger is another after World War II.

of the physicist is to arrive at those universal elementary laws from which the cosmos can be built up by pure deduction."[38] And in his 1949 "Autobiographical notes" he was even more explicit: [39]

> I would like to state a theorem which at present can not be based upon anything more than upon a faith in the simplicity, i.e., intelligibility, of nature:...nature is so constituted that it is possible logically to lay down such strongly determined laws that within these laws only rationally completely determined constants occur (not constants, therefore, whose numerical value could be changed without destroying the theory).

Oppenheimer's philosophy was very different. It could be characterized as more Bohrian than Niels Bohr himself. In 1957 he had been invited to deliver the William James Lectures at Harvard, and he invested a great deal of time in their preparation. What he hoped to accomplish in them can best be understood as trying to make sense of, and come to terms with the constantly accelerated, seemingly exponential growth of knowledge, and finding meaning in this new world. The lectures were Oppenheimer's attempt to attain, he said, "as unified, consistent and complete an outlook" as was possible, given what he had learned from recent archeological and historical studies; from some of the social sciences and economics; from the psychology of perception and learning; from advances in logic, i.e., from Gödel's incompleteness theorem; and from physics when it had reconciled, on the basis of Bohr's complementarity principle, the seemingly contradictory aspects of the description of the atomic world.

In particular, Oppenheimer wanted to make clear our cognitive predicament: as he described it, [40]

> the fact that there is a great deal to know; that it is not well ordered; that it does not follow from a few comprehensible principles; that it is offered to us in diverse tongues; that it is shared by rather small groups of very specialized people; that it is growing apace; and that we do not know very much of it.

He did believe that there was an order, a unity to the world, but it is not a "closed, or global, or architectonic unity, but rather, [as William James had put it] that of a network, a reticular unity."[41] And he made explicit his high regard for James: "[I] admire

38. "Motive des Forschens, " E. Warburg, M. v. Laue, A. Sommerfeld, and A. Einstein, with a reply by M. Planck, *Zu Max Plancks sechzigstem Geburtstag: Ansprachen, gehalten am 26. April 1918 in der Deutschen Physikalischen Gesellschaft* (Karlsruhe, 1918), 29-32; translated as "Principles of research," *Ideas and opinions* (ref. 28), 224-227.

39. Einstein, "Autobiographical notes" (ref. 19), 63.

40. JRO, The William James Lectures, JROP, box 259. A photocopy is available in the Harvard University Archives. These do not include Oppenheimer's preparatory notes, however. Lecture 1, "A pluralistic world."

41. In his 1956 talk at a symposium on "Science and the modern mind" at the American Academy of Arts and Sciences, Oppenheimer had spoken of a "reticular unity" as well. See

James," he said, "because he fought against limitation and preconception; against the applicability of ideas from which specifics could be deduced; against certitude, against doctrine, against system."[42]

The fundamental point that Oppenheimer wanted to make, what he desired his lectures to accomplish, was something he had learned through his interactions with Jerome Bruner, Ruth Tolman, and the other psychologists with whom he had conferred at the Institute, and more recently by reading the manuscript of Bruner, Goodnow, and Austin's *A study of thinking*:[43]

> No man ever notices more than a minute fraction of those things which are for him to notice; and that is the price for him noticing anything at all....It means that the potential that may be known is very greater than what is really known. It means that the potential always transcends the real....The things that are cut out are possibilities that are lost in order that one may perceive at all. [44]

Thus "[i]n order to know at all, man must act and choose; in doing that he loses the opportunity of other knowledge."[45] This, Oppenheimer believed to be a generalized formulation of the notion of complementarity that Bohr had introduced. It was to be the principle that would help unravel the cognitive chaos, and characterize and help understand the order and unity of the new open world that Oppenheimer was talking about.

This point came at the end of Oppenheimer's first lecture, which had lasted for over two hours—and was undoubtedly lost on the majority of the audience, held spellbound by his lyrical delivery. And those who understood the point probably thought that it was too inconsequential and superficial an instrument to resolve the difficult problems of the modern, open world. And yet Oppenheimer's closing words in his seventh and last lecture were normative and hortatory: [46]

> We have, I think, a dual duty: a duty of faithfulness and firmness and steadfastness in the things which by accident in our own time, in our own place, are our knowledge, our skill, our arts; [and] we have a duty of great openness to others...learning to

---

JRO, "The growth of science and the structure of culture," *Daedalus* (Winter 1958), 67-76. A few months later he came across the powerful passage in James's lectures on pragmatism: "Loosely speaking, and in general, it may be said that all things cohere and adhere to each other *somehow*, and that the universe exists practically in reticulated or concatenated forms which make of it a continuous or 'integrated' affair." William James, *Pragmatism*, Bruce Kuklick, ed. (Indianapolis, [1907] 1981), 64.

42. JRO, The William James Lectures (ref. 40), Lecture 1, "A pluralistic world."

43. Oppenheimer wrote a lengthy review of the book. JRO, "A study of thinking," *The Sewanee review, 46* (1958), 481-490.

44. JRO, The William James Lectures (ref. 40), Lecture 1, "A pluralistic world."

45. Ibid.

46. Ibid., Lecture 8, "The hope for order."

> welcome the strange and being glad to learn something that is new, and that we have not suspected, and that does not fit, that is different and in some sense alien. And both duties are part of any love of order and of truth that can exist today. Both are helped when we not only love other truth but each other who purvey it.

Oppenheimer's James Lectures presented a viewpoint that was at variance with the views of his colleagues in physics, one held or defended, moreover, by but very few philosophers. The pragmatism of Dewey and James, and the neo-pragmatism that Oppenheimer was trying to elaborate, was considered soft and not rigorous. It was more concerned with issues having biological foundations—and largely ignored at the time. Nor was the Cold War conducive to endorsing uncertainty and fallibility.

Oppenheimer was more successful in later talks at making clear his views. In those later settings he made much more explicit, using simpler language, how history, contextual factors, and culture-specific values and motives play a role in making sense of one's world and one's moral horizon. He became much more forceful in expressing his distrust of any order which is hierarchical in the sense that "it says that some things are more important than others—that some things are so important that you can derive everything else from them."[47] And he would compellingly avow that as far as science was concerned, "[n]o part of science follows, really, from any other in any usable form. I suppose nothing in chemistry or in biology is in any kind of contradiction with the laws of physics, but they are not branches of physics. One is dealing with a wholly different order of nature."[48]

Surely, Oppenheimer must have felt that Einstein during the last twenty-five years of his life had shirked the second duty of the James Lectures, "learning to welcome the strange and being glad to learn something that is new, and that we have not suspected, and that does not fit." Oppenheimer's pragmatic stance was at odds with Einstein's search for a unitary theory. It also put his expectations skew to those of many of his own colleagues, who followed Einstein's vision of unity, if not Einstein's version of its realization. In April 1961, for instance, the Massachusetts Institute of Technology celebrated its hundredth anniversary with a week-long conference on scientific and engineering education. The morning session of the last day of the gathering was devoted to a panel discussion on "The future of the physical sciences." Speaking for many, Rudolf Peierls, who at the time was professor of mathematical physics at Birmingham University, suggested that "if we think about the foundations, i.e., the front line of physics...we shall be looking for connections which will unify the very broad range of basic concepts and facts which we now have to use to describe what we know in phys-

47. JRO, *Some reflections on science and culture* (Chapel Hill, NC, [1959] 1960), 39.
48. JRO, "Tradition and discovery" (1960), N. Metropolis, Gian-Carlo Rota, and David Sharp, eds., *Uncommon sense* (Boston, 1984), 106. In another lecture from this period he asserted: "There are no synapses, no gaps, no gulfs in the picture of the various orders and parts of nature; but they are not derivable from one another. They are branches,...each with its own order, each with its own concern, each with its own vocabulary." JRO (ref. 47), 15. "The metaphor of a tree is invoked," he said, "because it is a growing thing; it is a tree, not a temple": ibid., 17.

ics." Although Peierls felt that there had to be "some new kind of connection which interprets the many fields and couplings [of the then known elementary particles] as different aspects of some common principle," he thought that "we should not expect, as we expected at one time [during the 1930s], that the unification will come with one go....There will be many stages of partial clarification before everything fits together, and, of course...the last stage may never be reached."[49]

At issue was indeed the future of the physical sciences. In the early 1960s a group of high energy physicists in the United States was pursuing intensive studies on the design of a several-hundred-GeV particle accelerator and the experimental program that could be carried out with it. The American high energy community strongly supported the project, but it "sensed the apparent existence of some misunderstanding of the objectives of high energy physics, not only among the general public but also among the scientific community as a whole."[50] It was therefore felt that some statements addressed to both the scientific community and the general public regarding the aims of high energy physics were in order. Luke Yuan, at Brookhaven National Laboratory, was asked to solicit statements from some thirty theoretical physicists, which together would "provide a comprehensive basis for a better understanding of the fundamental importance and great depth of high energy physics."[51] The statements were gathered, and Oppenheimer agreed to contribute the foreword to the volume.

In his remarks, Oppenheimer affirmed his belief that the techniques of high energy physics—experimental, instrumental, computational, mathematical—would prove of great value throughout the sciences, and in technology. But for him the justification a new accelerator rested on the following: [52]

> [T]he last centuries of science have been marked by an unabating struggle to describe and comprehend the nature of matter, its regularities, its laws, and the language that makes it intelligible. The successes in this struggle, from the Sixteenth Century until our own day, have inspired the whole scientific enterprise, and lighted the world of technology, and the whole of man's life. They have informed the education and the devotion of young people. They have played an ineluctable part in the growth, the health, the spirit, and the nature of science. We are now, despite tempting and brilliant topical successes, deep in the agony of this struggle. This volume attests the conviction of those who are in it that, without further penetration into the real of the very small, the agony may this time not end in a triumph of human reason. That is what is at stake; that is why the book is written.

49. The session took place in the MIT Armory on 8 Apr 1961. The records of the conference are available in the MIT Archives. I thank the archivist of the MIT Library for permission to quote from the transcripts.
50. Luke C.L. Yuan, *Nature of matter: Purposes of high energy physics* (Upton, NY, 1965), preface (n.p.).
51. Ibid.
52. Ibid., 5.

What is striking in Oppenheimer's remarks is that he did not claim, as did most of the other contributors, that, for instance (to quote Robert Sachs), "[h]igh energy physics is the essence of today's science of physics...and is the domain [in which] the fundamental laws of nature are now being discovered." The consensus—expressed in the words of Hans Bethe—was that high energy physics was "indeed the most basic field of knowledge in the physical world...no other field will give us such deep understanding....[P]article physics deserves the greatest support among all the branches of our science because it gives the most fundamental insights." It provided, so Henry Primakoff in turn put it, "in *some sense* the fundamental building blocks...out of which the material universe is constructed."[53] So although Oppenheimer certainly agreed that seeking to understand the structure of matter had been astoundingly intellectually fruitful in the past, he refrained from giving his assent to Abraham Pais's conviction that the very goal of high energy physics was to find a fundamental, unitary theory of the interactions between elementary particles. To Pais's mind, on the contrary, an "interruption of the pursuit toward high energy machines would be disastrous" for just that reason, as "[w]e may be stranded without their synthesis....We may stop short of...integrating such disjointed forces into a new unification."[54]

Epilogue

It was at Los Alamos, building weapons of mass destruction, that Oppenheimer experienced to the fullest a sense of community and of communal living. As the director of the Institute for Advanced Study he would expend enormous energies and efforts to try to make it an intellectual community. In his 1954 report to the Board of Trustees, Oppenheimer commented, perhaps too optimistically and not very objectively, that the members of the Institute formed a community because[55]

> many members have...a ranging and wide understanding and interest, and some substantial knowledge outside their own field of specialization. They are a community because close friendships contribute to mutual understanding and common interest. The fact that most of the members of the Institute live in the same apartments, eat in the same restaurant, share the same common room and the same library helps to bring them together.

Yet as time passed, George Kennan observed that Oppenheimer was often discouraged, in the end deeply disillusioned by the failure of community—the fact that[56]

53. Ibid., 20 (Sachs), 9 (Bethe), 18 (Primakoff). The sentiment is echoed by Weisskopf (26), Wick (29), and Dalitz (41).
54. Ibid., 16-17.
55. Institute for Advanced Study, "Report of the director," 1954, 26, JROP, box 233.
56. George F. Kennan, *Memoirs 1950–1963, 2* (Boston, 1972), 19.

> the members of the faculty of the Institute were often not able to bring to each other, as a concomitant of the respect they entertained for each other's scholarly attainments, the sort of affection, and almost reverence, which he himself thought these qualities ought naturally to command. His fondest dream had been one of a certain rich and harmonious fellowship of the mind. He had hoped to create this at the Institute for Advanced Study; and it did come into being, to a certain extent, within the individual disciplines. But very little could be created from discipline to discipline; and the fact that this was so—the fact that mathematicians and historians continued to seek their own tables in the cafeteria, and that he himself remained so largely alone in his ability to bridge in a single inner world those wholly disparate workings of the human intellect—this was for him a source of profound bewilderment and disappointment.

The contrast could not be greater with Einstein's message to the Italian Society for the Advancement of Science in 1950:[57]

> [A]part from the knowledge which is offered by accumulated experience and from the rules of logical thinking, there exists in principle for the man in science no authority whose decision and statements could have in themselves a claim to "Truth." This leads to the paradoxical situation that a person who devotes all his strength to objective matters will develop, from a social point of view, into an extreme individualist who, at least in principle, has faith in nothing but his own judgment. It is quite possible to assert that intellectual individualism and scientific eras emerged simultaneously in history and have remained inseparable ever since.

Einstein and Oppenheimer differed sharply in their views of individualism and collectivism, of creativity, and of the growth of knowledge. Nonetheless I believe that both would agree that in our times there should still be institutions and places within universities which, as Oppenheimer himself put it, "[encourage] scientific work, and [treasure] the knowledge and power which are its fruits"—where the creation of knowledge remains an autonomous activity and where some of its producers are free to set their own intellectual agenda, are able to pursue it, and are supported to do so.

57. Einstein, *Ideas and opinions* (ref. 28), 357.

CHAPTER EIGHTEEN

Oppenheimer in Film

A Transcript

Preface

In books, plays, films, and now an opera, Oppenheimer's life has given rise to a panoply of representations. These portraits carry the figure of Oppenheimer to the broadest of audiences. Naturally, the variety of public Oppenheimers has sparked reflections on historical fiction, documentary film, and scholarly history. It has simultaneously raised the question of the strengths and weaknesses of those genres in conveying the story of the nuclear age.

One session at the conference was devoted to the possibilities of filmic portrayals. The session began as a discussion between two distinguished participants in the venture: Jon Else, a professor of journalism and producer of *The day after Trinity: J. Robert Oppenheimer and the atomic bomb*, and Peter Galison, a historian of twentieth-century physics who co-produced (with Pamela Hogan) *The ultimate weapon: The H-bomb debate*. Else and Galison began their dialogue with short presentations. After several exchanges, they opened the floor to questions and comments. The transcript preserves the conversational flavor.

Jon Else

Looking out at the audience here is like looking at the bookshelf in my office. I'm not a scholar, nor a scientist, but a television producer, someone who makes people like Oppenheimer public. So I make public Oppenheimers. And I'm not one of those television producers who see academics as our natural enemy. Quite the contrary, I depend utterly on scholars in doing my work—to keep us honest, frankly, to make sure that we get it as right as it can possibly be.

I chose a long time ago to work prime-time television. That was not because I like prime-time television. In fact I don't like it: it's sort of like working in hell, an awful landscape into which to throw yourself. But in prime-time television, we can get ten million people to pay attention, all at once, on one night. That is no small thing. I've stuck at it. And you, the scholars, are sort of stuck with us. We are, for better and for worse, the popularizers of what you do. We try to do it with some humility, and we try to come as close as we can to maintaining the rigor and complexity that you bring to the work. But the greatest strengths of broadcast television, particularly the documentary, are also its greatest weaknesses. Its single greatest strength is that it is superbly powerful in conveying a few ideas. The bad news is then that it can only convey a few ideas. Documentaries, and television in general, are lousy vehicles for complexity, for nuance. They're lousy vehicles for encyclopedic chronicles of history.

That is then one reason Oppenheimer is so difficult. Oppenheimer is so very complex, such a ball of paradoxes. His life spanned so much of the century and so much of its moral and ethical and scientific and cultural history. Today I want to talk about two Oppenheimer documentaries: one that I did, as hard as it is to believe it, a quarter-century ago—for we started to work on *The day after Trinity* in 1977; and then a film that I'm just beginning—on which we just started shooting this month, in fact.

The topic of this panel is "The public Oppenheimer." I'm not sure there is a public Oppenheimer in any mass-culture sense today—in the sense that we have a public Einstein, or even a public Teller. Oppenheimer has not as a public image become a shorthand for something else. We have, for instance, McCarthyism—but we do not have Oppenheimerism. In the whole world of public images from that age, the mushroom cloud trumps everything else. Oppenheimer certainly comes nowhere near that, nor even does Teller.

To the extent that we *do* have a public Oppenheimer, we have in fact a cascade of many public Oppenheimers. We get about two a generation. Fortunately, right now we are getting more than two a generation in this wonderful flood of new books coming out. Each of those public Oppenheimers is a construction, constructed by someone: sometimes me, sometimes, usually, one of you. Public Oppenheimers will be constructed in the future by people who haven't even been born yet: this will go on for another thousand years, I hope. And each of those constructions differs slightly from the one before, just as it differs slightly also from the unmediated, factual, world historical Oppenheimer. That is both a good and bad thing.

So far in the series of public Oppenheimers, we first had the post-Hiroshima, pre-hearing Oppenheimer. That was almost the original *Revenge of the nerds*, seeing him so confident on the cover of *Life* magazine. Then we had the post-hearing, pre-scholarship Oppenheimer; we had the Oppenheimer of what I call the movie years, the late 1960s through the '80s, when the BBC and Hollywood (and I) were doing films. We now have the Oppenheimer of a new wave of scholarship. We are about to have the opera Oppenheimer, which I will talk about later. And there is an infinite series of wonderfully permutating Oppenheimers yet to come. It reminds me of *Being John Malkovich*: as I look into the future I see more and more variations.

For those of us who create public portraits, Oppenheimer is irresistible, like catnip. There is so much going on that we just cannot stay away. Before talking about the films, let me talk about the raw ingredients. First of all, there are his paradoxes. There is the conscience-stricken geek who apparently won the war. There is the misfit with militarism who created the greatest military weapon in history. There is the Berkeley radical who resolutely refused to spy, so far as we know, against his country. There is the frail aesthete who is also the tough horseman, riding in Jicarilla Apache country in the rain, at night, with the thunder rolling around him, one presumes while reciting Baudelaire.[1] There is the romantic recluse who becomes the Washington insider, comfortable in the halls of power—who is then brought to his knees, a whole rolling cascade of transformations.

Then, too, his story is attractive because it is laced with a sort of lurid mystery: all the FBI agents skulking around Jean Tatlock's house. There's a fascination of forbidden knowledge that underlies all of it. For those of us who do biographies, Oppenheimer is also the great anti-villain compared with, say, Curtis LeMay, Teller, or McCarthy. At least on the surface, to biographers these figures appear much more one-dimensional than Oppenheimer. There's not a whole lot of self-doubt and sniveling about having blood on one's hands and knowing sin.

Finally, Oppenheimer is attractive because of the bombs. There is simply a straight-ahead fascination with atomic bombs. When we did *The day after Trinity*, you couldn't say "fuck" on public television, and we talked to more than a few scientists and historians who made explicit connections to eroticism and the sexuality of the bombs. One scientist who had worked at length on all the iterations of the hydrogen bomb said to us, "If you love to fuck, you will love hydrogen bombs." I think he was half-joking, but half-not joking, and that underlies, I think, part of the attraction. If Oppenheimer had in fact been the prime mover in radar or had developed DARPAnet, I don't think that he would hold that fascination for biographers.[2]

Let's go back twenty-five years to producing *The day after Trinity*. That was done in its own historical context, as all television programs, as all biographies are. It was, believe it or not, still in the afterglow of World War II. World War II was all around us, through our parents, our uncles. It was the war in which we had whipped the pants off fascism. Also I began *Trinity* in the dark shadow of Vietnam, which was seen at the time as a very unjust war. There was nuclear testing going on underground in Nevada. Atmospheric testing had stopped, but every month or so they'd fire a shot at the Nevada Test Site.

Our original intent when we started *The day after Trinity* was to do a very bold, almost opaque, avant-garde film about the conundrums and paradoxes of Robert Oppenheimer and these weapons. We quickly abandoned that; for it became clear that very few people in the country knew anything at all about Oppenheimer. We had to back off. The revised intent of *The day after Trinity* was threefold. It was to do, very simply, a primer of why and how that first weapon was built and used; and then

1. I never knew Robert, but I knew his brother Frank fairly well in the later years of his life; and even in old age he was one tough cowboy. Knowing him was a little window into the Oppenheimer brothers. He was a tough guy; I presume Robert was also.

2. Before I leave behind the fascination of hydrogen bombs, there is now a weird mini-industry of "nuke porn" on the internet. It involves compilation DVD discs that are just lots and lots and lots of detonations for atmospheric tests. They're packaged very adroitly, they have wonderful music, but they are essentially just ninety-minute compilations of one mushroom cloud and one military effects test in Nevada after another. Very odd.

it was to convey a couple of very blunt, broad ideas. One was that no matter how abhorrent you found the weapon, if you looked at Los Alamos in 1943, you had to say, "That could be me—there's a good chance that I would have been on that next train to Lamy, ready to go up to the Magic Mountain and devote years of my life to making this weapon." And the other blunt, broad idea was to open up the thought that there are unintended consequences of good ideas.

Digging around in storage, I found an opening for *The day after Trinity* that we in fact abandoned, a fossil from twenty-five years ago. So this was our first stab at the first minutes of the film:

> ***Freeman Dyson***
> I never heard Oppenheimer speak about nuclear weapons directly. He avoided the subject, obviously quite deliberately. But I can imagine what that must have meant to him. I felt it myself—the glitter of nuclear weapons. It is irresistible, if you come to them as a scientist, to feel it's there in your hands to release this energy that fuels the stars, to let it do your bidding, to perform these miracles, to lift a million tons of rock into the sky. It is something that gives people an illusion of illimitable power and it is...in some ways...responsible for all of our troubles, I would say. This, what you might call technical arrogance that overcomes people when they see what they can do with their minds. Oppenheimer had this wonderful phrase, "technically sweet," which he used on several occasions to describe what he had done, what other people had done. When something is technically sweet, it is like climbing a great mountain. It beckons you and you climb it, just because it's there. And you only think about the consequences afterwards.
>
> ***Francis Fergusson***
> Well, he had various appearances. When I first knew him, he struck me as a rather odd-looking guy, because he had a habit of walking with his feet turned way out to the side, and he would walk often very fast, leaning forward and spreading these feet.
>
> ***Dorothy McKibbin***
> He seemed taller than he was. He's six feet, I'm sure, six-one, something like that. He was very, very slender and he walked sort of on the balls of his feet and held himself a little stiffly.
>
> ***Francis Fergusson***
> I remember meeting him once, on the street, just after he had been seeing his psychoanalyst. And he was standing on one foot bent sharply to the side, no hat, and his hair standing up in all directions and looking absolutely crazy. He wasn't, though. When I came along, he began talking and walking and we went on just as we had ever since I first knew him.
>
> ***Freeman Dyson***
> He was very thin, nervous, jittery. He constantly moved around, he couldn't sit still for five seconds. You had the impression of somebody who was tremendously ill at ease. He smoked all the time. He looked, probably, much more delicate than he really was. I think he must have been very tough, really, but you had the impression of somebody very frail, even when he was quite young.

Dorothy McKibbin

He had the bluest eyes I've ever seen. They're very blue, clear blue. If you have a seen a flower called gentian, they grow wild here—it was gentian blue.

As I said, we abandoned that. In this version we were setting out to do something like a moral and physical description of the man. In the final version, a story gets set in motion right away. We begin with Haakon Chevalier reading his letter addressed to Oppenheimer after the war, and then we go into the thick of the action right after Hiroshima, and the film unfolds as a chronicle.

Here two very important strands came together and have in fact been with television biography ever since. They have to do with the people that I worked with on *The day after Trinity*. There were two constellations. One group was the historians. It was the first time I had ever worked with historians, in this case Marty Sherwin and Bart Bernstein. On the other side, I was working with a brilliant young editor and screenwriter named David Webb Peoples and his wife Janet. There was a wonderful, fantastic tension between those two camps. At the same time that David Webb Peoples was working on *The day after Trinity*, he was writing his first two screenplays: first of all a Clint Eastwood film called *Unforgiven*, a Western about an old gunfighter who does the wrong thing for the right reasons; and also *Blade runner*. And Bart Bernstein and Marty Sherwin were working on their scholarly research.

I was caught in the middle, and it was a great place to be caught because we public biographers have our own Faustian bargain. We take real, actual people who lived in the world, walked on the earth, and we bend and spin them to fit into existing myths. We bend Oppenheimer to behave like Hamlet. I did a film in which I shamelessly bent Henry Ford to behave like King Lear. So there are templates. And the reason we do it is because it works. When I speak of getting ten million people to watch the film in one night, they're not going to watch if the story isn't moderately recognizable. Now whether or not Robert Oppenheimer fits into the Prometheus algorithm, or the Faust algorithm, or the Hamlet algorithm, or the King Lear algorithm, whether or not Kitty Oppenheimer is Lady Macbeth is in a way beside the point. The point is that we cannot do this honorably without someone to rein us in. We don't have terribly good circuit breakers, especially those of us who work in prime-time television. Ever since then, I have made it my business to solicit the help of scholars to try to get these things right.

Now let me hop ahead. I am currently doing a documentary in which, for the next eighteen months, I am going to follow the composer John Adams and the opera director Peter Sellars. They are doing an opera about Oppenheimer and Teller and the war years at Los Alamos. Interestingly enough, it is commissioned as part of a series by the San Francisco Opera based on the legend of Faust. Without telling you too much, the characters, including Oppenheimer, sing a libretto which is made up of the poetry of Baudelaire and John Donne and the *Bhagavad-Gita*. There are characters who sing from Robert Serber's *Los Alamos primer*, from Szilard's letters, and from War Department memos. Kitty sings the poetry of Muriel Rukeyser, a wonderful American poet from the 1940s. There are several characters who sing parts of unedited transcripts from *The day after Trinity*, out-takes that we did not use.

Their artistic creation is only the latest in a line of what I see as what I see as two diverging and converging public Oppenheimers. The one is the artistic, and the other is the scholarly. As long as those two keep talking to each other, I think we'll be all right.

Peter Galison

I too have worked on these projects with many of you, so I am speaking about things that you know very well. For over twenty years I have been in discussion with Bart Bernstein about the nuclear weapons projects and Oppenheimer in particular; in the course of the film *Ultimate weapon: The H-bomb debate*, with Marty Sherwin and Bart Bernstein, who both appear in it, and with Dan Kevles, who is on the board. For much of my involvement with this first film, I was above all inspired by Jon Else's *The day after Trinity*, which, however modestly he describes it as adapted to work on the television screen, is a remarkably allusive, even poetic exploration of a conflicted physicist in the midst of a catastrophic change in the politics and military posture of the United States and the Soviet Union.

Today I would like to speak about a subject that I have been thinking about for some time. That is: What does film open up that print obscures, what does print open up that film obscures, and how should we think about the process? For quite a number of years I have been teaching a course with Robb Moss, a filmmaker at Harvard. We launch many of our discussions with a contrasting pair—for example, Stephen Hawking's book *A brief history of time*, compared with Errol Morris's treatment of it in film. The book explains very clearly and in short compass how to think about the entropy of black holes. But that kind of explanation, with its diagrams and short-form arguments, would not work well at all in a film.

On the other hand, the film is able to do things that are quite remarkable in the way that it multiplies meanings. For instance, there is a spectacular moment in the film when Hawking is talking about a result he expected to show that under certain circumstances time would go backwards. His student says (approximately), "I kept coming back to Hawking with the results of the calculation and saying, 'No, Dr. Hawking, it won't do that.'" And Hawking keeps replying with each iteration, "Do it again." All of this is said in the visual and auditory presence of Hawking's physical disability, of his wheelchair, struggling over the course of the film with the physical reality of his machine-aided speech. In this visual and emotional context, the push and pull over a "routine" calculation gains a significance both for Hawking's physics and for his life. Most importantly, the simultaneity of meaning can be conveyed without having to lay it out, because, unlike print, the reality of Hawking's physical situation is present at each moment.

The terrific film anthropologist David MacDougall puts this very nicely: when we watch anthropological film, one thing that *doesn't* have to be constantly repeated is that the people we are writing about are *people*. At every moment the humanity of the subject is reinstantiated—in their presence, their gestures, their smiles, in exchanges, even in (subtitled) language where we cannot understand the original. The kind of generalization that this makes—that is, the generalization across cultures that occurs through the specificity of the constant physical presence of things and people—moves in a different way from the generalization and abstraction that take place in writing.

I've had two experiences working on topics very closely tied to Oppenheimer, one in print, one in film. First, I was interested, in collaboration with Bart Bernstein, in trying to understand the flip-flops of Oppenheimer and his colleagues, back and forth, for and against building the hydrogen bomb. Oppenheimer was for it during World War II, or at least willing to countenance it; then against it after Hiroshima; then briefly for it, and then against it, and then for it. That was true for many of the participants. When Bernstein and I wrote an article about this, we did what historians do: we periodized and structured these arguments in such a way that Oppenheimer fit into a wider frame.[3]

H-Bomb Periodization

- Stage I. WW II (1942-45): H-bomb assimilated to A-bomb.
- Stage II. Postwar Lull (1945-47): H-bomb nearly halted, pressure to internationalize.
- Stage III. Cold War (1947-49): H-bomb assimilated to nuclear weapons development; H-bomb medium to low priority.
- Stage IV. Joe 1 (August 1949): Radical polarization of scientific community.
- Stage V. Truman Decision (January 1950): Debate public.
- Stage VI. Teller-Ulam; Test (January 1951 – October 1952): Techno-political inevitability.

This is the kind of thing that can be done in print. It makes an argument about (in this case) periodization crystal clear. As it turns out, Oppenheimer was not at all unique in pitching back and forth—and from this chart we can understand the larger historical context that conditioned these changes in viewpoint. In stage I, the H-bomb was thought of as a "natural" continuation of the A-bomb, and the A-bomb easily assimilated to big conventional bombs like the "blockbuster." In that moment, JRO as well as many of his colleagues simply did not distinguish (morally) between blockbusters, A-bombs, and H-bombs. And so one can go down the list— stage II set nearly everyone against the H-bomb, stage III for it with a mid-level commitment, stage IV precipitated a polarization, et cetera.

Periodization in print is easy to compose once the hard historical work has been done in charts, in sections, in chapters. Easy as it sounds, such a division does not work well *at all* in film. Nor would reorganizing it (unless you do film as a kind of

3. The tables that follow are adapted from Peter Galison and Barton Bernstein, "In any light: Scientists and the decision to build the Superbomb, 1942-54," *Historical studies in the physical and biological sciences, 19:2* (1989), 267-347.

slide show) into a more analytic structure lining up the arguments in synchronic, not diachronic fashion. In this way one could put the pro-H-bomb arguments on the left and the anti-H-bomb arguments on the right. For example, pro-H-bomb advocates would say that the thermonuclear weapon gave more "bang per buck," whereas their opponents claimed that the weapon put resources needed elsewhere out of kilter. Pro: scientists had an intrinsic and political technological imperative to build what they could; Anti: scientists by their expertise had a particular moral obligation to oppose just such technological slippage. Pro: the H-bomb would give the U.S. a strategic advantage; Anti: the U.S. by its geography was *more* vulnerable to the weapon because most of its cities were on the coasts. And so on.

Analytic Structure

Pro-H-Bomb	**Anti-H-Bomb**
Power per Dollar	Resource Allocation
Strategic Asset	Coastal Vulnerability
Diplomatic/Psychological	Moral Standing
Principle Double Effect	Means/Ends
Continuum A-bomb/H-bomb	If A Bad, Why Worsen?
Best Weapon	Overkill—Genocide
Technological Imperative	Scientists' Responsibility

Again this short-form chart can be opened up by print, but it is hard to do without becoming pedantic in film.

Film, on the other hand, can give life to the people and the way they present their views. Just to illustrate this, let me show a short clip:

> ***John Wheeler***
> For ideological reasons, they didn't want to. In the end, we ended up with about thirty-five mostly younger people.
>
> ***Narrator***
> Among this new generation of physicists at Los Alamos is Ted Taylor, whose wife isn't thrilled about being there.
>
> ***Carol Taylor***
> My understanding when we left Berkeley was that he would be working on some kind of research. I was quite naïve; I think he may have been also a little naïve about what he'd be doing.
>
> ***Ted Taylor***
> I was a twenty-four-year-old non-entity on the scale of these big deliberations of conscience and everything else. I had my own immediate matter of ethics to deal with and pretty well disposed of it. I can rationalize, as Carol has said, almost anything that I really want to do. And I found within a few days I really wanted to work on these things. Set aside from people being killed, and people dying of radiation sickness, just the pure, physical thing of big explosions—I've always loved them. You used to take

us to fireworks on the 4th of July. It was your...You pushed that as much as I did. But I've always had the feeling that big bangs are things you just don't like.

Carol Taylor
You don't go to the fireworks to hear the bang.

Ted Taylor
Well, I do. [Laughter.]

Narrator
June 1950: South Korea is invaded by the North. The threat of a wider war, one perhaps fought with atomic weapons, dramatically intensifies the sense of urgency at Los Alamos. But the added pressure does not change the fact that no one knows how to make Teller's vision for a superbomb a reality.

Herbert York
I remember sitting in Ernest Lawrence's office, and Lawrence asked Teller how things were going, and Teller said that, "Well, we really still don't know how to make one." At which point Lawrence said, "Well, where's the progress? You didn't know that six months ago." And Teller said, "Oh, yes, but now we don't know it on much better grounds."

Ted Taylor
One would see, walking down the hall, John Reitz, or especially Marshall Rosenbluth, who was not only a very good physicist, but tended to be very innovative, just sitting there, thinking. What are you thinking about? "Well, how we can make this Super work?"

Emil Konopinski
The essential idea wasn't coming along, and as far as we knew, would never come along.

Narrator
Among those working on the problem is a Polish émigré mathematician named Stanislaw Ulam. To young Americans like the Taylors, Ulam and his wife Françoise are the very picture of a cultured European couple.

Carol Taylor
The music and the books and the way of doing things and the way of talking and the courtesy to each other—they had what seemed to be a civilized life, the two of them.

Ted Taylor
I found myself really spellbound by him. He had fun with his mind.

Narrator
Ulam invents a complex mathematical technique to study Teller's notion of the H-bomb. With it, he discovers fundamental flaws in the design.

Hans Bethe
I was at Los Alamos in January of '51 and found great depression in the whole laboratory. The most depressed person was Edward Teller himself. He was really desperate.

Narrator
That desperation lingers through the winter. Then Stan Ulam has a moment of inspiration.

Françoise Ulam
At lunch one day, he looked very preoccupied; and he was staring out of the window and when I came home from work, found him there and he said, I think I've found a way to make the Super work. And that rather appalled me.

Peter Galison
Maybe it's the somewhat separate, somewhat distanced position that Ulam takes relative to everybody who's in there struggling to build this weapon that gives him just that separation necessary to rethink the weapon in an entirely different way.

Narrator
Ulam's idea unleashes a wave of creativity from Teller.

Peter Galison
Together they begin to start to work on what they call their two-bomb paper, in which they develop a notion of how a hydrogen bomb could be built, utterly different from anything they've ever considered over the previous nine years.

Carson Mark
It looked as if it would work. And in that respect it was completely different from the classical Super, which might work, but how could you tell? It was going to take an awful lot of work to find out. This thing looked as if it ought to work if you did it right.

Narrator
The H-bomb team turns to some of the world's first electronic computers to confirm that the Teller-Ulam scheme will work—their self-imposed deadline, a GAC meeting, just weeks away.

John Wheeler
These young people that I had recruited were working, sometimes 36 hours at a stretch, to get out a calculation.

Ken Ford
What we were calculating was the burning process, obviously I can't go into detail. We wanted to show that the thermonuclear fuel, once ignited, would continue to burn. And by the time I finished this job, the GAC meeting at the Institute had already

gotten underway. It was, of course, highly classified. I was not authorized to enter the room in which the meeting was taking place, so I signaled to Professor Wheeler from a window, it was a first floor meeting room. He came over to the window, opened the window. I passed the graphs through to him.

John Wheeler
And I can recall the look on people's faces when they saw this convincing evidence that the thing would work.

Narrator
To Robert Oppenheimer and the GAC, it was now clear: the H-bomb puzzle would surely be solved by both sides. The simple, elegant Teller-Ulam solution had trumped their ethical objections.

Cyril Smith
We couldn't as a technical advisory committee *not* vote for it at that stage. On the other hand, I resigned from the committee not much afterwards, because I wanted to return to my university research, so I said. I've been rather ashamed of myself that I didn't publicly say why I resigned, which was just that I wanted nothing whatever to do with this kind of business.

The goal of both the written work and the film work, for myself and for my collaborators, was to do something complementary to the biographical and to embed Oppenheimer into a wider set of developments. On the written side, we did this through integration into a synchronic and diachronic understanding—diachronically of the periodization of the larger political-military developments of the time, and synchronically in terms of the argument structure. On the filmic side, we were able to use people who were not just the principals, to bring this away from the death struggle between Oppenheimer and Teller into a wider sphere of characters. In film this cannot be handled with a cast of thousands, as you can do much more easily in print with quotations from one person after another. Rather, we could take people like Ted Taylor and his wife and use them as a way of alluding to a larger set of people and developments, giving them a kind of humanity and engagement that comes out in film much more than it does in written form.

Discussion

Jon Else

Peter, in my history of documentary class, I use that clip of Taylor and his wife, together with a similar clip of Robert Wilson and his wife Jane with the same sort of back-and-forth. If you were to transcribe the conversation and put it on the page, it might be a bit flat, not quite as rich. It's the non-verbal stuff that gives it the oompf, the weird sort of moral whistling in the cemetery.

Peter Galison

Yes, I thought of that scene with the Wilsons from *The day after Trinity*—especially when Carol Taylor turns to her husband with both a look of both affection and complete incomprehension at what he's saying about fireworks and nuclear bombs, at words that are so discordant with what she understands of the situation. And he carries on with this rationalizing —after telling us how he can rationalize anything—with this fantastical, bizarre story about the fireworks, as if fireworks and nuclear weapons were an easy slide one to the other.

In your film, Jon, it's a long time before we actually see Oppenheimer; the film opens with people talking about him. It's almost as if you created an Oppenheimer in relief, a kind of negative space, by all the different ways that people describe him—physically, his gestures, his background; the narrator tells us about the conundrum of this poetically-inclined young man who ends up building the most destructive military weapon. In different ways, both of us had to face a problem: we had many other people who knew Oppenheimer, but Oppenheimer himself was dead, so there wasn't the possibility of interviewing him. The question is how to handle the absence of somebody in film. When you were making the film, how did you think about that absence and how you were going to handle it?

Jon Else

I knew so little about Oppenheimer, I actually thought I was going to be making a film about a man who was alive. It's an interesting litmus test for the public Oppenheimer. I undertook this project in '76-'77, and I had been pretty well-educated at some high-priced universities and some good public universities (including this one). And I really knew nothing at all. I think I had seen the cover of Heinar Kipphardt's play *In the matter of J. Robert Oppenheimer*. I knew it was about a man who had something to do with the atomic bomb, and there was a woman involved, and that was about all I knew. I didn't know he was dead. Then I sat down to begin writing a concept, and when I did the research I remember stumbling across the *New York Times* obituary and thinking, "Awww..."—because I thought I was going to interview him; I thought I was going to be able to ask him, was he a member of the Party?

The issue of negative space is interesting. In film, probably more than in print, materializing and evoking the presence of something is easy. But evoking the absence of something is difficult. In *The day after Trinity*, Oppenheimer is absent from the screen for a very long time. You don't see him in live action, actually speaking a word, until after Hiroshima, in fact. And there are two reasons for this. One is a sort of deal with the devil: there is a long-established tradition that works well in Hollywood films, having to do with method actors like Marlon Brando. Think of any Marlon Brando film: you don't see Brando; you hear about him, you learn about him from a distance. The most obvious example is *Apocalypse now*, in which you spend an hour and twenty minutes going up the river looking at Kurtz's dossier before you finally meet him. *Apocalypse now* was in fact released right around the time we were working. We had all read *Heart of darkness*. So there was a conscious storytelling impetus to build mystery around the man.

The other reason was simply a practical concern. I did not want to use Oppenheimer too much out of chronology. And there is no live action footage of Oppenheimer available until post-Hiroshima. In fact, there was some footage shot before the war with Lawrence and Oppenheimer at Lawrence's cyclotron up on the Hill here. I went and pulled it out of a vault in New York in '77; it had turned to cornflakes, and it was gone. That was it—the original negative in my hands, the only footage of Lawrence and Oppenheimer together.

So Oppenheimer's absence was partly a practical concern and partly a storytelling trick. That's also why Oppenheimer's famous remembered line, reciting from the *Bhagavad-Gita*, comes at the very end of the film, and not in the middle, at the Trinity test. It has purely to do with storytelling.

Peter, let me ask you a question. You teach the history of science, and you work with Robb Moss, a filmmaker. One thing struck me when reading Richard Rhodes's book *The making of the atomic bomb*: in these films we don't do the science very well. Somehow that is impervious to us. I'm curious if you think it's possible to do better. In this film, you do stay away from the science to a good extent, just as I do in *Trinity*. Can it be done?

Peter Galison

I think there are two issues at stake. I'm optimistic about it, if it's done in a particular way. As much as I admire *Nova* for what it's done over the years, that's not what I'm thinking of. In fact, one of the things that Robb Moss and I have done as an experiment in our class is to have each student go into a laboratory and start by just filming a single process, the equivalent of making a pot in visual anthropology: classifying a plant that comes to the botanical gardens, or looking at a particular run of a physics experiment, or getting someone set for tumor treatment at a medical cyclotron. The key is then to build out from the process. Starting with process is not the way that science film and science documentary has classically been structured. A typical science-popularizing film generally starts with the result and asks, "How do we make this result easily accessible?" But the more interesting parts of history of science, over the last fifteen or twenty years, have been emphasizing the sets of practices that are constitutive of the result, rather than starting with final results and making them simple.

The other issue, and one very close to my motivations for making this film, is that we do not have a good forum for public debate about the moral-ethical dimension of scientific issues. What interested me about the building of the hydrogen bomb—unlike the atomic bomb—was that, because the debate occurred in the Cold War, and not in the secrecy of World War II, the thermonuclear weapon was the subject of intense moral scrutiny before the weapon even existed. It seemed to me worth trying as a kind of existence proof: trying to make a film that would bring the moral and political issues front and center, if not handle the details of radiation implosion.[4] But

4. In fact, we talked to the people at *Nova* about this. They were excited by the possibilities: "We'll do the animation of radiation implosion, and we'll show the rays coming out from the

I also wanted to do it in a way that didn't simply make Oppenheimer into the fallen hero and Teller the villain, or the inverse. Rather, the goal of the film was also to give an account of the conflict so that both sides were truly compelling. I wanted to make vivid the position of Lawrence, Nitze, Teller, and others—to show that they were as worried that the end of the world would come by *not building* the H-bomb as Bethe, Conant, and Oppenheimer were that the end of the world would come by *building* it. So I wanted to engender a kind of conundrum for people. We don't have this for any of the big issues that now face us—for instance, in the European debates over genetically modified crops or the American one about stem cell research. Television or film could handle such issues, but we're not set up to do it through the standard exposition techniques of popular science.

Jon Else

We could do it if we get away from films on television. The issues that you describe require a tension. They require people to actually sit for more than—what is it between commercials?—seven minutes at a time, without the phone ringing, the girlfriend coming to the door, the kids crying. And where is that happening? In theaters. The example before us right now is Errol Morris's film *The fog of war*, a fascinating biography of former Secretary of Defense Robert McNamara. The film involves a lot of the same moral and policy and philosophical issues that revolve around Oppenheimer. It was very deliberately made to be shown in theaters. It is also a film in which Errol Morris, to his everlasting credit, did not succumb to the temptation to remove the complexity from McNamara's life, his positions, and his role in the Vietnam War. And it's a film in which you pay your seven dollars and sit down and watch it, uninterrupted, in the dark for the whole time.

I've chosen to work in prime-time television, but in fact I think prime-time television is part of the problem here. It's becoming increasingly difficult to work in because of the pace of imagery: the cascade of images and sounds that we can't just seem to free ourselves from. The question is who would pay for a feature documentary film about the debate over the hydrogen bomb. If you have some investors, let me know.

Questions

John Heilbron

To follow up on this famous line you mentioned, Jon, from the *Bhagavad-Gita*, which comes at the end of your film: I once heard Frank Oppenheimer on this subject. Somebody with due piety asked whether JRO quoted it at the Trinity test. Frank said,

---

atomic bomb and causing the plasma, and...." My fear was that animation of this explanatory type would suggest a purely scientific triumph, a technical solution that would get away from issues I wanted to spotlight.

"That's not what he said at all. What did he say? 'Thank God it worked.'" It seems to me that the complexity of character that these two different utterances suggest is very difficult to bring out with the help of people who knew him, especially if the standard story is that of a refined, ascetic man somehow in the wrong pew. So even if he did revert to the Hindu scriptures, it would probably be a setup of the kind to which his posing often brought him.

Thus my question: is it possible at all, if the hero or victim of the piece is not living, to obtain the richness of character that might be wanted in a film for presentation in a theater, or in a play in which the character is allowed to develop at some length? As an example, we might take Michael Frayn's recent play *Copenhagen*, portraying a conversation among the ghosts of Niels and Margrethe Bohr and Werner Heisenberg.

Jon Else

In *The day after Trinity* we did wrestle with the issue. After I finished the film, I swore I would never do another one about a dead person—and then I did half a dozen more. The grammar of biography and the grammar of film change over the years. In a way, I am even re-doing *The day after Trinity* right now, making a film about two men who are cooking up a, to great degree, fictionalized Robert Oppenheimer; they are going to give life to their own creation.

If we were to do *The day after Trinity* now, we would probably address that issue directly. We would probably more closely intercut Frank's recollection, which does appear in the film and which is clearly in opposition to Robert's recollection. So it would be very clear to a lay audience: wait a minute, something's fishy here—somebody is writing his own script, or maybe somebody is unwriting someone else's script. As the film is now constructed, those two recollections of the moment after Trinity are widely separated. A lot of books now raise the question: just when did the lightbulb go on for Oppenheimer that the bomb was like Krishna as he rose up in his multi-armed form?

Finally, on the question of giving life to people who are not with us, the answer has a lot to do with how much material there is. Henry Ford, for instance, was easy, because Henry Ford had a private motion picture company of his own that filmed his every moment.

Dan Kevles

How is it different to make a film about a group of people, *some* of whom are dead, compared to a film about a group of people, *all* of whom are dead? Much historical documentary on television is about the latter group—the Civil War, for example. So is there something special about the enterprise that the two of you are engaged in?

What you are doing, as an aside, is a very valuable thing to do. All of us who work in recent or contemporary history are doing so partly to write an account with an interpretation of the events—but we do it also to provide material for the future. That is because future historians won't be able to interview the people now alive. We can: we record it and make it available, in the context of what we are doing. But in the long run, what we are writing about becomes, so to speak, the history of the dead.

Then, with regard to the particular issues you just raised, is there something special about doing contemporary history on film, as distinct from doing contemporary history in print?

Peter Galison

This question and John Heilbron's question are closely related. When I started the film, I thought that it would use only the actors—not fictional actors, but people who were themselves participants. I had the idea of no match cuts, no privileged views, long shots—to try and make it in a *vérité* kind of style, which was supposed to be the style that wasn't a style. But when we started to put it together, we found that it felt hermetic, enclosed, airless. There was also, on the flip side, the problem of being able to step out of the participants' collective memory, which is the danger that historians who do oral history have known about forever.

So we decided to have a narrator, who could also set scenes and talk about the larger historical frame. But it still felt that there was a gap. And there were interpretive moves that I didn't just want to ascribe to the omniscient narrator. That was when we got Bart Bernstein and Marty Sherwin and myself to speak in the film, to take responsibility for our positions. So four things are used: archival material, a narrator, the actors, and historians.

Of course, the balance changes as the actors die out; but they were very much a part of my interest in making the film. Pam Hogan and I started the project in 1988, and it took twelve years. And already along the way, Luis Alvarez, whom I had figured on filming and talked to a great deal, had lost muscle control over part of his face; and he didn't want to be seen that way, so he didn't appear in the film. Likewise, I.I. Rabi was dying—we did film him, but his voice was not clear enough.[5] That said, when you are filming a close-knit group of people—as both Jon and I were—there is something that emerges from the juxtaposition of voices and memories that goes beyond isolated interviews.

Jon Else

In a perfect world, we would make films entirely with living people in them. Then the next best thing to a living person is someone who knew the living person, who can talk about the living person. We're also acutely aware that our filmed interviews become part of the record, including the out-takes that never end up in the film. For instance, for *Eyes on the prize*, all of the material is available and archived at Washington University for anyone doing research on the civil rights movement. A lot of the material for *The day after Trinity* is here in the Bancroft Library at UC Berkeley.

5. Part of my experience, having to make all possible mistakes starting in film, was to see how different a filmed interview was from an oral history interview, as I was used to in doing contemporary history. I did an interview with one of the pilots who found the evidence for the Russian bomb. I got back and my co-producer, Pam Hogan, said, "We're not going to be able to use this interview—the refrigerator switches on during it." The refrigerator! So I learned to do refrigerators.

And while we don't do our work with the precise methodology of historians, we are aware that our material does become part of the record. That is one of the arguments in favor of finding the real people.

You mentioned the Civil War. We can't learn anything from President Lincoln's facial expressions in the Civil War. But we learn a lot from Ted Taylor's facial expressions and his body language with his wife in Peter's film. I'd rather make that film, if I could, than make a film about Ted Taylor a hundred years from now.

Jessica Wang

It's somewhat ironic, isn't it, that of the two of you it's the historian who is uncomfortable with an omniscient narrator, who feels the need to bring in talking heads and present different viewpoints. When we historians do written narrative, there's always an omniscient narrator who rarely puts him- or herself forward—that is, the writer. Yet that approach is much more like what Jon Else did in *The day after Trinity*. In *The day after Trinity* you have a narrator at points, but you don't use talking heads. Instead you concentrate on interviews and getting people's inner states in their own words.

I was curious about the decision whether or not to bring in the talking heads. What do you think are the drawbacks and virtues of that approach to historical documentary?

Peter Galison

I would not say it prescriptively: there's not a rule or a better or worse. Of course, it's painful to see myself onscreen; but that aside, seeing Bart Bernstein and Marty Sherwin is a pleasure—because they have thought a lot about the matter, and their interventions give a sense of that depth. And it gains from their physical presence, as opposed to its translation back into narrative lines. The original idea was actually that we would integrate the historians into the background narrative. But for all the same reasons that the participants are interesting to see, the historians who have been thinking about it for a long time can express a lot, too. So I think it's an option to give another kind of dimensionality to the storytelling. In the balance of things, I found it helpful for this particular film; but I don't think that that's necessarily a stricture for other films, or that I'd do it again in the future.

Charles Thorpe

It's interesting that Jon said his first encounter with Oppenheimer was Heinar Kipphardt's play. To what extent has our public Oppenheimer, the Oppenheimer most people know, even the Oppenheimer that has influenced historical accounts, been created by Kipphardt's play and Robert Jungk's book *Brighter than a thousand suns*? I believe Jungk was in fact the main influence on Kipphardt and the way Kipphardt framed his story.

As for the quotation from the *Bhagavad-Gita*, I think it is Jungk's book that made that quote into the iconic Oppenheimer quote. Oppenheimer mentioned it in a *Time* magazine interview in 1948. But it did not become iconic at that point; James Hijiya goes into this in his article "The *Gita* of J. Robert Oppenheimer." My understanding is that very little was made of the quote until Jungk took it up and made it central, and then Kipphardt arranged his play around the Faustian story that Jungk developed. And that really set the model for the public image of Oppenheimer.

Jon Else

I agree. Is that the earliest public utterance of that quote, '48? That's three and a half years after the fact.

John Heilbron

Plenty of time to compose it.

Marty Sherwin

About fifteen years ago, I started to teach a course, "The camera and the Cold War: Documentary films as history." Of course, one of the reasons for this initiative was my experience as an advisor to *The day after Trinity*. But the main reason I began to use documentary films in class was an experience I had at WGBH, the public television station in Boston. I was an advisor to the film series *America in the nuclear age*. I had a debate with the executive producer over how to present the decision to drop the bomb, and I said, "You can't do it that way. Look at all this new information, this document and this document and this document...." He said, "Marty, there are a thousand ways to interpret this. One is as good as the other. We're doing it this way." I went away fuming because, if there's anything that all historians can agree on, it is that there may be numerous ways to interpret any issue, but one is definitely not as good as the other. So I began to stew about this conversation. A few years later, I decided that documentary films were too important to be left to filmmakers and that historians had to take films seriously. So I changed my lecture course on the Cold War into a course that compared history on film with written history.

There is something missing from our discussion: we're talking about documentary films as if there was just one kind of documentary film. But there are many types. If you have seen *Atomic café*, for instance, you will see immediately that it is different from anything else you've seen. It is a historian's film in that it is based completely on archival footage. It has no voice-over and no "talking heads." It is an extraordinary and innovative film with a strong interpretive message. Then there is the mixed-element film that Peter made, with contemporaries as well as historians discussing the hydrogen bomb decision. In Jon's film, people who knew Oppenheimer discuss him interlaced with contemporary footage, but no historians appear. And then there are films like the ABC documentary on the decision to drop the atomic bomb, *Hiroshima: Why the bomb was dropped*, which features historians discussing the event, interlaced with contemporary footage and a strong narrative voice-over (Peter Jennings) that guides the viewer through the issue.

We have different kinds of books, we have different kinds of histories, and we have different kind of documentary films. In a classroom setting, the important thing, I've found, is to put all of these together in the context of studying a particular subject. But it is absolutely critical to never leave out the written history. That is because students have to understand the relationship between what they learn from the film and what they learn from their books. This is a topic that this panel has taken up: what is the difference between visual and written history? One very important difference, which has shaped the way I teach history in any environment, is that film conveys the emotional content of history. You can never get this emotion from history books. As hard as we may try, we cannot recreate the same sense of the human experience as writers as can be conveyed in a well-made film. Written history is obviously better for analysis. But history on film is a superior medium for conveying how history felt.

Jon Else

A wonderful filmmaker on this campus, Marlon Riggs, who passed away several years ago, used to describe his documentary filmmaking method as "promiscuous." It's an excellent term, and I've adopted that method: do whatever it takes to get the story told in the best possible way. If it takes historians onscreen, great: I love historians. If it takes no historians, if it takes all cinema *vérité* footage, if it takes animation, if it takes cartoons, if it takes ephemeral flat art, great. Let's do whatever it takes.

If it appears to you (and to me, as it often does) that a particular "cookie-cutter" kind of historical documentary has been canonized in the last decades, that is true. It comes from the tyranny of the funding mechanisms within public television. That is another discussion for a different time. But it's not entirely our choice.

Keay Davidson

In *The fog of war*, Errol Morris does something very unusual and controversial: he tries to draw "lessons learned" from the life of McNamara ("empathize with your enemy," "maximize efficiency," and so forth). Since you are chroniclers of Oppenheimer and nuclear war via film, do you think that that approach, attempting to draw broader lessons learned, is possible or desirable for Oppenheimer?

Jon Else

I would say two things. First, if there were not lessons to be drawn here, we wouldn't be making these films. They are so hard to make, and they ruin your life. If I didn't think someone was going to learn something from it, I wouldn't be doing it.

Second, the explicit chapterized lessons in *Fog of war* are in fact parallel to explicit chapterized lessons in McNamara's book. The lessons onscreen in *Fog of war* are Errol Morris's list of lessons; they are not Robert McNamara's list of lessons; and in fact McNamara does not like the lessons that are there.

For myself, I wouldn't go so far as Errol Morris did in manifesting those lessons; I would somewhat shy away from it. I think he did a good job of it, but I wouldn't have a crack at it.

Peter Galison

What I find most interesting are not so much lessons learned, as conflicts encountered. In Jon's film, the conflicts stem from the biographical conflicts he mentioned:

What are the different directions in which Oppenheimer is torn? That is how the film is set up and pursued throughout. In my film it was more the political conflict that I wanted to bring out. I wanted to give enough weight to both sides so that you could feel it to be divisive in yourself, so that it would keep you from sailing too smoothly, too easily into saying, "The obvious answer is X." So, yes, I want people to learn something; but the thing learned, I hope, would be a conflict, rather than an after-action report à la McNamara.

Sam Schweber

Jon, you made two very striking statements: the mushroom cloud trumps everything, and there is no public Oppenheimer. The former doesn't quite explain the latter. From your perspective, why isn't there a public Oppenheimer?

Jon Else

What I meant by saying that there was no public Oppenheimer was that there wasn't one in the sense of our almost cartoonish, one-dimensional public McCarthy—or to some degree a cartoonish public Teller. There are a lot of reasons for it. Among others, there is the silly, simple reason that Oppenheimer is not much to look at. It may sound trivial, but it has a lot to do with making images in our culture.

My sense, though, is that the overriding reason we don't have a public Oppenheimer is that you cannot fit him into meaning one thing. McCarthyism is a shorthand for just one thing. But the very fact that Oppenheimer is such a wonderful bundle of contradictions and conundrums is what makes it very hard for us to have a single public image. If we're going to be crass, let's think about Madonna: Madonna reinvents her own, very public image about every three years. If you took all of those Madonnas and had them existing simultaneously, then that is the problem we have with Oppenheimer.

Peter Galison

There's also the fact that in a certain sense McCarthy won and Oppenheimer lost. The story we like to tell is that McCarthyism was defeated at the Army-McCarthy hearings and all bad things had a good end. I'm not sure that's true. If Oppenheimer's legacy had been that of the hero, the number one atomic thinker, as it says on the cover of *Life* magazine, and not the man whose security clearance was revoked at the hearing, it could have looked differently to us. And so we are not just in the shadow of the bomb, but in the shadow of the AEC hearing and all that it represented about his exclusion from the canon of importance and distinction.

In Europe, of course, he does play a different role. Foucault, as you know, asks what it is that will replace the dreaded Zola-to-Sartre generalized intellectual. He says that it is the specific intellectual, and Oppenheimer is his name.

Mitchell Ash

Except for rather vague, allusive references, we haven't yet discussed audience and how it plays into the "how" of making films. One aspect of this is demographics. For whom are these films being made? In the history of science popularization, we know that there is no such thing as a single public. That was already true in the nineteenth century, and yet people persisted in speaking in an undifferentiated way of "we" and "they," as though there were such a single public.

A second aspect is cultures. Although Oppenheimer is an American topic, Peter's allusions have encouraged me to think about cultural differences, while Jon referred to Heinar Kipphardt, and Charles Thorpe made the suggestion that Jungk and Kipphardt were the initiators, in some respects, of the public Oppenheimer. If you look at historical documentary films in Germany and Austria, they are made differently and presuppose different kinds of viewers. Pacing is a big issue here: documentary films (both European and American) always seem to be slower; and yet in the German and Austrian context, the slowness is less egregiously obvious because even regular entertainment programming is slower.

Jon Else

I have a simple answer to the question of the demographic. We talk about high school AP history classes as our audience.

Jim Hershberg

I'm wrestling with the issue of print versus film for dealing with Oppenheimer and some other documentary projects on Cold War history. One reason why film Oppenheimers have been so unsatisfactory—I'm thinking of the BBC/PBS miniseries *Oppenheimer* in 1980 and *Fat Man and Little Boy*—is that they have attempted to represent the extremely articulate, complex Oppenheimer through occasional delphic utterances, long silences, and the physicality of the gestures, the pipe, the hat.

Jon, you spoke of the difficulties of conveying Oppenheimer in a documentary (though I think you did a remarkable job). What do you think of the more fictional possibilities? Are there ways to get around the problems? There is Kipphardt's play, but there are also these film versions. What about something like Michael Frayn's play *Copenhagen*?

Jon Else

I think Sellars and Adams's opera *Doctor Atomic* will bring to the stage a strange and powerful Oppenheimer like nothing we've seen before, but we won't know for sure until a year from now. (I think parts of it may be like *Copenhagen* on acid or on a lot of caffeine—at least I hope so.) In film so far we've had some less-than-Faustian, somewhat smaller-than-life Oppenheimers—in *Fat Man and Little Boy,* and in the BBC series. It's partly because the scripts have been a bit pedestrian, and partly because we haven't yet had a really great actor in the role. (I'm waiting for Johnny Depp to turn 39.) I'm thinking of someone on the scale of Jack Nicholson or Marlon Brando—because you need a great actor as well as a great script. And the difficulty goes back to the same reasons why he doesn't fit into an easy iconic slot: there's too much there, and no one has been able to handle it. I don't know how we get documentaries to go further. We just need more money and more time and more material.

There's another irony with Oppenheimer. We talk about his being so charismatic and bristling with attraction. But if you look at the real Oppenheimer on film—of which there's not very much, mainly newsreels from '46 to '52 or '53—he's just not very charismatic on film, suprisingly. He's not very interesting. The affect is sort of flat. So newsreel footage is obviously not the answer. And that is one of the reasons we kept him off screen, by the way. He wasn't quite ready for prime time.

Peter Galison

In this context I think not so much of Michael Frayn's *Copenhagen*, but of Tom Stoppard's *Arcadia*. Stoppard's play has a very dense, complex reasoning that nonetheless moves along and has a kind of poetic quality to it. If I were fantasizing about a model for a fictional Oppenheimer, one that would be a big step up from Kipphardt, it would be a Stoppard character of that caliber.

Bart Bernstein

Jon, I've watched your film probably about twenty-four times now. What has struck me more and more is how Freeman Dyson, if anyone, gets to carry the interpretive line. He is presented in the guise of a man who knew Oppenheimer. Of course, we know (offscreen more than onscreen) that he knew Oppenheimer *after* Trinity. And yet it's he who gets to promulgate the Faustian bargain. When you put the film together, I was so wonderfully naïve in multiple ways that I never really focused upon this. But over the years, I've come to wonder: if you were doing it again, would you give Dyson the role of insider, observer, omniscient interpreter—I think he has all these roles, for he gets to be what in Peter's film is Peter, Marty, myself, and all the other folks. Or would you play it another way?

Jon Else

I've thought a lot about this, too. Dyson is the only outsider from Los Alamos in the film. And I knew that when we interviewed him. I had just read his book *Disturbing the universe* and was really taken with it. If I were to do it all over again, I probably would keep Dyson in his role in the film. At the same time, I would turn cartwheels to make it clear that he was not at Los Alamos, that he is speaking as the one person in that film who has more distance than the others. So I'll stick by the decision to keep him in talking about the Faustian bargain. It's one of many bargains that Oppenheimer made, and to make the film without it would not have been a good idea.

Alexei Kojevnikov

In some sense, Dyson is also trying to impersonate Oppenheimer: Dyson is trying to fill the space of the public intellectual after Oppenheimer stepped off the scene.

Jon Else

I will say that sitting across from Dyson at his office at the Institute for Advanced

Study, I felt the closest that I ever did to being in the presence of someone who was as smart as Oppenheimer. I remember literally getting white knuckles hanging onto the chair as he spoke. I was young; we were all much, much younger, and it was my first film. I had never done a biography before, and I'd never been in the presence of these sorts of people.

Peter Galison

I've also looked at your film many times over the years, and it's interesting how it changes from one time to the next. Now when I see Dyson, I also see him very much in terms of his own conflicts—his own war work in World War II and how he thought about it, his own Faustian bargain. You mention that your film came out of thinking about World War II and Vietnam. The film that I made was very much in the light of the late '80s and the big escalation in nuclear armaments then. And *The fog of war* is being viewed by everyone in light of Iraq, thinking about that conflict and precision weapons and munitions and so on. Film isn't different from print in that way. But it does mean that we're constantly reinterpreting and rewriting our understanding, not only of the film's subject matter, but of the witnesses as well.

Jon Else

And these things that we've made, whether they're print or film, themselves become documents. They become snapshots of how we viewed history in 1977.

AFTERWORD

David A. Hollinger

Whatever else J. Robert Oppenheimer may have been, he was a uniquely prominent site for the emergence and often violent action of forces that largely defined the historical epoch in which he lived. We usually do not speak of a person as a "site." We use this word for geographic locations, such as "Trinity Site," where the first atomic bomb was exploded on July 16, 1945. To speak of a person as a "site" might seem to deny agency, individuality, or dignity. But not necessarily. To speak of an individual human being as a site can help us to recognize that some individuals compel our attention because of the magnitude of what happens to them as well as the magnitude of what they accomplish. Our attention is compelled all the more if the individual registers deep awareness of his dual role as the recipient and the agent of multiple and portentous fates, as Oppenheimer's eyes seem to do in many photographs, and in the Ben Shahn drawing reproduced on the cover of this book.

Oppenheimer was one of very few people to have experienced, with extraordinary intensity, "both the illusion of being history's conqueror and the fact of being its victim," wrote Murray Kempton in one of the most sensitive essays ever written about this commanding but enigmatic presence in our recent past.[1] Kempton, like most of those who have tried to assess Oppenheimer's life in the round, focuses first on the triumph of building the bomb and its resulting prestige and authority, and second on what is often called the tragedy of destruction at the hands of the government Oppenheimer had served. Yet Kempton also confronts a host of his subject's decisions in relation to the forces that swirled around him. He does this through the medium of a conversation that reportedly took place between Oppenheimer and a visitor shortly after his public humiliation by the Atomic Energy Commission. Kempton's searching meditation on Oppenheimer's mixture of agency and passivity is worth following for a moment. "What bothers me," the battered Oppenheimer declared to his visitor, "is the complicity."

1. Murray Kempton, "The ambivalence of J. Robert Oppenheimer," Kempton, *Rebellions, perversities, and main events* (New York, 1994), 118-135, esp. 134. This essay was first published in *Esquire* in December 1983.

But where, the visitor wondered, did the complicity lie? Complicity in having been closer to Stalinism than he later wished he had been? Complicity in his own humiliation by Strauss? Complicity in the development of an arms race? Kempton speculates about a number of the possibilities:

> Was it in making the atomic bomb or resisting the hydrogen bomb, in denouncing Bernard Peters or defending Bernard Peters, in leaving Jean Tatlock or returning to her for a few skimpy hours in the night, in hurting Haakon Chevalier or trying to help him, in serving his government too unquestioningly or not questioningly enough—in all the mess we make of life and life makes of us, just where do we locate and house the complicity?

Where did the complicity lie? As Kempton tells the story, Oppenheimer answered his visitor: "In all of it," he said.[2]

Being "complicit" in so many happenings, and not being able to sort out with conviction where one's decisions had been honorable or wise, and where not, makes sense if we do indeed approach Oppenheimer's life as a field on which a variety of historic trajectories confronted one another and sometimes collided. Some of these trajectories obviously derived from the weapons-relevant discoveries of physicists and from the interests of the military establishment of the United States in the early Cold War, but others derived from certainties and uncertainties about Stalin and Hitler, and still others derived from the escalating violence of the Pacific War that created the immediate context for Oppenheimer's bomb-related decisions in the late spring and summer of 1945.

Oppenheimer was the point-person, so to speak, for his society in dealing with the rapidly appearing sequence of novelties now central to what historians study when they address the 1940s and 1950s. If Oppenheimer the person is hard to find in this welter of events, it is partly because his personality was articulated in a dialectic between an inventory of traits he brought to government service in 1942 and the political and intellectual objects of varying shapes and textures and weights thrown at him with different velocities by Leslie Groves, Hans Bethe, Lewis Strauss, Edward Teller, James Conant, Haakon Chevalier, Ernest Lawrence, Harry Truman, and a host of other contemporaries who were also, like Oppenheimer himself, agents as well as recipients of fate. What is so special about Oppenheimer, then, is the concentration in a single individual's life of so many of his era's most important pressures and novelties.

Only rarely does this central truth about Oppenheimer surface in *Doctor Atomic*, the John Adams opera organized around the bomb test at Trinity Site. But understanding this truth can make us appreciate all the more that opera's most powerful moment. At the end of the first act Adams's Oppenheimer is drawn back and forth, like a yo-yo, from the bomb to the front of the stage and back again, while singing

2. Ibid., 134.

John Donne's words, "Batter my heart, three-person'd God."[3] In his attraction to that sonnet we can sense Oppenheimer's yearning for some authority outside himself to take control, to assure him that things are all right. Oppenheimer was not a religious believer, and certainly not a Christian, but he was greatly attracted to this poem, first called to his attention not by some devout friend, but by his Communist sometime lover, Jean Tatlock. The poem is about grace.

"Holy Sonnet XIV" is spoken in the voice of a lost soul pleading with the deity to grant him the power to believe, to cleave to God. You "as yet but knocke," Donne's lost soul complains to God, rather than apply the "force" that can "breake, blowe, burn and make me new." Oppenheimer knew he did not have such an outside authority: he knew he was on his own in the force fields of the 1940s and 1950s. He was battered, but not by the God of John Donne. But his well-known affection for this verse, now rendered with unforgettable ferocity in *Doctor Atomic*, enables us to grasp Oppenheimer's predicament. He found himself the site for the engagement of a complex of human elements spinning in and out of one another, just as the expanse of New Mexico desert he named Trinity was a site for a complex of physical elements actually exploding.

A Hindu scripture Oppenheimer quoted in translation two days before the Trinity test bespeaks this same yearning, exactly as it invokes the battering by earthly forces and admits to deep self-doubt.

> In battle, in the forest, at the precipice in the mountains
> On the dark great sea, in the midst of javelins and arrows,
> In sleep, in confusion, in the depths of shame,
> The good deeds a man has done before defend him.

Here, in lines Oppenheimer actually recited to Vannevar Bush,[4] and addressed within this volume by John Heilbron, we can see the builder of the bomb assaying with a measure of hope his own prospects in the uncertainties of a post-bomb world.

Oppenheimer's flashes of self-knowledge about his predicament contribute to our ongoing engagement with him, and our determination, it would seem, to treat him as our moral and political representative within the historic domains in which he was an actor. I attribute this determination to Oppenheimer's appraisers because so much of the literature on Oppenheimer during the entire half-century since his humiliation by the Atomic Energy Commission has displayed a special, earnest eagerness to hold him accountable, to decide just how well he met the multiple, successive challenges he faced.

3. John Adams, *Doctor Atomic*, was premiered by the San Francisco Opera on October 1, 2005. John Donne's "Holy Sonnet XIV," written in 1633, is widely available, e.g. in M.H. Abrams, et al., eds., *The Norton anthology of English literature* (New York, 1962), 374-375.
4. Vannevar Bush, *Pieces of the action* (New York, 1970), 148.

Hence there is in the literature a persistent "almostness." Everyone agrees that Oppenheimer was a great man in some way or another, yet we learn from each other's books and articles and opinion pieces that Oppenheimer was almost—but not quite—a remarkable number of things. He was almost, but not quite, a genius, Abraham Pais is often quoted as saying. Others in the same mode have concluded that Oppenheimer was almost a great physical theorist, almost a convincing analyst of the relations of science to culture and politics, almost an interesting philosopher, almost an effective giant in the liberal intelligentsia of his day, almost a modern Faust, almost a blameless victim of McCarthyism, and, as Robert P. Crease explains with great perspicacity in this volume, almost tragic.[5] Some commentators, to be sure, have insisted that Oppenheimer became all of these things in the full, but the syndrome to which I want to call attention is the holding of Oppenheimer to extraordinarily demanding standards of human achievement.

Perhaps this persistent almostness in our assessments of this man is a metaphor for the failure of even the most talented of us to ever become complete, and thus a moving emblem for our own petty lives, for our own failures to achieve tragic depth? Whatever the measure of truth in that speculation, I believe it is fair to say that the reputation of J. Robert Oppenheimer has been long held hostage to our own yearning for an intellectual hero who met the challenges of the 1940s and 1950s era more impressively than the bulk of that generation did. Oppenheimer is our best retrospective shot, and we, as a community of interpreters, professional and lay, just cannot seem to forgive him for not getting it right.

This is the case even with Kempton. In words that seem brutal coming from someone who admires Oppenheimer as much as Kempton did, Kempton concluded that Oppenheimer proved to be the physicist the government of 1954 wanted: "someone who could be bullied."[6] What Kempton means is that our hero, for all the confidence that sometimes lapsed into arrogance, backed down on a series of important issues when he was in a position to put up more of a struggle. Oppenheimer was passive in the extreme when on the stage of his legendary security hearing, passing up opportunities to challenge the charade, allowing himself a self-representation as an idiot, and placing trust in officials whom he was sophisticated enough to know should not be trusted. And earlier, when he alerted security officers of the dangers of espionage, he was not capable of concealing the name of his indisputably dear friend Haakon Chevalier when the opportunity was clearly there to bypass the kitchen conversation at One Eagle Hill and to simply name the figure behind it, George Eltenton. Oppenheimer surrendered too easily, Kempton laments, to government authorities. When he had his audience with President Truman, and perhaps a unique opportunity to exercise his influence, he blew it badly. And what about the Superbomb itself; was not Oppenheimer awfully

5. The applicability of the concept of tragedy to Oppenheimer's fate at the hands of the Atomic Energy Commission is called into question in the important new book by Patricia J. McMillan, *The ruin of J. Robert Oppenheimer and the birth of the modern arms race* (New York, 2005).
6. Kempton (ref. 1), 130.

quick to fall in behind it after Truman approved it? But what most nags Kempton is Oppenheimer's passive acceptance of the judgment against him, and the role that passivity played in strengthening the power of men like Lewis Strauss.

Kempton's broodings about Oppenheimer's passivity are given point by Richard Polenberg's scrutiny, within this volume, of the transcript of the 1954 hearings. Had Strauss been shrewd enough to renew Oppenheimer's security clearance while releasing the transcripts, might not Oppenheimer have been even more fully destroyed than he was? Might he have forfeited much of the sympathy that has sustained his reputation down to the present day? Might the crude logic of the Faust legend, so often applied uncritically to Oppenheimer's making of the bomb, have been applied in a fashion to a man who named names in order to save his security clearance?

In the end, Kempton relents a bit, and remarks that his own analyses are somewhat hushed when he looks at Oppenheimer's "eyes in the old photographs," where we confront "the inexplicable nobility of stuff so adamant that it can be passed through every variety of the ignoble, do the worst and have the worst done to itself, and somehow shine through all its trash and its trashing."[7] And so it is with Polenberg, and with most of the analyses contained in this volume: the tone is more relaxed, not because the issues have ceased to matter but because our community of interpreters is now less inclined to make Oppenheimer do our own moral and political work for us. The reappraisals and reflections offered by this group of scholars on the occasion of the Oppenheimer Centennial try to liberate Oppenheimer from the moral and political needs of his successors. That liberation, and the deep respect it carries with it, is best achieved if we view Oppenheimer as a site on which the history of a generation of American intellectuals was the most importantly enacted.

7. Ibid., 135.

Guide to Further Reading

Online

Information from the Oppenheimer Centennial at Berkeley, including web links, may be found at http://ohst.berkeley.edu/oppenheimer/.

The award-winning exhibit "J. Robert Oppenheimer: A Life" is located at http://ohst.berkeley.edu/oppenheimer/exhibit/. PDF versions of the exhibit (high- and low-resolution) are available for download or classroom use.

Book-Length Treatments

Bernstein, Jeremy. *Oppenheimer: Portrait of an enigma*. Chicago, 2004.

Bird, Kai, and Martin J. Sherwin. *American Prometheus: The triumph and tragedy of J. Robert Oppenheimer*. New York, 2005.

Cassidy, David. *J. Robert Oppenheimer and the American century*. New York, 2005.

Chevalier, Haakon. *Oppenheimer: The story of a friendship*. New York, 1965.

Davis, Nuel Pharr. *Lawrence and Oppenheimer*. New York, 1968.

Else, Jon, David Webb Peoples, Janet Peoples, Pyramid Films & Video, KTEH-TV. *The day after Trinity: J. Robert Oppenheimer and the atomic bomb*. San Jose, CA, 1981.

Goodchild, Peter. *J. Robert Oppenheimer: Shatterer of worlds*. London, 1980.

Herken, Gregg. *Brotherhood of the bomb: The tangled lives and loyalties of Robert Oppenheimer, Ernest Lawrence, and Edward Teller*. New York, 2002.

Holloway, Rachel L. *In the matter of J. Robert Oppenheimer: Politics, rhetoric, and self-defense*. Westport, CT, 1993.

Kipphardt, Heinar. *In the matter of J. Robert Oppenheimer: A play freely adapted on the basis of the documents*. New York, 1968.

Kunetka, James W. *Oppenheimer: The years of risk*. Englewood Cliffs, NJ, 1982.

McMillan, Priscilla. *The ruin of J. Robert Oppenheimer and the birth of the modern arms race*. New York, 2005.

Michelmore, Peter. *The swift years: The Robert Oppenheimer story*. New York, 1969.

Pais, Abraham, with Robert P. Crease. *J. Robert Oppenheimer: A life*. New York, 2006.

Polenberg, Richard, ed. *In the matter of J. Robert Oppenheimer: The security clearance hearing*. Ithaca, NY, 2002.

Rabi, I.I., et al. *Oppenheimer*. New York, 1969.

Schweber, S.S. *In the shadow of the bomb: Bethe, Oppenheimer, and the moral responsibility of the scientist*. Princeton, 2000.

Smith, Alice Kimball, and Charles Weiner, eds. *Robert Oppenheimer: Letters and recollections*. Cambridge, 1980. Stanford, 1995.

Stern, Philip M., with Harold P. Green. *The Oppenheimer case: Security on trial*. New York, 1969.

Thorpe, Charles. *Oppenheimer: The tragic intellect*. Chicago, 2006.

United States Atomic Energy Commission. *In the matter of J. Robert Oppenheimer: Transcript of hearing before Personnel Security Board and texts of principal documents and letters*. Cambridge, 1971.

York, Herbert F. *The advisors: Oppenheimer, Teller, and the Superbomb*. San Francisco, 1976.

Articles and Essays

Bernstein, Barton. "The Oppenheimer loyalty-security case reconsidered." *Stanford law review, 42* (1990), 1383-1484.

Bernstein, Barton J. "In the matter of J. Robert Oppenheimer." *Historical studies in the physical sciences, 12:1* (1982), 195-252.

Bethe, H.A. "J. Robert Oppenheimer." National Academy of Sciences, *Biographical memoirs, 71* (1997), 174-219.

Day, Michael A. "Oppenheimer on the nature of science." *Centaurus, 43* (2001), 73-112.

Galison, Peter, and Barton Bernstein. "In any light: Scientists and the decision to build the Superbomb, 1942-54." *Historical studies in the physical and biological sciences, 19:2* (1989), 267-347.

Hijiya, James A. "The Gita of J. Robert Oppenheimer." American Philosophical Society, *Proceedings, 144* (2000), 123-167.

Holton, Gerald. "'Success sanctifies the means': Heisenberg, Oppenheimer, and the transition to modern physics." *The advancement of science, and its burdens*, 141-162. Cambridge, 1986.

Kempton, Murray. "The ambivalence of J. Robert Oppenheimer." *Rebellions, perversities, and main events*, 118-135. New York, 1994.

Rieff, Philip. "The case of Dr. Oppenheimer." Philip Rieff, ed., *On intellectuals: Theoretical studies/case studies*, 341-369. New York, 1970.

Schweber, Silvan S. "J. Robert Oppenheimer: Proteus unbound." *Science in context, 16* (2003), 219-242.

Taylor, Brian C. "The politics of the nuclear text: Reading Robert Oppenheimer's *Letters and recollections*." *Quarterly journal of speech, 78* (1992), 429-449.

Thorpe, Charles Robert. "Disciplining experts: Scientific authority and liberal democracy in the Oppenheimer case." *Social studies of science, 32* (2002), 525-562.

Thorpe, Charles, and Steven Shapin. "Who was J. Robert Oppenheimer? Charisma and complex organization." *Social studies of science, 30* (2000), 545-590.

Additional Background

Badash, Lawrence. "Science and McCarthyism." *Minerva, 38* (2000), 53-80.

Badash, Lawrence, Joseph O. Hirschfelder, and Herbert P. Broida, eds. *Reminiscences of Los Alamos, 1943-1945*. Dordrecht, 1980.

Butler, Loren. "Robert S. Mulliken and the politics of science and scientists, 1939-1946." *Historical studies in the physical and biological sciences, 25:1* (1994), 25-45.

Cassidy, David C. "Cosmic ray showers, high energy physics, and quantum field theories: Programmatic interactions in the 1930s." *Historical studies in the physical sciences, 12:1* (1981), 1-39.

Collins, Martin J. *Cold War laboratory: RAND, the Air Force, and the American state, 1945-1950*. Washington, DC, 2002.

Conant, Jennet. *109 East Palace: Robert Oppenheimer and the secret city of Los Alamos*. New York, 2005.

Ghamari-Tabrizi, Sharon. *The worlds of Herman Kahn: The intuitive science of thermonuclear war.* Cambridge, 2005.

Heilbron, J.L., and Robert W. Seidel. *Lawrence and his laboratory: A history of the Lawrence Berkeley Laboratory*, vol. 1. Berkeley, 1989.

Herken, Gregg. *Cardinal choices: Presidential science advising from the atomic bomb to SDI.* New York, 1992.

Hershberg, James. *James B. Conant: Harvard to Hiroshima and the making of the nuclear age*. New York, 1993.

Hewlett, Richard G., and Oscar E. Anderson, Jr. *The new world: A history of the United States Atomic Energy Commission*, vol. 1, *1939-1946*. Berkeley, 1990.

Hoddeson, Lillian, Paul W. Henriksen, Roger A. Meade, and Catherine Westfall. *Critical assembly: A technical history of Los Alamos during the Oppenheimer years, 1943-1945*. Cambridge, 1993.

Hodes, Elizabeth. "Precedents for social responsibility among scientists: The American Association of Scientific Workers and the Federation of American Scientists, 1938-1948." Ph.D. dissertation, History, University of California, Santa Barbara, 1982.

Holloway, David. *Stalin and the bomb: The Soviet Union and atomic energy, 1939-1956*. New Haven, CT, 1994.

Hughes, Jeff. *The Manhattan Project: Big science and the atom bomb.* New York, 2003.

Hunner, Jon. *Inventing Los Alamos: The growth of an atomic community.* Norman, OK, 2004.

Jewett, Andrew John. "To make America scientific: Science and democracy in American public culture, 1900-1950." Ph.D. dissertation, History, University of California, Berkeley, 2002.

Jungk, Robert. *Brighter than a thousand suns.* New York, 1958.

Kaiser, David. "Cold War requisitions, scientific manpower, and the production of American physicists after World War II." *Historical studies in the physical and biological sciences, 33:1* (2002), 131-159.

Kevles, Daniel J. *The physicists: The history of a scientific community in modern America.* Cambridge, 1977, 1995.

Kojevnikov, Alexei. *Stalin's great science: The times and adventures of Soviet physicists.* London, 2004.

Kuznick, Peter J. *Beyond the laboratory: Scientists as political activists in 1930s America.* Chicago, 1987.

Needell, Allan A. *Science, Cold War and the American state: Lloyd V. Berkner and the balance of professional ideals.* Amsterdam, 2000.

Norris, Robert S. *Racing for the bomb: General Leslie R. Groves, the Manhattan Project's indispensable man.* South Royalton, VT, 2002.

Rhodes, Richard. *The making of the atomic bomb.* New York, 1986.

Rudolph, John L. *Scientists in the classroom: The Cold War reconstruction of American science education.* New York, 2002.

Schweber, S.S. "The empiricist temper regnant: Theoretical physics in the United States 1920-1950." *Historical studies in the physical and biological sciences, 17:1* (1986), 55-98.

Serber, Robert. *The Los Alamos primer: The first lectures on how to build an atomic bomb.* Berkeley, 1992.

Serber, Robert, with Robert P. Crease. *Peace and war: Reminiscences of a life on the frontiers of science.* New York, 1998.

Sherwin, Martin J. *A world destroyed: Hiroshima and its legacies,* 3rd ed. Stanford, 2003.

Smyth, Henry DeWolf. *Atomic energy for military purposes*. Princeton, 1945.

Taylor, Brian C. "Organizing the 'unknown subject': Los Alamos, espionage, and the politics of biography." *Quarterly journal of speech, 88* (2002), 33-49.

Walters, Ronald G., ed. *Scientific authority and twentieth-century America*. Baltimore, 1997.

Wang, Jessica. *American science in an age of anxiety: Scientists, anticommunism, and the Cold War*. Chapel Hill, NC, 1999.

Wang, Jessica. "Scientists and the problem of the public in Cold War America, 1945-1960." *Osiris, 17, Science and civil society*, Lynn K. Nyhart and Thomas H. Broman, eds., 323-347. Chicago, 2002.

Wang, Zuoyue. *In Sputnik's shadow: The President's Science Advisory Committee and Cold War America*. New York, 2006.

Weart, Spencer. *Nuclear fear: A history of images*. Cambridge, 1988.

Westwick, Peter. *The national labs: Science in an American system, 1947-1974*. Cambridge, 2003.

Oppenheimer's Published Works

Oppenheimer, J. Robert. *Atom and void: Essays on science and community*. Princeton, 1989.

Oppenheimer, J. Robert. *The flying trapeze: Three crises for physicists*. New York, 1964.

Oppenheimer, J. Robert. *The open mind*. New York, 1955.

Oppenheimer, J. Robert. *Science and the common understanding*. New York, 1954.

Oppenheimer, J. Robert. *Uncommon sense*. Eds. N. Metropolis, Gian-Carlo Rota, and David Sharp. Cambridge, 1984.

A bibliography of Oppenheimer's scientific papers may be found in *Robert Oppenheimer: Letters and recollections*, edited by Alice Kimball Smith and Charles Weiner, cited above.

Contributors

BARTON J. BERNSTEIN is Professor of American History at Stanford University. He teaches and researches on American political and diplomatic history since 1939, with a special interest in the history of the nuclear age and the figure of J. Robert Oppenheimer.

KAI BIRD is the author of *The chairman: John J. McCloy and the making of the American establishment* (1992) and *The color of truth: McGeorge Bundy and William Bundy, brothers in arms* (1998). He is also co-editor with Lawrence Lifschultz of *Hiroshima's shadow: Writings on the denial of history and the Smithsonian controversy* (1998). A Contributing Editor of *The nation*, he is the recipient of fellowships from the John Simon Guggenheim Foundation, the Alicia Patterson Journalism Fellowship, the John D. and Catherine T. MacArthur Foundation, the Rockefeller Foundation, the Thomas J. Watson Foundation, the German Marshall Fund, and the Woodrow Wilson International Center for Scholars.

CATHRYN CARSON is Associate Professor of History and Director of the Office for History of Science and Technology at the University of California, Berkeley. She has worked on the history of quantum field theory and on the public role of the scientist after the Second World War, focusing on Werner Heisenberg. Her current project is a history of the science behind nuclear waste management, while her teaching centers on the history of physics, science in Germany and the United States, and nuclear history.

DAVID C. CASSIDY is Professor of Natural Science at Hofstra University in Hempstead, New York. A former associate editor of the Albert Einstein Papers Project, he is the author of *Uncertainty: The life and science of Werner Heisenberg* and *Einstein and our world.* His biographical study *J. Robert Oppenheimer and the American century* was published in 2005.

ROBERT P. CREASE is Professor of Philosophy at Stony Brook University, historian at Brookhaven National Laboratory, and author of a monthly column entitled "Critical

Point" on the social dimensions of science for the magazine *Physics world*. He assisted in publication of *J. Robert Oppenheimer: A life*, by Abraham Pais with supplemental material by Robert P. Crease. His other books include *The prism and the pendulum: The ten most beautiful experiments in science* (1993); *Making physics: A biography of Brookhaven National Laboratory* (1999); *Peace and war: Reminiscences of a life on the frontiers of science* (1998, with Robert Serber); *The play of nature: Experimentation as performance* (1993); and *The second creation: Makers of the revolution in twentieth century physics* (1986, with Charles C. Mann).

Jon Else is a documentary filmmaker and Professor of Journalism at the University of California, Berkeley. He is best known for his *The day after Trinity: J. Robert Oppenheimer and the atomic bomb* (1980), *Cadillac desert: Water and the transformation of nature* (1997), and *Sing faster: The stagehands' Ring Cycle* (2000). He was series producer and cinematographer for *Eyes on the prize: America's civil rights years* (1987). He has served as cinematographer on hundreds of documentaries for PBS, the BBC, ABC, MTV and HBO, including the BBC/PBS *History of rock and roll* (1993), *Who are the DeBolts* (Academy Award 1976), the Paramount/MTV feature documentary *Tupac: Resurrection* (Sundance 2003), *Alice Waters* for *American masters*, and several independent feature films, as well as numerous commercials and music videos. He is currently directing *Wonders are many: The making of Doctor Atomic*, a documentary about the creation of John Adams and Peter Sellars's opera about Oppenheimer.

Peter Galison is Mallinckrodt Professor of the History of Science and of Physics at Harvard University. In 1997, he was named a John D. and Catherine T. MacArthur Foundation Fellow; in 1999, he was the recipient of the Max Planck Research Award. His books include *How experiments end* (1987); *Image and logic: A material culture of microphysics* (1997), which won the Pfizer Award from the History of Science Society; and *Einstein's clocks, Poincaré's maps* (2003). With Pamela Hogan he co-produced the film *Ultimate weapon: The H-bomb debate* and is now, with Rob Moss, making a film about science, security, and secrecy.

J.L. Heilbron is Professor of History and The Vice Chancellor Emeritus at the University of California, Berkeley. His most recent books are *Geometry civilized: History, culture, and technique* (1998); *The sun in the church: Cathedrals as solar observatories* (1999); *Rutherford and the explosion of atoms* (2003); and, as editor in chief, *The Oxford companion to the history of modern science* (2003). His current research interests include relations between science and religion in the seventeenth and eighteenth centuries, physics and its institutions in the twentieth century, and the use of history of science in the teaching of science.

Gregg Herken is Professor of History and a founding faculty member of the University of California, Merced. Between 1988 and 2003 he was a senior Historian and Curator of Military Space at the Smithsonian Institution's National Air and Space Museum in

Washington, D.C. He is the author of four books: *The winning weapon: The atomic bomb in the Cold War* (1981), *Counsels of war* (1985), *Cardinal choices: Presidential science advising from the atomic bomb to SDI* (1992), and *Brotherhood of the bomb: The tangled lives and loyalties of Robert Oppenheimer, Ernest Lawrence, and Edward Teller* (2002), which was a finalist for the 2003 *Los Angeles times* Book Prize.

James G. Hershberg is Associate Professor of History and International Affairs at George Washington University. He authored *James B. Conant: Harvard to Hiroshima and the Making of the Nuclear Age*, published in 1993, and numerous articles on cold war and nuclear history. He directed the Woodrow Wilson Center's Cold War International History Project and currently edits the CWIHP Book Series, co-published by the Stanford University and Wilson Center Presses.

David A. Hollinger is Department Chair and Preston Hotchkis Professor of History at the University of California, Berkeley. His books include *Cosmopolitanism and solidarity: Studies in group affiliation in the United States* (Wisconsin, 2006) and *Science, Jews, and secular culture* (Princeton, 1996). He is also the editor of *The humanities and dynamics of inclusion after World War II* (Johns Hopkins, 2006). His recent articles have appeared in *Church history*, *Harper's*, *Modern intellectual history*, and *The Cambridge history of science*.

David Holloway is the Raymond A. Spruance Professor in International History, Professor of Political Science, and Senior Fellow at the Institute for International Studies at Stanford University. Since coming to Stanford in 1986, he has served as Co-Director of the Center for International Security and Arms Control (1991-97), as Associate Dean in the School of Humanities and Sciences (1997-98), and as Director of the Stanford Institute for International Studies (1998-2003). His research has focused on the impact of nuclear weapons on international relations, the history of the Cold War, and science and technology in the Soviet Union. His book *Stalin and the bomb: The Soviet Union and atomic energy, 1939-1956* was published by Yale University Press in 1994 and has appeared in several languages, including Russian. His current research focuses on the history of nuclear weapons.

Karl Hufbauer, who retired from the University of California, Irvine in 1999, has written on the formation of the German chemical community in the eighteenth century, the development of solar science since Galileo, and the emergence of what has come to be called nuclear astrophysics. He hopes to write a book about research on the stellar-energy problem during the first half of the twentieth century. He now resides in Seattle, where he is affiliated with the University of Washington's Department of History and is active as a stone sculptor.

David Kaiser is Associate Professor of the History of Science in MIT's Program in Science, Technology, and Society, and a Lecturer in MIT's Department of Physics. His historical research focuses on changes in American physics during and after World War II, looking in particular at how the postwar generation of graduate students was trained. His book *Drawing theories apart: The dispersion of Feynman diagrams in postwar physics* was published in 2005 by the University of Chicago Press. He has also recently edited a book, *Pedagogy and the practice of science: Historical and contemporary perspectives*, which appeared in 2005 from MIT Press.

DANIEL J. KEVLES is the author most recently of *The Baltimore case: A trial of politics, science, and character* and a coauthor of *Inventing America: A history of the United States* (2nd edition, 2006). His previous books include *In the name of eugenics: Genetics and the uses of human heredity* and *The physicists: The history of a scientific community in modern America*. From 1964 to 2001, he taught at the California Institute of Technology, and since 2001 he has been the Stanley Woodward Professor of History at Yale University, where he chairs the program in History of Science and Medicine. In 2001 he received the History of Science Society's George Sarton Medal for career achievement.

ALEXEI KOJEVNIKOV is Associate Professor of History at the University of Georgia, Athens. He has worked on the history of quantum physics, the relationship between science and ideology, and the history of Soviet science. He is the author of the book *Stalin's great science: The times and adventures of Soviet physicists*, published in 2004 by Imperial College Press in London.

W. PATRICK MCCRAY is Associate Professor of History at the University of California, Santa Barbara. He researches and teaches the history of science and technology in the twentieth century and is, most recently, the author of *Giant telescopes: Astronomical ambition and the promise of technology* (Harvard University Press, 2004).

RICHARD POLENBERG has taught at Cornell University since 1966 and is now Goldwin Smith Professor of American History and Stephen H. Weiss Presidential Fellow. He is the author of *Fighting faiths: The Abrams case, the Supreme Court, and free speech* (1987), which received the American Bar Association's Silver Gavel Award, and *The world of Benjamin Cardozo* (1997). He has recently published an edited version of the 1954 AEC hearing, *In the matter of J. Robert Oppenheimer* (2002).

S.S. SCHWEBER is currently a Senior Research Associate at the Dibner Institute. His publications include *QED and the men who made it: Dyson, Feynman, Schwinger, and Tomonaga* (1994) and *In the shadow of the bomb: Oppenheimer, Bethe, and the moral responsibility of the scientist* (2000). He is at work on a biography of Hans Bethe.

MARTIN J. SHERWIN is the Walter S. Dickson Professor of English and American History at Tufts University, where he founded the university's Nuclear Age History and Humanity Center. The author of *A world destroyed: Hiroshima and its legacies* (1975, 1987, 2003), he is the recipient of the Bernath Prize, awarded by the Society for Historians of American Foreign Relations, as well as the American History Book Prize awarded by the National Historical Society. He has served as a historical advisor for several documentary films and is co-executive producer of a documentary on the life of Igor Kurchatev. He has received fellowships from the John D. and Catherine T. MacArthur Foundation, the John Simon Guggenheim Foundation, the American Academy of Arts and Sciences, the National Endowment for the Humanities, and the Rockefeller Foundation.

CHARLES THORPE is Lecturer in the Department of Science and Technology Studies at University College London. He was trained in sociology and science studies at the University of California, San Diego. His sociological biography *Oppenheimer: The tragic intellect* will be published by the University of Chicago Press in 2006.

STEPHANIE YOUNG is a doctoral student in the Department of History at the University of California, Berkeley. Her interests focus on federal science policy in the United States after the Second World War. She is particularly interested in scientists and the development of military strategy in the early Cold War.

Index